AF631258

ENCYCLOPEDIA OF GENOMICS

ENCYCLOPEDIA OF GENOMICS

Vol. 3

GENOMIC EVOLUTION

By

Dr. Shiv Kant Prasad

School of Studies of Zoology & Biotechnology

Vikram University

Ujjain

(India)

D P H

DISCOVERY PUBLISHING HOUSE PVT. LTD.

NEW DELHI-110 002

Published by:
Tilak Wasan
DISCOVERY PUBLISHING HOUSE PVT. LTD.
4383/4A, Ansari Road, Darya Ganj
New Delhi-110 002 (India)
Phone: +91-11-23279245, 43596064-65
Fax : +91-11-23253475
E-mail: parul.wasan@gmail.com
discoverypublishinghouse@gmail.com
web : www.discoverypublishinggroup.com

***First Edition:* 2012**

ISBN: 978-93-5056-032-7 (Set)
978-93-5056-033-4 (Vol. 1)
978-93-5056-034-1 (Vol. 2)
978-93-5056-035-8 (Vol. 3)
978-93-5056-036-5 (Vol. 4)
978-93-5056-037-2 (Vol. 5)

Encyclopedia of Genomics

© 2012, Author

All rights reserved. No part of this publication should be reproduced, stored in a retrieval system, or transmitted in any form or by any means: electronic, mechanical, photocopying, recording or otherwise, without the prior written permission of the author and the publisher.

This book has been published in good faith that the material provided by authors is original. Every effort is made to ensure accuracy of material, but the publisher and printer will not be held responsible for any inadvertent error(s). In case of any dispute, all legal matters are to be settled under Delhi jurisdiction only.

Printed at:
Shree Balaji Art Press
Delhi

Preface

Progress in any scientific discipline is dependent on the availability of techniques and methods that extend the range and sophistication of experiments which may be performed. Over the last 30 years or so this has been demonstrated in a spectacular way by the emergence of genetic engineering. This field has grown rapidly to the point where, in many laboratories around the world, it is now routine practice to isolate a specific DNA fragment from the genome of an organism, determine its base sequence, and assess its function. The technology is also now used in many other applications, including pharmaceutical analysis of samples, paternity disputes, medical diagnosis, genomic and the biotechnology industry.

The present title **"Encyclopedia of Genomics"** is a fast growing area of research for undergraduate and post-graduate students in life sciences, chemistry, engineering and medicine. The overall objective of this publication is to provide a professional level reference work with comprehensive coverage of the molecular basis of genome and the application of that knowledge in genetics, evolution, medicine, and agriculture. It deals with the life processes at a molecular level genetic disease diagnosis and genetic therapy; the theory and techniques for understanding manipulating, and synthesizing biological molecules and their aggregates; and the application of biological processes to make or modify products to improve plants or animals or to develop microorganisms for specific uses.

The development of techniques for genomic science is based on an understanding of the molecular aspect of genomic processes occurring in nature. These techniques have, in turn, enabled expansion of our knowledge and understanding of how genes are organized and expressed in living cells and has provided major impetus to the development of biotechnological enterprises.

The present title has been written to provide readers with a thorough understanding of the techniques which underpin the developments in the field of genomic science. By gaining this understanding, readers will be able to apply the techniques of genomic manipulation to tackle specific problems or to achieve desired objectives in a wide variety of fields.

There can be no claim to originality except in the manner or treatment and much of the information has been obtained from the books and scientific journals available in the different libraries.

The author express his thanks to his friends and colleagues whose continue inspirations have initiated him to bring out this book.

The author expresses her gratitude to Mr. Wasan and staff of M/s Discovery Publishing House Pvt. Ltd. for their whole hearted co-operation in the publication of this book.

S.K. Prasad

Contents

Chapter 1

Introduction

Centromeres are required for normal chromosome *segregation* in mitosis and meiosis, and a substantial proportion of human *pathology* stems from *abnormalities* of *chromosome* segregation, the *underlying* genomic basis and mechanism(s) of which are largely unknown. Human centromeres consist of megabases of a-satellite DNA, a tandemly repeated DNA family whose *genomic organization*, evolution, and function is *increasingly* well *understood*.

The study of normal, abnormal, and engineered human chromosomes is providing *insights* into the nature of human centromeres and their mechanism of action, as well as enabling *comparison* with *centromeres* of other *eukaryotic* organisms and the *identification* of genomic elements required for normal centromere function. In all eukaryotic organisms including humans the centromere plays a *critical role* in each step of chromosome segregation in mitosis and meiosis.

As the site of the proteinaceous kinetochore, the *centromere* is responsible for attaching chromosomes to spindle *microtubules* that then align the chromosomes at the metaphase plate. Proteins localized to the *centromere* are also involved in the *metaphase* to *anaphase checkpoint* and signal the attachment of all chromosomes to *spindle microtubules* before allowing the cell to progress into *anaphase*.

Finally, sister *chromatid cohesion* must be resolved at the centromere, and proper *timing* of the removal of cohesion is of *vital* importance for *segregating* chromatids. The centromere is defined by specific genomic DNA sequences, as well as by a *specialized* chromatin structure. Although centromere proteins are highly conserved among a number of eukaryotic organisms, DNA sequences at the centromere are not well conserved. In fact, centromeres range in size and complexity from the 125-bp *centromere* found in *budding* yeast to the human centromere that spans several *megabases*. Like *centromeres* in most complex organisms, the human centromere is made up of repetitive DNA. α-satellite DNA, a tandemly repeated DNA family based on a fundamental unit length of approx 171 bp, has been found at all human centromeres and has served as a model to study the genomics, *evolutionary dynamics*, and general potential function(s) of *satellite* DNAs. The focus of this chapter is the structure and function of centromeric regions in the human genome. We discuss the genomic organization and evolution of α-satellite DNA, current concepts of centromeric chromatin, and efforts to assess normal centromere function in human chromosomes, utilizing both structurally abnormal chromosomes *derived* from patient material and *engineered* human *artificial chromosomes.*

STATUS OF CENTROMERE

In Eukaryotes

The sequences that make up the *centromeres* of diverse organisms are extremely variable. Most well-characterized centromeres contain repetitive DNA with an AT-richness greater than that of the genome average. However, *individual* organisms have evolved different genomic structures to create a locus capable of chromosome segregation. The simplest centromere organization is found in the chromosomes of the yeast, *Saccharomyces cerevisiae.* Only approx 125 by is required for centromere function in the *budding* yeast, and this sequence motif is shared among the centromeres of all 16 chromosomes.

Unlike other characterized centromeres, the budding yeast centromere consists of largely unique DNA. In contrast to the simple centromere of the budding yeast, the fission yeast centromere is more similar to the *centromeres* of higher *eukaryotes* in its size and *complexity. Schizo-saccharomyces pombe* centromeres are made up of inner and outer inverted repeats flanking a *nonrepetitive* central core, and each of these regions is AT-rich.

Among the three *S. pombe* chromosomes, the centromeres are similar, but not identical in organization. Overall, *S. pombe* centromeres are 35-110 kb in size, spanning a few percent of the linear length of each chromosome. The inner repeats and central core are necessary for *centromere* function and bind spindle *microtubules,* whereas the outer repeats recruit *heterochromatin* proteins and are more likely responsible for functions such as *heterochromatin* formation and sister *chromatid cohesion.* Centromeres in several other organisms are characterized by long *stretches* of so-called "*satellite DNA.*"

The *Drosophila* centromere has been defined by a 420-kb region of a minichromosome that is required for chromosome *transmission.* The fly *centromeric* region consists of two adjacent blocks of short *microsatellites,* based on AATAT and AAGAG repeats, that are interspersed with *transposons* as well as AT-rich DNA.

Normal fly centromeres have not been fully sequenced, owing to the difficulty of sequencing and assembling highly *heterochromatic* regions of the *genome.* However, the chromatin *environment* of endogenous *Drosophila* centromeres has been very well characterized. Plant centromeres are very similar to the satellite- and transposon-rich fly centromeres. The major component of *the Arabidopsis thaliana* centromere is an AT-rich 180-bp repeat unit, spanning 400 kb to 1.4 Mb among *chromosomes.*

The *Arabidopsis* centromere is also enriched for *retrotransposons* not usually found on chromosome arms. A similar picture has emerged for the rice *(Oryza sativa)* centromere, which is comprised predominantly of a 155-bp tandem repeat unit arranged in arrays ranging from 65 kb to 2 Mb among the 12 rice chromosomes.

These arrays are interspersed with gypsy-class *retrotransposons.* The *Arabidopsis* and rice centromeres have recently been defined at the level of chromatin, and *chromatin immuno*

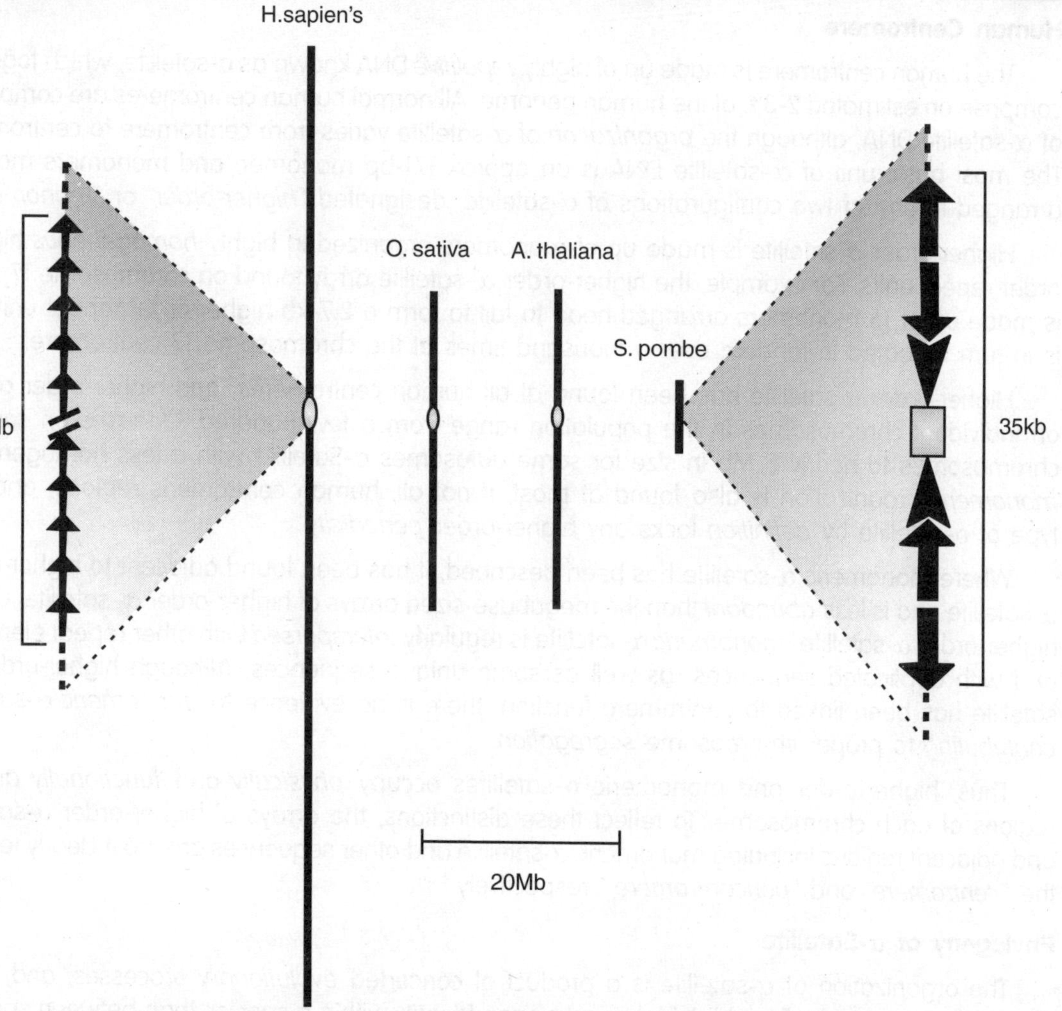

Figure 1.1: Schematic representation of genomic organization of centromeres.

precipitation experiments using *antibodies* to proteins required for centromere function have been *conducted* in both species. As expected, centromere proteins are associated with satellite repeats in both *Arabidopsis* and rice.

However, within the functional domain of the smallest rice centromere, there are also four expressed genes. This finding is *surprising* because centromeres are *classically* thought of as heterochromatic regions resistant to gene expression. As opposed to all other normal centromeres described, the centromeres of *Caenorhabditis elegans* appear to be completely *sequence-independent*.

C. elegans chromosomes are *holocentric*, meaning that many sites along the chromosome act as a centromere, capable of *recruiting* centromere proteins necessary for segregation. Holocentric chromosomes are a curious contrast to the *monocentric* chromosomes found in most other species *typically* containing *repetitive* AT-rich DNA at the centromeres.

Human Centromere

The human centromere is made up of highly *repetitive* DNA known as a-satellite, which together comprise an estimated 2-3% of the human genome. All normal human centromeres are comprised of α-satellite DNA, although the *organization* of α-satellite varies from centromere to centromere. The most basic unit of α–satellite DNA is an approx 171-bp monomer, and monomers may be arranged in one of two configurations of α-satellite, designated "*higher-order*" or "*monomeric*".

Higher-order α-satellite is made up of monomers organized in highly *homogeneous* higher-order repeat units. For example, the higher-order α–satellite array found on chromosome 17, DIM, is made up of 16 monomers arranged head to tail to form a 2.7-kb higher-order repeat unit that is in turn repeated in tandem over a thousand times at the chromosome 17 centromere.

Higher-order α-satellite has been found at all human centromeres, and higher-order arrays on individual chromosomes in the population range from a few hundred *kilobases* on some Y chromosomes to nearly 5 Mb in size for some autosomes α-Satellite with a less homogeneous *monomeric* organization is also found at most, if not all, human centromeric regions, and this type of α-satellite by *definition* lacks any higher-order *periodicity.*

Where monomeric α-satellite has been described, it has been found adjacent to higher-order α-satellite and is less *abundant* than the megabase-sized arrays of higher-order α–satellite. Unlike higher-order α-satellite, *monomeric* α-satellite is regularly *interspersed* with other repeat elements and with duplicated sequences, as well as some unique sequences. Although higher-order α-satellite has been linked to centromere function, there is no evidence for *monomeric* α-satellite *contributing* to proper chromosome *segregation.*

Thus, higher-order and monomeric α-satellites occupy *physically* and *functionally* distinct regions of each chromosome. To reflect these distinctions, the arrays of higher-order α-satellite and adjacent regions including monomeric α-satellite and other sequences are most clearly terrned the "*centromere*" and "*pericentromere,*" respectively.

Phylogeny of α-Satellite

The organization of α-satellite is a product of concerted *evolutionary* processes, and, thus, these sequences typically exhibit higher sequence *identity* within a species than between species. Although α-satellite has been found at all primate centromeres studied, the organization and types of α-satellite vary among species and have begun to inform hypotheses about centromere evolution.

Higher-order α-satellite has been found at some of the centromeres of chimpanzees, gorillas, and orangutans. Notably, higher-order α-satellite has not been found in more distant primates; indeed, only *monomeric* α–satellite has been found in Old World monkeys, New World *monkeys,* and *prosimians.*

As the centromeres from these monkeys have not been fully analyzed, however, the apparent absence of higher-order α-satellite should be *interpreted* with caution. Nonetheless, these findings are consistent with a model of α-satellite evolution in which *higher-order evolved* relatively recently from monomeric α-satellite.

Although a number of processes, collectively referred to as "*molecular drive*" and including

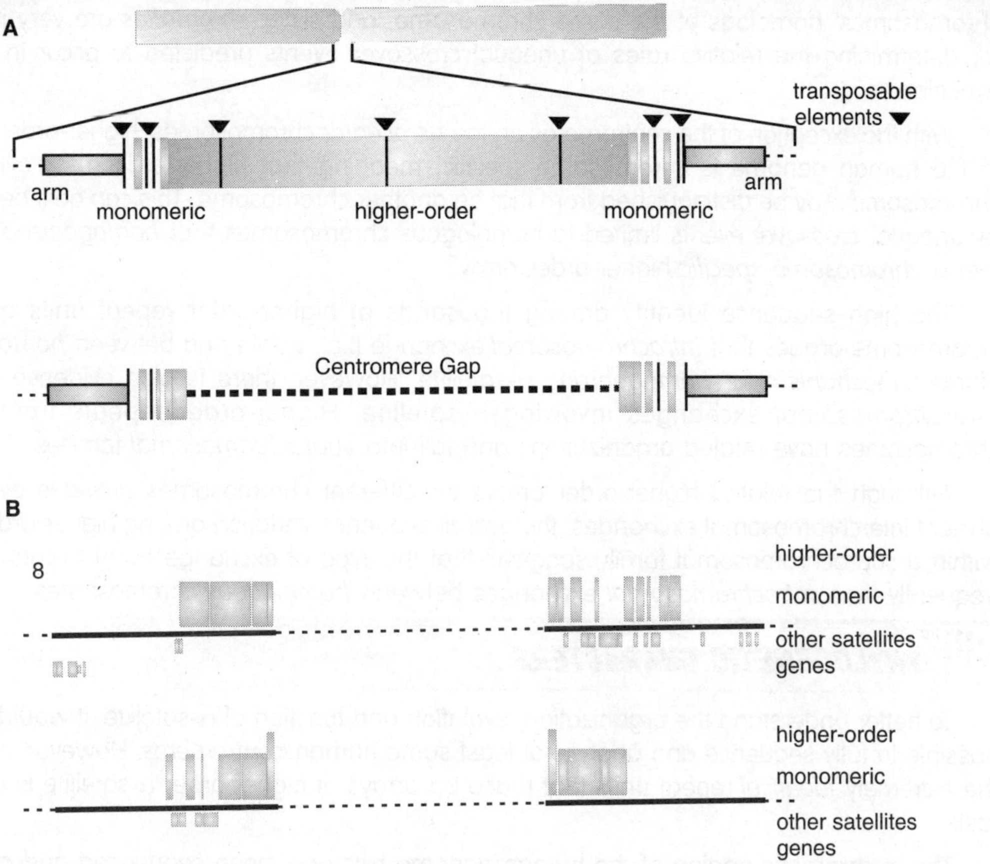

Figure 1.2: Genomic organization and annotation of human centromeres. (A) General model of centromeric region. The centromere itself is contained within a large array of higher-ordera-satellite (light gray) that spans several megabases and is only rarely interrupted by nonsatellite sequences such as transposable elements. The array is flanked by shorter segments of monomeric a-satellite, which is interspersed with other satellite sequences, duplicated sequence, a high frequency of transposable elements, and occasional single-copy sequences before transitioning into the euchromatin of the chromosome arms. (B) Because the human genome sequence covers mostly euchromatic sequence, the available contigs do not contain much (and in most cases, any) of the higher-order repeat arrays on individual chromosomes. Thus, by comparison to the model in (A), the current map and sequence of each chromosome has a large centromere gap, whose size can only be approximated. Two of the most complete annotated maps are shown for chromosomes 8 and the X, indicating the location of known higher-order repeat α-satellite, monomeric α-satellite, other satellite sequences, and genes close to the centromere.

mechanisms such as unequal crossover, gene conversion, and *transposition*, may be participating in α-satellite evolution to some extent,-the homogenization of α–satellite can largely be accounted for by unequal *crossover*. Recurring rounds of crossovers will homogenize tandem repeats, leading to nearly identical repeat units.

This process can explain not only the emergence of α-satellite DNA approx 30-50 millions of years ago, but also the initial homogenization of subsets of monomeric α-satellite to form the higher-order repeat units that subsequently expanded to make up the *megabase-sized* arrays currently present on human *centromeres*. The *relationships* among α-satellite on different chromosomes, homologs of the same chromosome, and sister *chromatids* are very informative for determining the relative rates of unequal crossover events predicted to occur in α-satellite evolution.

With the exception of the centromeres on the *acrocentric* chromosomes, higher-order α-satellite in the human genome is chromosome-specific, meaning that higher-order α-satellite on one chromosome may be distinguished from that on another chromosome. This can best be explained by unequal *crossover* events limited to homologous chromosomes that *homogenized* α-satellite into a *chromosome-specific* higher-order array.

The high sequence identity among thousands of higher-order repeat units on a given chromosome argues that *intrachromosomal* exchange (i.e., within and between *homologs*) is an efficient *mechanism* for *homogenizing* α-satellite. However, there is also evidence of ancient *interchromosomal* exchanges involving α-satellite. Higher-order repeats from different chromosomes have related organizations and fall into *suprachromosomal* families.

Although the related higher-order arrays on different chromosomes provides evidence for ancient interchromosomal exchanges, the overall sequence variation among higher-order repeats within a suprachromosomal family suggests that this type of exchange event occurs much less frequently than *intrachromosomal* exchanges between *homologous* chromosomes.

PHYLOGENETIC CHARACTERS

To better understand the organization, evolution and function of α-satellite, it would ideally be possible to fully sequence and analyze at least some human *centromeres*. However, *assembling* the extremely *identical* repeat units that make up arrays of higher-order α-satellite is a *daunting* task.

The euchromatic portion of the human genome has now been sequenced and assembled, yet, the centromere regions were *intentionally* omitted from the human genome project, and thus no human centromere has been *completely* assembled. *Notwithstanding* the incomplete nature of centromere sequences, several of the contigs of chromosome arms in the now-completed sequence off the eukaryotic portion of the human genome extend into higher-order arrays of α-satellite.

The next step in sequencing human centromeric regions would logically focus on further connecting existing chromosome arm contigs to higher-order α-satellite and then developing a strategy to sequence across the highly *homogeneous* arrays of higher-order α-satellite. The most *challenging* part of sequencing across *megabases* of higher-order α–satellite is not the *sequencing per se,* but the assembly process.

The assembly of sequences within genomic clones from these arrays is expected to be far more *complicated* than assembling typical genomic DNA, as higher-order repeat units are up to 100% identical. Although the merits of *attempting* to sequence across several megabases of nearly identical repeats are open to debate, in the meantime it is important to recognize and acknowledge that the existing *chromosome arm contigs* end some *distance-substantial* for some chromosomes-from the functional centromere itself.

Thus, from a practical standpoint, analysis and especially interpretation of human chromosome defects involving the centromere or *pericentromeric* region on the basis of the available genome sequences and resources must necessarily be made with caution.

CENTROMERIC ORGANIZATION

Although, as outlined above, the organization of centromeric DNA varies widely among *eukaryotic* organisms, the chromatin modifications and proteins involved in centromere function are very well conserved from yeast to humans. The likely primary *epigenetic* mark of the functional centromere is the histone H3 variant CENP-A, also known as CenH3.

CENP-A is found at active centromeres in every organism studied. Depleting CENP-A causes chromosome segregation defects and also has downstream effects on the localization of other centromere proteins, *supporting* its role as the primary *epigenetic* mark. CENP-A and histone H3 nucleosomes are interspersed at the centromeres of flies and humans, and CENP-A can substitute for histone H3 in reconstituted nucleosomes in vitro. Given the fact that CENP-A is a histone variant, it is tempting to posit that it sets up the centromere-specific *chromatin* conformation that then recruits other centromere proteins.

Such proteins include a number of DNA-binding proteins, as well as motor proteins, that are part of the *kinetochore.* Centromere protein B (CENP-B) is a DNA-binding protein found at the centromeres of diverse organisms, and it recognizes a 17 by sequence known as the *"CENP-B box"* in mouse minor satellite and human higher-order α-satellite.

The CENP-B box sequence has also been found at the centromeres of the great apes, but not in Old World monkeys, New World monkeys, or prosimians. Despite its evolutionary conservation, the role of this protein in centromere function is *questionable.* Another DNA-binding protein, however, CENP-C, which localizes only to active centromeres, is clearly directly involved in centromere function, as its depletion causes chromosome *segregation* defects.

Other proteins such as dynein, MCAK, and CENP-E are also members of the kinetochore, playing a role in chromosome movement along the microtubules. Also, spindle *checkpoint* proteins such as Mad2 and Bubl are critical for chromosome *segregation* as they signal the start of anaphase once all kinetochores are attached to the spindle. Proper resolution of sister *chromatid cohesion* is also required for chromosome *segregation.* After proceeding into mitotic anaphase, sister *chromatids* completely lose cohesion and separate to opposite poles of the cell.

Loss of sister chromatid cohesion is a two-step process in meiosis; in the first meiotic division, cohesion is removed from chromosome arms but maintained at the centromere, and then in the second meiotic division cohesion is completely removed. In the absence of cohesins, chromosomes *missegregate, exhibiting* chromosome lag and premature sister *chromatid* separation. The relationships among centromeric *chromatin, kinetochore* formation, spindle checkpoints, and resolution of sister chromatid *cohesion* have been best described in model organisms.

Although the role of specific genomic elements in the various functions previously summarized and associated with proper chromosome segregation is unknown, the general picture to *emerge* from genetic and molecular studies of the *fission* yeast or fly centromere is that each can be divided into centromeric and *pericentromeric* chromatin domains, each of which is characterized by a distinct set of chromatin (and *heterochromatin*) proteins, but both of which are required for proper chromosome segregation. It is tempting to apply this domain model to the organization of the human centromere.

It may be that, although monomeric α-satellite cannot nucleate the site of the kinetochore on its own, it is required for setting up the pericentromeric chromatin state, similar to the kinetochore flanking sequences in *S. pombe* and *D. melanogaster.*

Future studies carefully dissecting the locations of centromere proteins, chromatin modifications, heterochromatin proteins, and cohesins at the human centromere will determine the unique nature of centromeric chromatin, its distinction from euchromatin and heterochromatin, and likely differences between the centromeric and pericentromeric domains.

GENETIC SYSTEM OF CENTROMERE

Although the requirements for centromeric and pericentromeric functions have been well-defined in model organisms, similar analyses of the human centromere lack the *tractable* genetic systems found in other organisms, making it difficult to test specific regions functionally for centromere activity.

The respective roles of genomic sequence and epigenetic modifications in specifying human centromere function area topic of substantial debate and have been much informed by two lines of investigation: first, the analysis of abnormal human chromosomes derived from patient material; second, the development of approaches to generate and *analyze* engineered human chromosomes suitable for testing specific *hypotheses.*

Dicentric Chromosomes

Dicentric human chromosomes have been known about for 30 years and contain two distinct arrays of α-satellite on the same chromosome, formed by chromosome breakage and fusion events. To *maintain* chromosome stability, there must be only one active centromere, because if the *chromosome* attaches to spindle microtubules at two sites it could be pulled to opposite poles of the cell, causing chromosome breakage or *anaphase bridging.*

The stability of *dicentric* human chromosomes can be *achieved* either by *inactivating* one centromere or by *coordinating* the activity of the two centromeres such that they orient to the same pole. Centromere *inactivation* involves the loss of the ability of that centromere to bind proteins involved in spindle *microtubule* attachment, most likely involving reversal of the epigenetic *modifications* that mark functional *centromeric* chromatin.

The fact that a region of otherwise functionally competent α-satellite can exist on a chromosome without conferring *centromeric* activity has led some to argue that α-satellite must not be sufficient for centromere function and that *centromere* function, therefore, must be sequence independent.

However, this argument seems misleading *sensu stricto; notwithstanding* the effects of epigenetic silencing, an inactive centromere is no less a "*centromere*" than an inactive gene (i.e., as a result of imprinting, X inactivation, or tissue differentiation) is still a "gene."

Neocentromeres

The existence of neocentromeres also provides an argument for the sequence independence of centromere function. Neocentromeres are regions of chromosomes that do not contain typical centromeric DNA, but that have been modified epigenetically to act as a centromere and segregate the chromosome faithfully. Neocentromeres were first described in *maize* and have also been engineered in flies. At least several dozen human neocentromeres are found on marker chromosomes detected in patient material and appear to derive from chromosome breakage events in which a previously acentric fragment acquires *centromere* activity.

Neocentromeres have been extensively characterized to determine the molecular structure and epigenetic modifications responsible for centromere activity. There appear to be *"hotspots"* for neocentromere formation as certain regions of the genome have been involved in neocentromere formation multiple times.

The simplest model that such regions share particular sequence characteristics has been challenged by a finding that three marker chromosomes derived from the same region of chromosome 13 all have different CENP-A *binding* domains. Thus, the regions of the genome from which marker chromosomes are derived may be hotspots for chromosome *breakage* events; however, based on current evidence, the acquisition of centromeric activity by *neocentromeres* is not obviously strictly sequence-dependent. Other *epigenetic* parameters such as centromere protein deposition, *replication timing*, and histone *acetylation* status have been implicated in neocentromere function.

These data clearly establish that α-satellite DNA is not always necessary for centromere function, although it should be emphasized both that *neocentromere* formation is an extremely rare event and that the existence of neocentromeres on abnormal chromosomes need not impugn the now well-established role of α-satellite in determining centromere identity in normal chromosomes.

Usual Centromeres

To complement the analysis of abnormal chromosomes, requirements for human centromere function can be addressed using normal human chromosomes. In one approach, antibodies to centromere proteins known to be present at functional *centromeres* are used to determine which DNA sequences *colocalize* with the active centromere. Such colocalization of centromere proteins and α-satellite DNA has been demonstrated in a number of studies. For example, antibodies to CENP-A bind to a portion of the α-satellite at human centromeres, at the site of the inner *kinetochore*.

Most recently, extended chromatin fiber experiments have *demonstrated* directly that both centromere proteins like CENP-A, as well as *centromere-specific* chromatin modifications, only characterize a portion of the α-satellite at human centromeres. Although, together, these data suggest that only a subset of α-satellite DNA is part of the functional centromere, they do not define the particular type of α-satellite participating in centromere function.

Chromatin immunoprecipitation experiments with antibodies to centromere proteins directly support a functional role for α-satellite DNA and identify the specific genomic DNA sequences involved in centromeric chromatin. Vafa and Sullivan first showed that α-satellite does in fact *immunoprecipitate* with antibodies to CENP-A and proposed a specialized phasing for CENP-A-containing *nucleosomes*.

In another study, antibodies to CENP-B and CENP-C as well as CENP-A were found to immunoprecipitate α–satellite-containing chromatin. After cloning and sequencing the chromatin immunoprecipitated DNA, the only type of α-satellite associated with centromere proteins contained CENP-B boxes.

As CENP-B recognition sites are found only in higher-order α–satellite and not *monomeric* α-satellite, these findings suggest that only higher-order α-satellite is part of the *kinetochore* complex at human centromeres. These two chromatin immunoprecipitation studies are consistent with numerous *cytological* centromere protein colocalization experiments and strongly support a role for higher-order α-satellite in normal centromere function.

Chromosomal Biotechnology

In addition to strategies that examine the DNA and protein composition at normal centromeres of endogenous chromosomes, studies of engineered chromosomes have explored the minimal requirements for centromere function. *Telomere* sequences can be introduced into human cell lines to truncate existing chromosomes into smaller "*minichromosomes*". Such minichromosomes can be mapped subsequently to determine the sequences that are present and responsible for *centromere* function on this minimal chromosome.

In contrast, human artificial chromosomes are derived from naked DNA *transfected* into tissue culture cells. *Artificial* chromosomes may be used as an assay to determine the types of sequences capable of forming *a de novo* centromere and are a *promising* tool for determining the sequence requirements for centromere function.

Minichromosomes

Farr et al. engineered *telomere* truncation chromosomes by introducing telomere repeats in a *nontargeted* fashion to truncate the human X chromosome at a number of locations along the q arm. These chromosomes were further truncated along the p arm to generate centric minichromosomes less than 2.4 Mb in size.

The minimal chromosome that retained mitotic stability was 1.4 Mb overall (compared to the approx 150 Mb normal X) with a 670 kb array of DXZ1 higher-order α-satellite (about a quarter of a typical DXZ1 array). Below this threshold, chromosomes with less DXZ1 or less *flanking* sequence on the p side of DXZ1 were *mitotically* unstable, suggesting that both higher-order α-satellite and neighboring *pericentromeric* sequence may be required for proper centromere function. Similar chromosome truncation studies have been conducted on the human Y chromosome.

The smallest Y chromosome-based minichromosome exhibiting faithful segregation was 1.8 Mb overall, with an approx 100 kb array of DYZ3 higher-order α-satellite. These data from X- and Y-based minichromosomes *demonstrate* that higher-order α-satellite is *capable* of maintaining centromere function after the original chromosome has been significantly *truncated* and/or rearranged.

The fact that the smallest minichromosomes are larger than just the higher-order α-satellite array may reflect a requirement for other *flanking* sequences for proper chromosome segregation, as also suggested by studies in model organisms.

Human Artificial Chromosomes

Human artificial chromosome studies address the requirements for centromere establishment as well as maintenance. Candidate DNA sequences are *transfected* into human tissue culture cells to test them for the ability to form an artificial chromosome with *a de novo* centromere derived from the input DNA.

Numerous artificial chromosome studies in the 7 years since development of this technology have tested α-satellite sequences, non-a-satellite sequences, and different types of α-satellite for centromere competence. The first human *artificial chromosome* study combined the principal components of chromosomes—*centromeres, telomeres,* and *genomic* DNA (presumably containing origins of replication)-to generate small linear artificial chromosomes. In an alternate approach, others engineered a yeast artificial chromosome construct containing α-satellite and telomere sequences on a single molecule.

In both approaches, artificial chromosomes were *mitotically* stable in the absence of drug selection and bound antibodies to centromere proteins, demonstrating the assembly of fully functional human centromeres. Since these two *original* studies, a number of constructs containing well-characterized higher-order α-satellite have been successful in *generating* artificial chromosomes.

Conversely, constructs containing *monomeric* α-satellite, the sequences comprising a neocentromere, as well as other non-α-satellite sequences have failed to form artificial chromosomes, thus clearly implicating higher-order α-satellite as a major genomic determinant of human *centromere* function. Higher-order α-satellite containing CENP-B boxes provides a functional *capability* that is absent from monomeric α-satellite. So what characteristic of higher-order α-satellite is responsible for conferring centromere function?

Is it the extremely *homogeneous* organization of higher-order repeats? Or the presence of CENP-B boxes in higher-order but not monomeric α-satellite? Or are specific basepairs present in higher-order repeats besides the CENP-B box responsible for centromere function? Expanding on the earlier study involving α-satellite from chromosome 21, Ohzeki et al. *generated* a number of constructs to begin to address the specific characteristics of higher-order α-satellite that nucleate centromere function.

A mutation was introduced into the CENP-B boxes in the higher-order repeat unit, *eliminating* the ability to bind CENP-B protein in a gel shift assay. Importantly, the mutated higher-order α-satellite was incapable of artificial *chromosome* formation.

These data suggest that the mutations in this construct are responsible for the absence of centromere function, but it remains to be determined if centromere function was abolished by the mutation specifically in the CENP-B box or if any mutation in higher-order α-satellite might similarly hinder *centromere* function. This remains an important area for future inquiry.

Chapter 2

Evolutionary Trees

This chapter discusses methods for analyzing and describing ancestor-descendant relationships among groups of organisms. These groups may be formalized groupings recognized by biological classification systems (species, genus, sub-family, etc.) or they may be different populations within a species. Recognized groups are called *taxa* (singular: *taxon*). The actual ancestor-descendant *relatiuriships* for a particular set of taxa are called a *phylogeny*.

The actual *phylogeny* ordinarily is not known because ancestral organisms may have become extinct long before modern humans evolved to *observe* them. Therefore, phylogeny is inferred from data derived from organisms alive today or represented in the fossil record. In this chapter, we assume we are using molecular data.

Phylogenetic relationships are of considerable historical and practical interest, and we first offer two specific examples that illustrate what phylogenies are and why they are important. Then we present an informal discussion of how to "*read*" trees before proceeding to quantitative methods for *inferring* them. There is some overlap with our discussion of clustering in enywhere in this book. In that chapter, we employed distances to create *hierarchical* clusters using either *morphological* or sequence data.

In the example of clustering of primates based upon sequence data, we employed the number of sites at which two DNA sequences differ (i.e.. the *Hamming distance*) to create a hier-archical cluster linking *human, chimpanzee, gorilla,* and *gibbon*. How does the construction of *phylogenetic* trees differ from *clustering*? First, ancestor-descendant relationships are explicitly invoked in tree-building. In contrast, points where OTUs join in *hierarchical clustering* need not correspond to *hypothetical ancestral* OTUs.

Second, evolutionary mechanisms suggest that, the distances between two OTUs in a tree should be divided between the two branches leading hack to their common ancestor. With hierarchical clustering, the junction point was mapped to the corresponding distance, and the distance was not divided between the two branches. Finally, with tree construction we *recognize* that evolutionary *processes* may not be *adequately* reflected in the distances that are directly measured, and accordingly appropriate corrections are applied.

Hierarchical clustering employs the distances calculated directly from the measured characters. Despite these differences, there are obvious operational and conceptual connections between *hierarchical clustering* and the building of trees.

Human Immunodeficiency Viruses

Hurnan *immunodeficiency* viruses HIV-1 and HIV-2 are the causative agents for the AIDS pandemic (HIV-1 being more prevalent). Because HIV-1 is a *retrovirus* (replicating by reverse *transcription* of an RNA molecule), it accumulates mutations more rapidly than does human *genomic* DNA.

The sequence variation that results can be revealed by sequencing DNA copies of the retroviral RNA molecules. Populations of the variant HIV-1 strains constitute the groups and sub-types of the virus found among human populations today. For example, group M, subtype B is common among *infected* persons in the United States, while group M, subtype C is more common in Southeast Asia. By examining the sequences of HIV *subtypes*, it is possible to track the pattern of transmission of the viruses around the world.

Figure elsewhere in this chapter shows the relationship between HIV strains and various simian *immunodeficiency* viruses (SIV). It is seen that HIV-1 types are more closely related to SIVcpz than they are to HIV-2, which itself is more closely related to SIVs isolated from the sooty *mangabey* and the *stump-tailed macaque*.

From these and similar data we can make the reasonable inference that the *ancestors* of the HIVs present today entered the human population by *transmission* of simian *immunodeficiency* viruses found in populations of *chimpanzees* and other primates. It is thus possible to infer the origins and date of inception of the pandemic (Hahn, 2000).

Finding Inter Relationships

We humans are keenly interested in our own origins. Although people from different regions of the world may differ substantially in appearance (e.g., a *Masai* from Africa compared with a Dane from Northern Europe), we all be-long to the same species. How and where did the various human populations originate? This question can be approached by *sequencing* DNA from individ-uals belonging to many human populations from all over the globe.

Sometimes chromosomal DNA (particularly X or Y chromosomal DNA) is employed, and sometimes *mitochondrial* DNA (mtDNA) is used. Mitchondrial DNA of humans is a circular molecule containing some 16,500 bp. Over time, this DNA accumulates mutations (base changes) within its genome, and these mutations are passed on to descendants by the mother.

(Mitochondrial DNA is *maternally* inherited.) Populations of humans that share particular sets of mutations are understood to be more closely related to each other than they are to populations lacking these mutations. One inferred set of relationships based upon mtDNA sequences is shown in Figure 2.2.

This tree suggests, for example, that native Australians are more closely related to East Asians (e.g., Japanese) than they are to the major groups of Africans. Of particular interest is the observation that populations from the rest of the world are a subset of African populations, *supporting* the anthropological hypothesis that modern humans in other parts of the world are

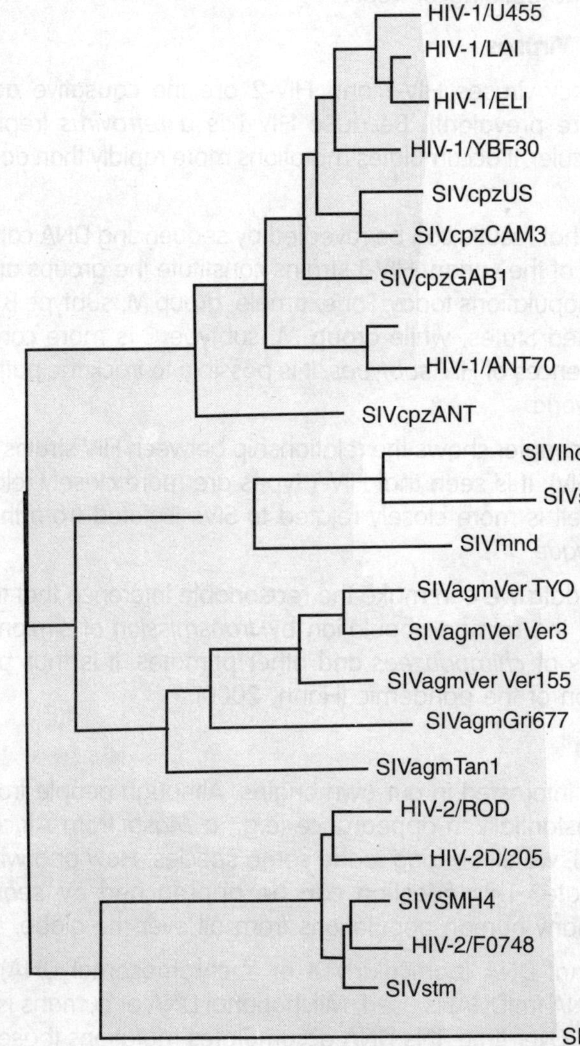

Figure 2.1: Examples of phylogenetic trees. Inferred relationships between human immunodeficiency virus (HIV) and simian immunodeficiency virus (SIV) strains isolated from humans and other primates: agm, African green monkey; cpz, chimpanzee; 1hoest, L'Hoest monkey; mnd, mandrill; sm, sooty mangabey; stm, stump-tailed macaque; sun, sun monkey; syk, Sykes monkey. The tree was constructed from viral sequence data but may be confounded by recombination between different virus strains.

descendants of migrants that originated from an African population (the "Out-of-Africa" hypothesis). Also noteworthy is the diversity of African populations. For example, the *Kikuyu* people differ more from the Effik than Europeans differ from Papua New Guinean *highlanders*.

Understanding the Evolutionary Flow

In both examples above, we stated conclusions based upon the phylogenetic trees presented in Figure's elsewhere in this chapter . Here we will be more explicit about how to interpret the *iconography*. We have already had some preparation for this because of the material presented in Chapter anywhere in this book.

The tree is seen to be a branching structure generated by *successive splittings* of prior branches. Ideally, the branches are formed by successive bifurcations (splits of one branch into two), but sometimes the branching order of several taxa cannot be determined, and in that case there are multiple branches *emanating* from a single branch. Eventually, branches end at tips (or "leaves"), and these represent extant taxa (e.g., strains of HIV or *contempo-rary* populations of humans).

We can trace back along branches leading to two different taxa (taxon A and taxon B, for instance) until these branches join. This junction represents the most recent common *ancestor*, C. The branch lengths connecting A and C are proportional to the distance between A and C (e.g., using a distance metric defined in other chapter). Under ideal *circumstances*, the distance between A and B is the sum of distances AC and BC.

For the examples shown in Figure's elsewhere in this chapter distances are depicted as horizontal lines: the vertical lines joining the *horizontal* ones are just connec-tors, and the lengths of the vertical connections are not related to distance. A set of all taxa derived from a particular common ancestor is called a *clade*, and the process of *branching* is sometimes called *cladogenesis*. A tree that depicts the order of branching of taxa without regard to the distances between them is called a *cladogram*.

To construct a tree, we often must identify the ancestral state of the characters employed (e.g., the ancestral DNA sequence). To do this, we employ an *outgroup*, which is a taxon that is clearly more distantly related to the *taxa* of interest than any of them is to another of these taxa. In Figure elswhere in this chapter the outgroup is *chimpanzee*.

If we look at the branches of humans at the bottom, we see results for two San individuals, preceded above by branches for two Mbuti, and then above those (skipping the Hausa) we see the branch for a Kikuyu individual. Using the convention described above, we can see that the distances separating the two San is less than the distance separating either of the San from either of the Mbuti, indicating (as expected) that the San are more closely related to each other than either is to either of the Mbuti.

The depiction of the data should be distinguished from the quantitative aspects, which relate to branch lengths and branching topology. The top-to-bottom ordering of the branches in this representation is chosen for convenience and has no quantitative *significance*. If we transposed the top-to-bottom order of the branches leading to the San and the Mbuti+Hausa, the tree would have the same meaning.

Similarly, the chimpanzee branch could be "*flipped*" up to the top, to lie adjacent to the non-Africans, with the same meaning as before. If we look at the lower African branches, we notice

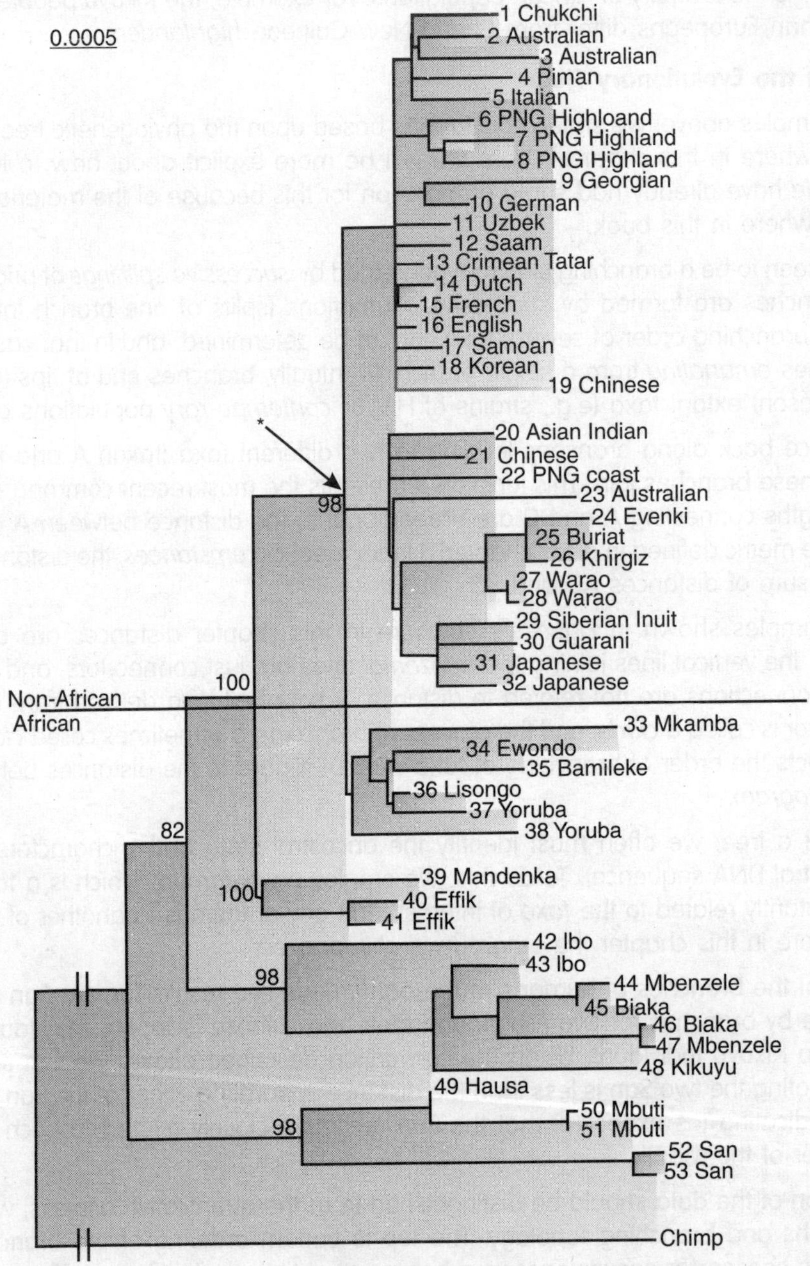

Figure 2.2: [This figure also appears in the color insert.] Phylogenetic relationships among representatives of human populations based upon complete mtDNA sequence data (D loop region excluded). The scale is in units of the number of nucleotide differences per site.

that branches emanate from longer branches, whereas *non-Africans* are related to each other through *shorter branches*. This implies that African populations *diverged* from each other earlier, which could produce greater genetic *diversity* among them. More sampling of African populations is needed to explore this point.

Trees help us to interpret the data. For example, notice the pat-tern of the ten taxa at the top of Figure elsewhere in this chapter. The first four branches of HIV-1 isolates (HIV-1/U455, /LA1, /ELI, /YBF30) emanate from one side of a *bifurcation*, the other of which contains SIVcpzUS and SIVcpzCAM3. This set of six viruses represents one of two branches, the other of which contains SIVcpzGABl. The four HIV-1 strains mentioned are said to form a *sister group* of SIVcpzUS and SIVcpzCAM3.

These six together form a sister group of SIVcpzGAB1, and by parsimony type arguments (see below) they share a common ancestor that was an SIVcpz. The appearance of HIV-1 strains among branches that are otherwise SIVcpz suggests that the first four HIV-1 strains *mentioned* are derived from SIVcpz as a result of a *cross-species* transfer event.

Considering the next connection back, we find two HIV-1 strains forming a sister group with respect to the first seven *mentioned* above and derived from an *ancestor* shared by SIVcpzANT.

Again, we can argue that this *ancestor* was an SIVcpz. Notice that there are two clades of HIV-1 within a larger Glade that is otherwise SIVcpz: the Glade formed from HIV-1 strains U455, LAI, ELI, and YBF30, and the Glade formed from strains MVP5180 and ANT70. This suggests that there were at least two *interspecies* transfer events involved in generating the first ten viruses.

Trees such as those shown in Figure's elswhere in this chapter are inferences based upon a particular data set and *employing* one of several possible methods of tree construction. Such trees may or may not represent the actual *phylogeny* or *evolutionary* tree.

There are underlying assumptions used in the construction of any tree, as will become clear in the following sections. It is possible to pour data thoughtlessly into a tree-building software application and get a *"result"* at the end. If the data are poor or represent inappropriate sampling, or the assumptions for building the tree don't match the data, then the result may be *misleading* or *worthless*.

NAMING THE GRAPHS

In the previous section, we discussed two particular examples of trees to illustrate why they are important. In that context, we have *already* introduced some *nomenclature*. Now we formalize the presentation. A tree is a particular kind of *graph*, which is defined as a set of *vertices* connected by *edges*.

Each *vertex* has a *degree*, which is the number of edges that emanate from that vertex. A *directed graph* has a defined ordering between *vertices* connected by one or more edges. A cycle in a graph is a collection $v_1, \ldots, v_r$ of vertices with edges between v_1 and v_2, v_2 and v_3, ..., v_{r-1} and v_r, and finally v_r and v_1. Trees are cycle-free graphs whose vertices correspond to current or ancestral species or populations. *Vertices* immediately below a vertex *v* and connected to it by edges are called the *children* of *v*. Similarly, a vertex *u* immediately above *v* and connected to it by an edge is called the *parent* of *v*.

Biological Entities

Trees relate species or other biological entities to each other by invoking *ancestor-descendant* relationships. In this discussion, we consider *relation-ships* between species, but remember that the related objects could be gene sequences or *populations* within species. The *observed* species (corresponding to the data) appear at the tips of the branches, and these are sometimes called *leaves*.

Leaves are vertices of degree 1. Vertices in the tree where leaves or branches join are also called internal *nodes*. For molecular sequences, the actual data correspond to the terminal vertices or leaves. Internal nodes correspond to ancestral species that are not part of the data. Ancestors *temporally* precede their descendants, and we can sometimes *infer* likely character states of these *ancestral* species. However, these states are not usually directly observed.

Binary trees are ones for which each internal node has two children. Internal nodes are connected by internal branches, and leaves are connected to the rest of the tree by external branches *emanating* from an *internal* node. The lengths of the branches connecting leaves to nodes and nodes to nodes correspond to distances between them. The trees discussed here are all *binary*, or *bifurcating*.

A tree is said to be rooted (or have a *root*) if there is a single ancestral node from which all other nodes descend, and in this case the root node is connected to two branches. Trees may be either rooted or unrooted, as illustrated in Figure elsewhere in this chapter. If the tree is rooted, direction is defined by the *evolutionary* time-scale, and we usually associate increasing time with the *downward* or *rightward* direction.

If there is no root, all we can say is that the leaves correspond to the ultimate descendants of some ancestor, but we do not know whether the internal node *adjacent* to any leaf is its ancestor, nor do we know the *ancestral* relationships of the *internal* nodes. Put another way, we don't know which direction along the edges corresponds to increasing time.

In the unrooted tree shown in Figure elsewhere in this chapter, OTU 7 may be an ancestor of 8, corre-sponding to placement of the root as indicated by arrow R1, or 8 might be an ancestor of 7, as would be the case if the root were placed as indicated by arrow R2. Once the root is placed, the *ancestor-descendant* orientation of all edges is established.

Figure elsewhere in this chapter shows the two different rooted trees corresponding to the two different placements of the root in Figure elsewhere in this chapter. We have so far been representing trees graphically, and we will continue to do so. However, note that there are other *representations*. For instance, if we construct a tree with branches like those illustrated in Figure elsewhere in this chapter, that tree could be represented as (((Ho,Pa),Go),Hy).

Methods of Constructing Trees

Before proceeding to a discussion of methods for constructing trees, we need to be aware of the magnitude of the problem. What we seek is the tree or collection of trees that best represents the ancestor-*descendant* relationships implicit in the data. We might *naively* imagine that we could draw all possible trees and see how well each fits the data. The *unsuitability* of this approach is evident if we calculate the possible trees that can be drawn relating *n* different species.

We first consider unrooted trees. The simplest unrooted tree is shown below. It connects three species, has a single internal node, and contains no internal branches. The number of internal branches is thus $n - 3 = 0$. Now consider the case $n = 4$ by adding one leaf or species (connected between the internal node and 1, for example), as shown in the diagram below.

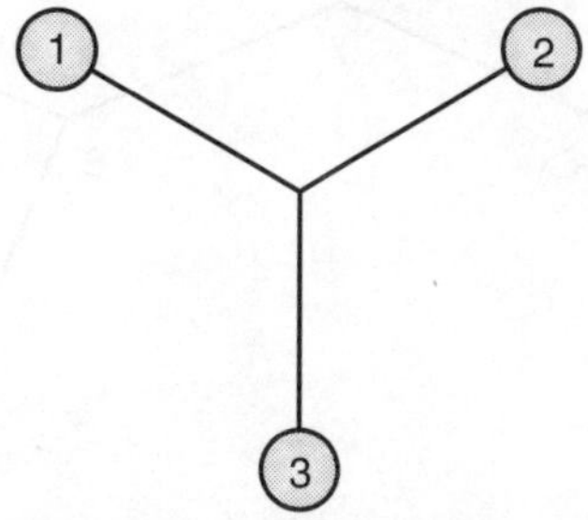

There is now one additional leaf and one *additional* node; there is one internal branch (the thick line). The number of *internal* branches is now $n - 3 = 4 - 3 = 1$. Repeating this step $n - 4$ additional times results in a tree with n leaves and $n - 4$ additional *internal* branches for a total of $n - 3$ *internal* branches.

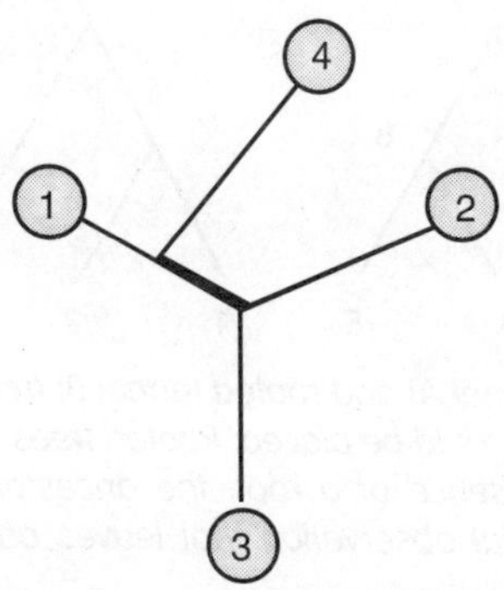

The total number of branches is therefore the sum of *n* external branches and *n – 3* internal branches. There are thus *2n – 3* branches in an unrooted tree with *n* leaves. We now find the number b_n of uprooted trees having '*n*, leaves; clearly $b_3 = 1$. We calculate this number by first asking how many ways there are to connect the last *leaf*; it is just the total number of branches for $n - 1$ leaves.

Then we ask how many ways there are to add branch $n - 1$ (which is the number of *branches* in a tree with $n - 2$ leaves) and so on until we reach $n = 1$. The total number of possible trees is the product of the numbers of ways of adding each *successive* branch. The number of ways of *adding* the last branch is

$$b_n = \left[2(n-1)-3\right]b_{n-1} = (2n-5)b_{n-1},\ n = 3,4,\dots.$$

The number of ways of adding the second-to-last branch is

$$b_{n-1} = \left[2(n-2)-3\right]b_{n-2} = (2n-7)b_{n-2},\ n = 4,5,\dots.$$

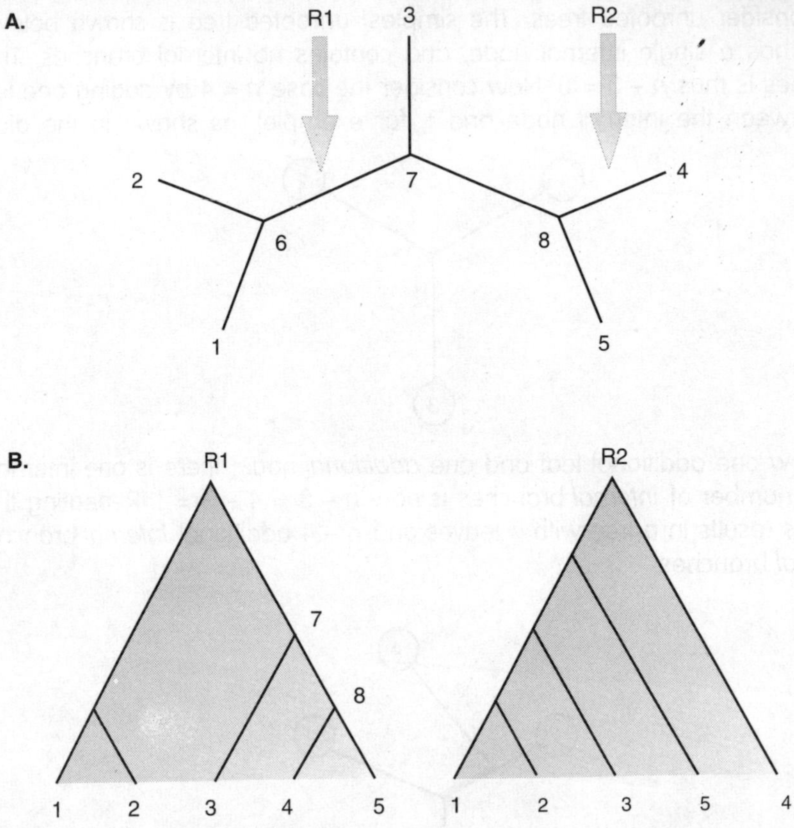

Figure 2.3: Distinctions between unrooted (panel A) and rooted (panel B) trees. In panel A, R1 and R2 represent two of the seven possible edges in which the root could be placed. Rooted trees corresponding to these two placements are shown in panel B. Note that in the absence of a root, the ancestor-descendant relationships among the vertices are not established beyond the trivial observation that leaves are not ancestors. R1 and R2 imply very different ancestors for most taxa.

The terms may now be calculated successively until reaching $b_3 = 1$. Hence

$$\begin{aligned} b_n &= (2n-5)\times(2n-7)\times\cdots\times 3\times 1 \\ &= \frac{(2n-5)!}{(n-3)!2^{n-3}}, n = 3, 4, \ldots. \end{aligned} \tag{2.1}$$

The proof of the last equality appears in the exercises.

The number of rooted trees with n leaves is closely related to the number of unrooted trees. All that is required is the *multiplicative* term for the number of ways of placing the root on one of the branches. That number is equal to the number of *branches,* which we saw above was $2n - 3$. Thus the number of rooted trees b'_n for n leaves is

$$\begin{aligned} b'_n &= (2n-3)\,b_n \\ &= \frac{(2n-3)!}{(n-3)!2^{n-2}}, n = 3, 4, \ldots. \end{aligned}$$

Now we are ready to calculate how many unrooted trees there are for several values of *n*. The result is shown in Table elsewhere in this chapter. For the example in Figure elsewhere in this chapter, there were 24 leaves on the *unrooted* tree. The number of possible trees in that case is then approximately 5.64×10^{26}. With the *naive* approach, we would have to search through all these trees to find the "*best*" one; this is currently a *computationally impossible* task.

BASIC METHODS

There are three basic methods for building trees from molecular data: *parsimony* methods, distance methods, and likelihood-based methods. In this section, we briefly introduce *parsimony* and distance methods.

Table 2.1: Numbers of possible unrooted trees b_n corresponding to different *numbers* of taxa or leaves, *n*.

n	3	4	5	6	7	8	9	10
b_n	1	3	15	105	954	10,395	135,135	2,027,025

Parsimony Methods

Finding a tree relating *n* species or sequences by parsimony employs the idea that the tree requiring the least number of mutations to relate those sequences is the preferred one. This approach is a specific application of Ockham's Razor: "*Plurality* should not be posited without *necessity*." We illustrate this with an example. The method tests *hypothetical* trees and then calculates the number of base changes needed to yield the observed data.

Suppose the observed data are:

Species:	***Site:***
	1 2 3 4 5
1	A C T T T
2	A C A T T
3	A A C G T
4	A A T G T
5	A A T T T

Expressions in the last section indicate that there are 105 possible rooted trees. We illustrate two of these alternative trees and evaluate which one is most *parsimonious*. First, we examine the tree (((3,4),5),(1,2)) with root at 9, diagrammed in Figure elsewhere in this chapter. What is the smallest number of changes required to produce the five sequences according to the species *relationships* implied by this particular tree? We have drawn a set of four mutations that explains the data given this tree. (We illustrate how to obtain these particular *mutations* a bit later.) Now let's consider the tree ((((1,2),3),4,),5) *diagrammed* in Figure 2.4.

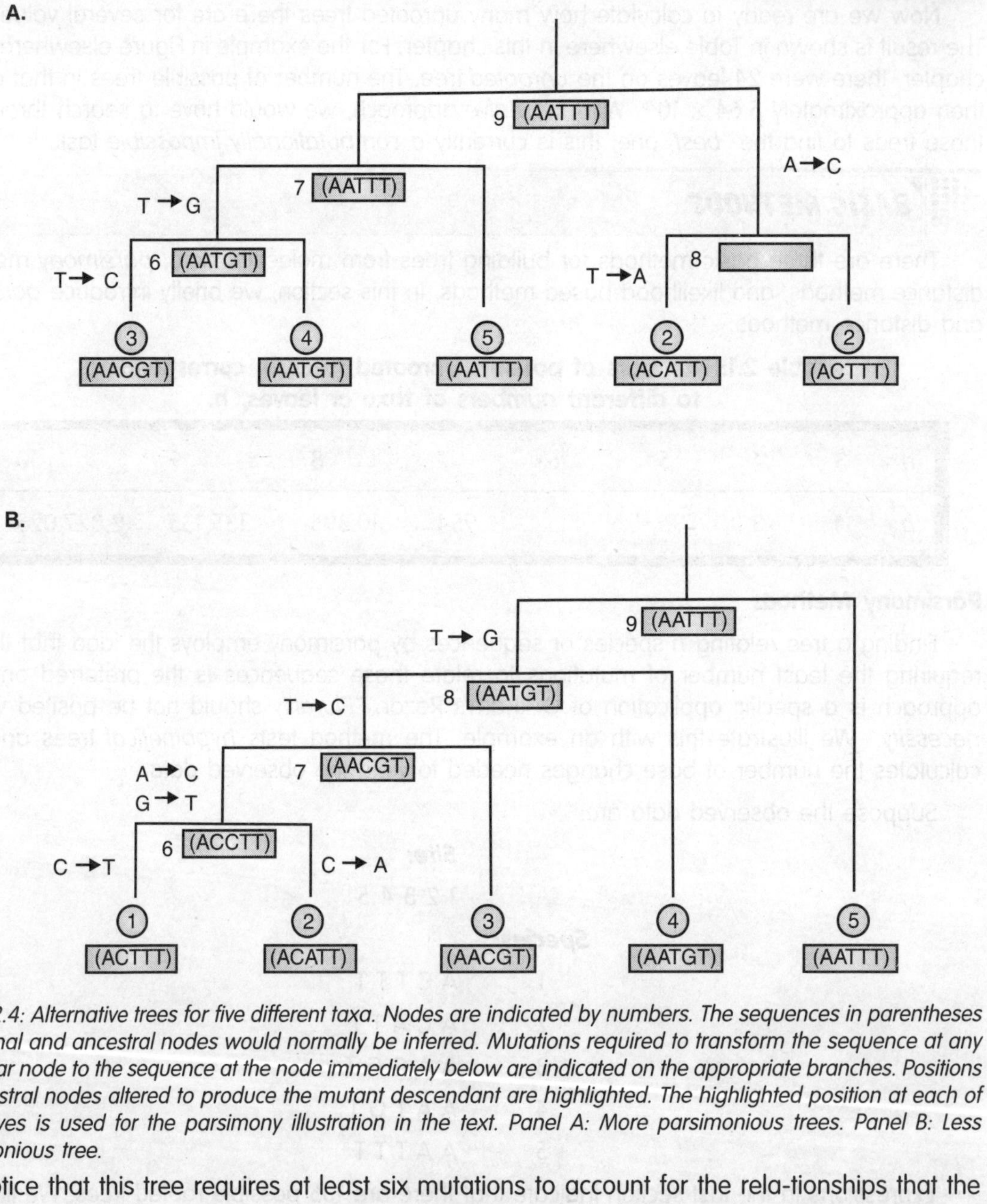

Figure 2.4: Alternative trees for five different taxa. Nodes are indicated by numbers. The sequences in parentheses at internal and ancestral nodes would normally be inferred. Mutations required to transform the sequence at any particular node to the sequence at the node immediately below are indicated on the appropriate branches. Positions at ancestral nodes altered to produce the mutant descendant are highlighted. The highlighted position at each of the leaves is used for the parsimony illustration in the text. Panel A: More parsimonious trees. Panel B: Less parsimonious tree.

Notice that this tree requires at least six mutations to account for the rela-tionships that the tree *implies*. Since the second tree requires a greater number of changes than does the first, it is less parsimonious. We therefore prefer the first tree. Is this tree the correct one? We actually don't know.

All we can say is that the probability of mutation at any position is usually very small over relatively short timescales (e.g., 1 to 100 million years) and that the *probability* of six independent *mutations* is less than the *probability* of four mutations. But how did we decide what the mutations should be and what the sequences at the *ancestral* nodes might have been? We focus on one position, position 3, and determine the smallest number of changes needed at that position to be *consistent* with the tree in Figure elsewhere in this chapter. (Bases at position 3 for the five leaves are indicated in *boldface* type.)

Start for example, with the join, corresponding to the letters {C, T}. The parent node, labeled 6, can therefore be either C or T. Next consider joining species 5, which has a T at that site, at the node labeled 7. The most parsimonious assignment is to put T at node 7. If we were to put a C in position 3 at node 7, we would then need to add another mutation leading to 5, which is less *parsimonious*. Now consider joining species 1 and 2 at node 8.

The sites are A, T, so that at node 8 we must have either A or T. Thus node 8 is {A, T} and node 7 is T. Thus the most parsimonious assignment is to put the T at node 8. This allows us to read off the assignment at each internal node: T (8), T (7), T (6). The resulting assignments of bases at position 3 imply two base changes over the tree. The same method is employed for the five other positions to determine the total number of changes for the tree as a whole.

We now *formalize* the approach a little. It is *convenient* to think of the tree as rooted, as for example in Figure 2.4 A. The root node is labeled $2n - 1$, which in Figure 2.4 A is $2 \times 5 - 1 = 9$. We are going to work our way up the tree in the following way. Let F_i denote the set of possible base *assignments* at a node *labeled i*, and let L_i denote the number of changes that have accumulated up to that node. We start with $L_i = 0$ if i is a leaf. We want to obtain values (F_3, L_3) at the next node up as a result of combining two nodes with values (F_1, L_2) and (F_2, L_2). The rules for doing this are as follows:

1. If $F_1 \cap F_2 = \phi$, then $L_3 = L_1 + L_2 + 1$, $F_3 = F_1 \cup F_2$.
2. If $F_1 \cap F_2 = \phi$, then $L_3 = L_1 + L_2$, $F_3 = F_1 \cap F_2$.

Reminder: For sets *A* and *B*,

A ∩ *B*, read as "*A* intersect *B*," denotes the elements in *A* and *B* that appear both in *A* and in *B*.

A ∪ *B*, read as "*A* union *B*," denotes the set of elements that appear in *A* or *B* or both.

ϕ, called the "empty set," is a set that contains no elements at all.

Let's try this out at position 4 using the tree in Figure 2.4 A. We start with $(F_3, L_3) = (\{G\}, 0)$ and $(F_4, L_4) = (\{G\}, 0)$, so that $(F_6, L_6) = (\{G\}, 0)$ as a result of applying rule 2. Moving onto node 7, we have $(F5, L_5) = (\{T\}, 0)$, so that $(F_7, L_7) = (\{G, T\}, 1)$ as a result of applying rule 1 for nodes 6 and 7. Next, since $(F1, L_1) = (\{T\}, 0)$ and $(F_2, L_2) = (\{T\}, 0)$, we have $(F_8, L_8) = (\{T\}, 0)$ (rule 2). To complete the assignments at node 9, we note that $(F_7, L_7) = (\{G, T\}, 1)$ and $(F_8, L_8) = (\{T\}, 0)$, which by rule 2 yields $(F_9, L_9) = (\{T\}, 1)$.

Now that we are at the top, we can read off the number of changes required in the tree from $L_9 = 1$. We have explored all possible paths to the root, and we then backtrack down the tree to construct the *assignments* to the *ancestral* states and the base changes that have occurred at each step. At position 4, the change was $T \rightarrow G$ after node 7, *leading* to node 6. We can perform this *calculation* for each site in the sequences related by the tree.

The total parsimony cost for a tree is the sum of the *parsimony* scores for all the sites. The method in its simplest form is to compute the parsimony cost of all possible trees, and choose the minimum-cost tree. *Reconstructing* the collection of best trees by this approach can be hard though!). For larger trees, a variety of heuristic search methods are used to attempt to identify the best ones without *examining* all of the trees. There are also versions of weighted parsimony in which different weights are given to different types of *substitutions*.

Distance Methods

Assume for the moment that we have a set of pairwise distances linking *n* sequences, and write d_{ij} for the distance between sequence *i* and sequence *j* . Let *D* be the distance matrix whose elements are d_{ij}. This is formally the same type of data employed for the *hierarchical clustering* described in previous Chapter in this book. In the evolutionary context, the distances meet the three criteria already listed.

Now that we are interpreting the data in the context of a tree, we need to recognize that biological *mechanisms* may impose additional conditions on the distances under particular circumstances.

These are described below. We call a tree *additive* if the distance between any pair of leaves is the sum of the distances between those leaves and the first node that they share in the tree. Furthermore, a rooted *additive tree* is called *ultrametric* (or clock-like) if the distances between any two leaves and their common ancestor are equal.

An example of an *ultrametric tree* is shown in Figure 2.5. There is an interesting and important biological concept related to the idea of *ultrametric* trees. In actuality, *mutations* may not occur at the same rate in different *branches* of the tree.

For example, mutations accumulate more slowly among primates than among *rodents*. If the ticking of the molecular clock is not at the same rate on two different branches, then the branch lengths for a given amount of time since the last common ancestor will differ.

With ultrametric trees, however, *clocks* "tick" at the same rate on both branches *emanating* from a *bifurcation*.

One natural question is: Given a distance matrix *D,* can we determine whether we can construct an additive or ultrametric tree corresponding to those distances? The answer is contained in the following result. Let *D* be the distance matrix.

(i) Three-point condition for ultrametric trees: *D* corresponds to an ultrametric tree if and only if for any three sequences *i, j, k,* the distances satisfy $d_{ij} \leq \max(d_{ik}, d_{kj})$.
(ii) Four-point condition for additive trees: *D* corresponds to an additive tree if and only if for any four sequences (labeled here 1, 2, 3, 4) two of the sums $d_{12} + d_{34}$, $d_{13} + d_{24}$, $d_{14} + d_{23}$ are equal and greater than or equal to the third.

The interpretation of part (ii) becomes evident when we sketch out trees with three or four leaves and examine the relevant sums. Note that if the tree is ultrametric, part (i) states that two distances must be the same and the third must be smaller. This is shown algebraically and graphically in Figure's 2.6 A and 2.6 B.

The hierarchical clustering method discussed in previous chapter is one way to calculate an *ultrametric* tree from a set of distances. When the distance between two clusters is defined by

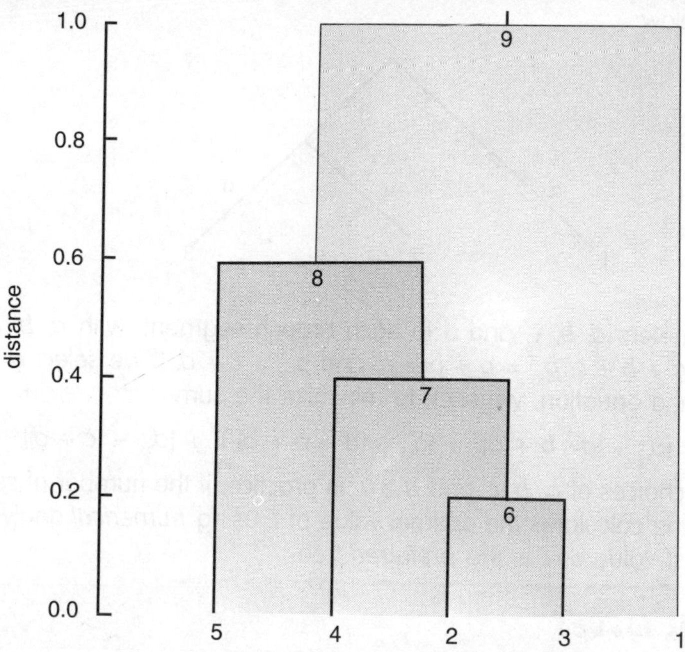

Figure 2.5: Example of an ultrametric tree. The distance from the leaves to node 6 = 0.2, to node 7 = 0.4, to node 8 = 0.6, and to node 9 = 1. Note the equal lengths of the branches from any node to all leaves derived from that node. This is a distinguishing feature of ultrametric trees.

group average linkage, the method produces what is called the UPGMA tree (UPGMA meaning "*unweighted* pair group method with *arithmetic* means").

Finding an additive tree is also an *interesting* problem, but in practice the distances estimated from molecular sequence data do not satisfy the conditions of the *theorem* and we are left with more work to do.

There are several approaches. For an approximately additive tree, a popular method uses Saitou and Nei's *neighbour* joining method, *implemented* for example in the software package MEGA.

Another approach uses weighted least squares to fit a best additive tree for which the distances in any tree to be evaluated are denoted by parameters p_{ij} and the observed pairwise distances d_{ij} as before. We find the *assignment* of sequences or species to nodes that minimizes the function E given by

$$E = \sum_i \sum_j w_{ij} \left| d_{ij} - p_{ij} \right|^{\alpha},$$

where α typically has the value 1 or 2 and the w_{ij} are weights. Choices of weights such as w_{ij} = 1, $d2_{ij}^{-1}$, or d^2_{ij} . are typical. This also allows for missing data by putting w_{ij} = 0 if there is no distance for pair (*i, j*).

An example serves to illustrate the rationale behind this approach. Consider the tree for three taxa shown below:

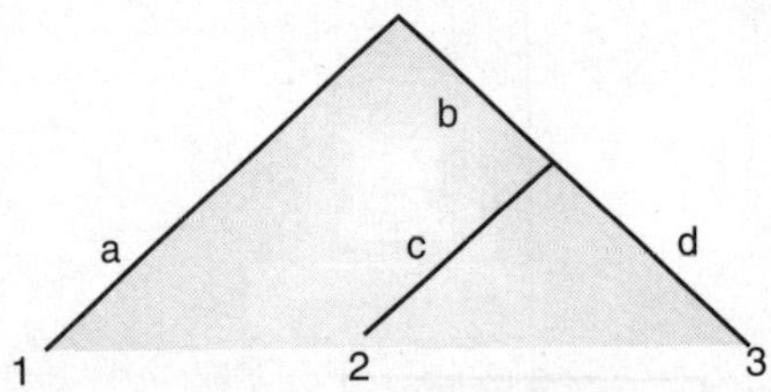

We assign parameters *a, b, c,* and d to each branch segment, with $a, b, c, d \geq 0$. For this particular tree, $p_{12} = a + b + c$, $p_{13} = a + b + d$, and $p_{23} = c + d$. If we select $\alpha = 2$ and let $w_{ij} = 1$, then according to the equation, we seek to minimize the sum

$$E = [d_{12} - (a + b + c)]^2 + [d_{13} - (a + b + d)]^2 + [d_{23} - (c + d)]^2$$

with the appropriate choices of $a, b, c,$ and $d \geq 0$. In practice, if the number of sequences or taxa is sufficiently small, one calculates the optimal value of *E* using *numerical* analysis methods. The tree having the lowest value of *E* is the *preferred* tree.

CONCEPTUAL ISSUES

There are two additional *conceptual* issues to be addressed before we proceed further. The first concept is modeling the *evolutionary* process, and the second is using this model to obtain the distance matrix, given that this evolutionary process is operating. It is indicated that a difference between hierarchical *clustering* and *construction* of trees is the inclusion of evolutionary models in the latter case.

Before proceeding to the actual construction of trees, we need to introduce a quantitative description of the evolutionary processes that influence the distances used for tree *construction*. An important aspect of this is the *introduction* of time and its relationship to distances. We use a simple *stochastic* model for the evolution of a DNA sequence through time.

Much of the *requisite* mathematical framework has already been presented in other chapters in this book. For *simplicity*, we assume that changes in the sequences occur through base substitutions only. Suppose that we have two aligned DNA sequences (in which any sites with gaps have been *removed*).

How can we measure the distance between the two sequences? We have already used the Hamming distance, calculated as the proportion of sites at which the two sequences differ. This measure is not *sufficient* for many evolutionary studies because it does not allow for the possibility of repeated substitutions at the same site.

For example, at a given position in the sequence, there might be an initial A $\rightarrow$ G transition that later reverts by *a* G $\rightarrow$ A transition, restoring the original state at this position. Because of this *phenomenon*, we might *underestimate* the true number of substitutions between our sequences. We treat the substitution process as a *stochastic* process and define the distance between the two sequences, measured on the path in a tree *separating* them, as the expected number of substitutions that change the base at that site.

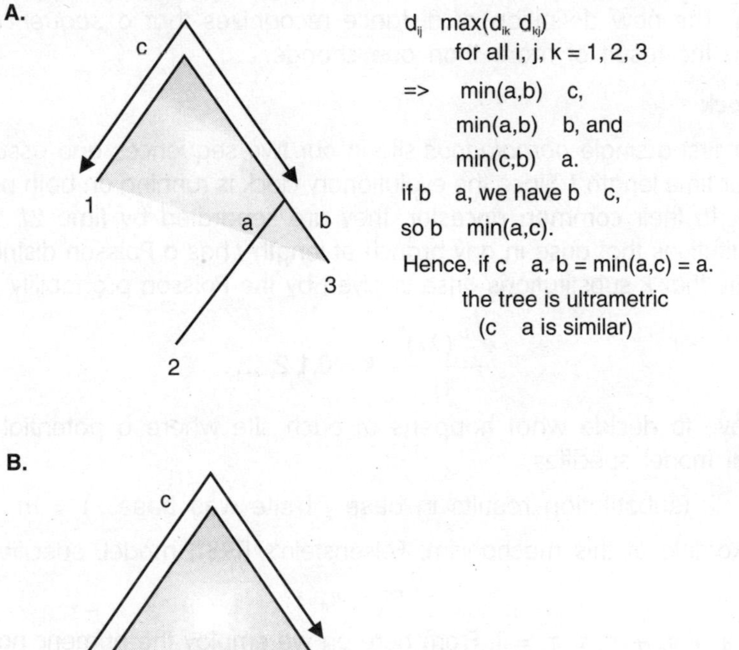

Figure 2.6: Illustrations of distance constraints imposed by the ultrametric requirement (panels A and B) and additive distances (panel C). Panel A: An arbitrary relationship between three taxa 1, 2, and 3 is shown with distances from the common node indicated. Assuming the three conditions shows the tree is ultrametric. Panel B: With taxa at the leaves of a rooted tree, the ultrametric condition $(a + b) \leq (a + c) = (b + c)$ can be observed directly. Panel C: Illustration of the four-point condition for additive trees (or a portion of a larger additive tree). The dashed line indicates a potential connection to a larger tree. As indicated earlier, two of the summed distances are equal to each other and larger than the third summed distance.

Note that the *Hamming* distance counts a difference between two sequences at a position as one change. The new definition of distance recognizes that a *sequence* difference at a position may be the result of more than one change.

Evolutionary Clock

We consider first a single *homologous* site in our two sequences and assume the two sites have *diverged* for time length *t*. Since the evolutionary clock is running on both paths from the two sequences back to their common ancestor, they are separated by time 2*t*. Suppose that the number of substitutions that arise in any branch of length *t* has a Poisson distribution with mean λt; the probability that *k* substitutions arise is given by the Poisson probability

$$\frac{e^{-\lambda l}(\lambda t)^k}{k!}, k = 0, 1, 2, \dots. \tag{2.2}$$

We then have to decide what happens at each site where a potential substitution can occur. A general model specifies

$$\mathrm{P}\ (\text{substitution results in base } j \mid \text{site was base } i) = m_{ij}.$$

A simple example of this mechanism, Felsenstein's (1981) model, specifies

$$m_{ij} - \pi_j, \tag{2.3}$$

with $\pi_i \geq 0$ and $\pi_1 + \pi_2 + \pi_3 + \pi_4 = 1$. From here on we employ the numeric notation introduced previously: 1, 2, 3, and 4 correspond to bases *A, G, C,* and *T*, respectively. This implies that the *substitution* that appears at a position in the *sequence* does not depend on the type of the base at that position when the substitution occurred. We also assume that the set of *probabilities* p*j is the same at every position in the sequence.

Next we calculate the probability q_{ij} (*t*) that a base that was *i* at time 0 has mutated to base *j* a time *t* later. For the model in (2.3), this is easy to do. We calculate the probability by *conditioning* on whether or not any mutations occurred in time *t*. Suppose first that $i = j$. If there were no *mutations* on that branch (probability $e^{-\lambda t}$), then the base time *t* later must still be *j*.

On the other hand, if there is at least one mutation, then the chance that the resulting base is *j* is just π_j; this follows because the mutation *mechanism* does not care about the type at a site when a mutation occurs. Summing over the two possibilities gives

$$q_{ij}\ (t) = e^{-\lambda t} + (1 - e^{-\lambda t})\pi_j \tag{2.4}$$

When $i \neq j$, the same argument shows that

$$q_{ij}(t) = (1 - e^{-\lambda t})\pi_j. \tag{2.5}$$

We need to assume something about the base that is the type of the most recent common ancestor of the two sites we are *considering*. The usual assumption is to make the evolution of base frequencies along a single branch *stationary*. By stationary we mean that the distribution of base frequencies is the same for every time *t*. Thus, if π^0_i denotes the chance that the *ancestral* base is of type *j*, then we want

$$\mathrm{P}\ (\text{base a time } t \text{ later} = j) = \pi^0_j$$

for any time *t*. Finding the *probability* distribution that makes this last equa-tion true requires some linear *algebra*. It turns out that we have to solve the *equations*

$$\pi_j^0 = \sum_i \pi_i^0 m_{ij} \tag{2.6}$$

subject to $\sum_i \pi_i^0 = 1$. It can be checked that for the model (2.3),

$$\pi_j^0 = \pi_j,\ j = 1, 2, 3, 4. \tag{2.7}$$

If the π^0_i do satisfy (2.6), then it can be shown that for any $t > 0$

$$\pi_j^o = \sum_i \pi_i^0 q_{ij}(t). \tag{2.8}$$

so that, indeed, the base frequencies are not evolving with time.

Mutation Mechanism

For building the tree, we usually employ distances that take into account the mutation *mechanism.* Here we present the mathematical background for *determining* the distance K for any position. The set of K_{ij} for a set of taxa or sequences forms the elements of the distance matrix, analogous to those distance *matrix* elements employed for *clustering* in previous Chapter in this book.

Site Substitutions

We know that an average of λt substitutions occur at a particular site on a branch of length t. However some of these substitutions result in a given base being *"replaced"* by the same base. In terms of the end product, this would not be detected as a substitution at all. Now consider a particular substitution. Because of the stationarity assumption, the base at this time is i with chance π_i. In the model in (2.3), the chance that a base of type i changes is just $1 - \pi_i$. Hence the chance that a mutation results in a change of base is

$$H = \sum_{i=1}^{4} \pi_i (1 - \pi_i) = 1 - \sum_{i=1}^{4} \pi_1^2, \tag{2.9}$$

and the average number of real substitutions in time t *is* therefore (average number of *mutations*) × (*proportion* that result in change) = $(\lambda t)H$. In the general case with *stationary ancestral frequencies,* this number can be calculated as

$$\lambda t \sum_i \pi_i (1 - m_{ii}).$$

If we assume the same mutation mechanism applies in the path from the common ancestor to each of its descendants then the expected number of real *substitutions* between the two present-day bases for the model in (2.3) is

$$K = 2\lambda t H. \tag{2.10}$$

We use K as our measure of distance.

Estimating *K*

We want to estimate K from the sequence data. To do this, we first have to calculate the chance that we observe an i in one species and a j in the other when they have diverged for time t. We denote this probability by $F_{ij}(t)$. By averaging over the possible ancestral *nucleotides*, we get

$$F_{ij}(t) = \sum_t \pi_l q_{li}(t) q_{lj}(t), \tag{2.11}$$

assuming that the two sites evolved independently of each other after splitting from their common ancestor.

The result in (2.11) can be simplified if the mutation process is *reversible* with respect to the stationary distribution π. This means that the *detailed balance equations*

$$\pi_i m_{ij} = \pi_j m_{ji} \text{ for all } i \neq j$$

hold. From this it can be shown that

$$\pi_i q_{ij}(t) = p_j q_{ji}(t) \text{ for all } i, j \text{ and } t > 0. \tag{2.12}$$

It is easy to check that the model in (2.3) is reversible. When the model is indeed reversible the result in (2.11) can be simplified, since

$$\begin{aligned} F_{ij}(t) &= \sum_l [\pi_l q_{li}(t)] q_{lj}(t) = \sum_l [\pi_i q_{il}(t)] q_{lj}(t) \\ &= \pi_i \sum_l q_{il}(t) q_{lj}(t) = \pi_i q_{ij}(2t). \end{aligned} \tag{2.13}$$

The second equality follows from reversibility and the final equality from elementary probability considerations.

Next we compute the probability $F = F(t)$ that the letters at a particular position in two immediate *descendants* from the same node are identical. Averaging over the possible *ancestral* bases gives

$$F = \sum_i \pi_i q_{ii}(2t)$$

for a reversible model, and for the model in (12.3), this reduces to

$$F = e^{-2\lambda t} + (1 - e^{-2\lambda t})(1 - H). \tag{2.14}$$

EVOLUTION OF SITES

We assume that sites evolve independently of one another, with identical *mutation* processes at each site. We discuss these assumptions later. We can now glue together the information from each site in the sequence by thinking in *coin-tossing* terms. Look at the aligned sequences (with no gaps), and define

$$X_i = \begin{cases} 1, \text{ if the } i\text{th pair of sites differs,} \\ 0, \text{ otherwise} \end{cases}$$

If there are s sites in all, then the *Xi* are independent and (for the model (2.3)) satisfy

$$\text{P}(X_i = 1) = 1 - F = (1 - e^{-2\lambda t}) H. \tag{2.15}$$

Then $D = X_1 + \ldots + X_s$, the number of mismatched pairs of bases, is a binomial random variable with parameters s and $1 - F$. (Note that D is the Hamming distance.)

Of course, F is unknown and we have to estimate it from the sequence data. Returning to the *coin-tossing* example, we know that a sensible *estimator* of the binomial success probability is the observed proportion of successes. Thus we can estimate $1 - F$ with the *quantity D/s.* We can then use this to estimate K by solving the equation

$$\frac{D}{s} = \left(1 - e^{-2\lambda t}\right)H,$$

obtaining (after taking logs and simplifying)

$$2\lambda t = -\log\left(1 - \frac{D}{sH}\right).$$

Recalling the definition of K in (12.10), we obtain

$$K = 2\lambda tH = -H\log\left(1 - \frac{D}{sH}\right). \tag{2.16}$$

This estimate of the distance K between the two sequences is known as the *Jukes- Cantor formula.*

If we knew H, we could estimate K using the Jukes-Cantor formula. Typically we do not know H, and we have to estimate this from the data also. The usual strategy is to estimate the base frequencies $\pi_A, \pi_C, \pi_G, \pi_T$ from the collection of homologous sequences being compared, and then calculate $H = 1 - \sum_i \pi_1^2$.

Pairwise distances such as K can be calculated for many other models of the evolutionary process. These often arise by changing the form of the mutation parameters m_{ij} in (2.3), and finding the transition probabilities $q_{ij}(t)$ for the new model. The subsequent analysis is simplified if the model is reversible.

DISTANCE CALCULATION

Now that we have seen how to calculate distances and to build trees based upon those distances, we move to the third of our methods, developing a *statistical* method for tree-building based upon a maximum *likelihood* approach. For illustration, we assume that we are trying to construct an *ultrametric*, or clock-like, tree.

Evolutionary Tree

A clock-like tree can be represented in many ways. Here is a simple one. First, we label then species 1, 2,..., n at the leaves and assign labels $n + 1, \ldots, 2n - 1$ to the $n - 2$ internal nodes. This should be done in such a way that any internal node has a label larger than any of its descendants. This labeling lists the names of the *descendants* of each internal node and hence the complete tree shape.

The other missing *ingredient* is the times of the joins in the tree. For an ultrametric tree, it is natural to record the distance to each internal node measured from the present time back into the past. For example, the tree *described* by

6	2	3	0.2
7	6	4	0.4
8	7	5	0.6
9	1	8	1.0

has n = 5 species. The left-hand column gives the *internal node labels;* these are followed by the labels of the two *direct descendants* and the *distance* from the internal node to the leaves. In the example, species 2 and 3 join first at node 6, 0.2 units back in time; node 6 is then joined to leaf 4 at node 7, the distance being 0.4 units. Leaf *5* joins the tree at node 8, a distance of 0.6 units from the present, and finally leaf 1 joins at the root node 9, a distance of 1 unit from the present. The *resulting* tree was shown in Figure elsewher in this chapter. Note that it is easy to represent other tree *topologies* using this data structure.

Comparative Sequences

As earlier, we assume that the sequences under *comparison* have been aligned and that any gaps have been removed. We also assume we know the *topology* of the *phylogenetic* tree *linking* the species. First, we calculate the probability $p(i_1, i_2, ..., in)$ that species 1, 2, ..., n have bases $i_1, i_2, \ldots, i_n$ at a given nucleotide site.

4	1	2	t_1
5	3	4	t_2

We get

$$p(i_1,i_2,i_3) = \sum_a \sum_b \pi_a q_{ai_3}(t_2) q_{ab}(t_2 - t_1) q_{bi_2}(t_1) q_{bi_1}(t_1). \tag{2.17}$$

To see what is involved, let's do the calculation for the case of three species with a tree of the form This formula arises by considering the possible assignments a for the root node 5 and b for the node 4 and summing over the *possibilities* for a and b. Here the π_a are the *stationary* base frequencies (i.e., base frequencies assuming the same distribution both for *ancestral* and *descendant* bases).

The $q_{ij}(t)$ give the probability of base j at a site given base i at that site a time t earlier. A formula of this type can be written down for any tree topology, but it can be very complicated. The standard way to arrange the *computations* is to use the *peeling algorithm,* due in this context to *Felsenstein* (1981).

This recursive algorithm goes through the tree calculating at each internal node the probability of the *subtree* data below that node as a function of the base at the node. This algorithm is routinely *implemented* in most *tree-building* software. The *principle* behind it is to "move all *summation* signs as far to the right as possible."

Likelihood of Data

We compute the likelihood of the data by *multiplying* the likelihood terms $p(i_1, i_2, \ldots, i_n)$ calculated using the peeling algorithm at each site in the data. This comes about because we assume that the sites are evolving independently of each other. The maximum likelihood estimator of the parameters of the mutation model and the branch lengths are found in two steps: first, maximize the likelihood over the model parameters for a given tree, and then repeat this maximization over all possible trees.

The maximum likelihood estimator we are looking for is the one that maximizes these maximized likelihoods. This recipe is manageable for small numbers of species but impractical for large numbers (bigger than 5, say). In such cases, a *variety* of *heuristic* search algorithms have been *devised* to explore tree space.

An important question to be addressed is: What can be estimated? It is conventional to reduce the number of parameters to be estimated by setting the *stationary* frequencies to be the observed proportions of each base in the entire data set. We do this, too. For a given *topology*, for the simple mutation model we are using, there is one parameter, λ, for the substitution rate, as well as $n - 1$ node heights that might be estimated.

However, looking at the form of the $q_{ij}(t)$ in (2.4) and (2.5) shows that times and the rate λ are *confounded;* that is, they cannot all be estimated separately. We then have two different (but equivalent) choices:

(i) We can fix the height of the tree (the distance to the root node) and estimate λ and $n - 2$ other node heights, or

(ii) We can fix parameter λ (at value 1, say) and estimate the $n - 1$ node heights.

The program used in the exercises allows for either possibility. This *observation* shows why we need external *calibration* of the tree height in order to obtain node *heights* in years and mutation rates in units of changes per year.

Mathematical Study

Often we want to use *molecular* data to test different *hypotheses* about the *phylogenetic* relationships between species or to estimate (and perhaps find confidence intervals for) parameters in the mutation model. A common *guiding heuristic* is to compare like-lihoods among different models. This is a subtle business, but a couple of examples should illustrate what can be done.

One problem concerns whether or not a given phylogeny explains a data set better than another phylogeny. We have already discussed this problem in our description of *parsimony* and distance methods. In the present context, we could compare the log-likelihoods of the data under each model; *presum-ably* the one with the highest log-likelihood provides a better description of the data.

This can be formalised into a *rigorous* statistical procedure, but we do not present the details here. Another problem concerns estimation of parameters in a mutation model given a particular phylogenetic tree topology. We can test (formally at least) whether a simpler model (the "reduced hypothesis" in what follows) provides an adequate fit to a data set, as opposed to a model with more parameters (the "general hypothesis"). We can use a χ^2 test as follows:

1. Calculate the maximized likelihood L_1 under the general hypothesis.
2. Calculate the corresponding likelihood L_0 under the reduced hypothesis.
3. Calculate $W = -2\ [\log L_0 - \log L_1]$.
4. Under the reduced hypothesis, the distribution of W is approximately χ^2 with degrees of

freedom given by the difference between the dimension of the general hypothesis and the dimension of the reduced one.

CONCLUDING REMARKS

There are numerous problems associated with tree-building using molecular data. We might wonder about the adequacy of a given mutation model as a description of the data. We have made a number of assumptions, among them

(a) Sites evolve independently of one another.

(b) Sites evolve according to the same stochastic model. (c) The tree is rooted.

(d) The sequences are aligned.

Point (c) is easy to deal with; there are likelihood methods for uprooted trees (assuming reversible models of *substitutions*) with the additive property. Point (b) is clearly violated in a coding region: it is well-known that third *positions* evolve faster than first positions, which in turn evolve faster than second positions. It is possible to fit models with a common tree while *allowing* for this type of rate *variation*.

It is sometimes useful to try a model of random rate variation, in which each site evolves independently but with a rate drawn *independently* across sites from a given distribution. Some of the computational and biological problems are described in Yang (1996). There have been some *attempts* to alleviate the difficulties of assumption (a), which can arise, for example, in sequences with *secondary structure*.

Point (d) has proved more difficult to address. Usually, gaps in aligned sequences are removed before tree-building. The effects of this are hard to assess, especially in distantly related species in which alignment is hard. Finally, we remark that there are *computational* problems with *exploring* tree space adequately. One way to address this issue is through the use of *Bayesian computation*, in particular the Markov chain Monte Carlo methods.

When the goal is to understand the evolutionary *relationships* between *organisms*, the distinction between a gene tree and a *species* tree becomes important. This distinction is illustrated in Figure 2.7. A *gene tree* is a tree drawn from DNA or protein sequences corresponding to a particular gene shared by a set of organisms.

Each different gene may (or may not) produce a different tree for the same set of organisms. A *species tree* is often produced from sets of macroscopic characters but also may be produced from sequence data. Generating a *consensus* species tree from a collection of gene trees is a, difficult problem that is beyond the scope of this book.

Chapter 3

Evolution of DNA Sequencing

The falling costs of DNA sequencing and the increasing abundance of genomic sequence data means that the ability to find functional elements from DNA sequences is an increasingly important part of molecular biology. This is especially true for functional noncoding DNA elements, as there are not clear rules for their sequence patterns, unlike *protein-coding* genes. One of the most powerful methods for finding functional elements from sequence data is *phylogenetic* footprinting.

The concept behind *phylogenetic* footprinting is simple: through time functional elements will undergo a slower rate of sequence change than nonfunctional elements, as they are functionally *constrained*. Therefore, if we compare sequence data between related species, we expect-functional elements to stand out as "*footprints*" of sequence conservation against the background sequence *variation* of the *nonfunctional* regions.

However, therein lies the trick: the key to successful phylogenetic footprinting is choosing species for comparison that have the appropriate level of relatedness to distinguish functional elements from the *background*. The number of studies employing *phylogenetic footprinting* has increased rapidly in the last few years, and these studies have used *phylogenetic* footprinting to reveal a variety of functional elements.

Most commonly, it has been used to identify elements involved in regulation of gene expression. However, it has also proved useful for detecting other functional elements, such as: finding genes that are difficult to detect by *conventional gene-prediction* tools; detecting functional elements not associated with gene coding (called NOCs), such as origins of *replication* and *replication* fork barrier (RFB) sites; and *identifying* elements that clearly must be functional but for which the functions are not known.

Furthermore, the availability of whole genome sequences for related organisms means these analyses can be performed on a genomic scale, without explicit regard as to the function. Finally, increasingly more sophisticated tools for *phylogenetic* footprinting are becoming available, often as web-based software tailored for specific purposes. Here, we *demonstrate* how to perform a basic phylogenetic footprinting analysis, involving *making* multiple sequence alignments, producing visual plots of the level of sequence *conservation* across these *alignments*, and some *subsequent analyses*.

This phylogenetic footprinting *methodology* is applicable to researchers wanting to identify functional elements in their sequences of interest. However, phylogenetic footprinting is also a powerful *confirmatory* tool: if the presence of a functional element is inferred through other means, *phylogenetic* footprinting can be used to see if this element is indeed conserved, and, thus, whether or not it is likely to be *biologically* relevant.

We provide a practical example of how phylogenetic *footprinting* can reveal *functional* elements by using the ribosomal RNA gene repeats (rDNA) from a number of *Saccharomyces* yeast species. The rDNA repeats are maintained in most *eukaryotes* as multiple tandem repeats.

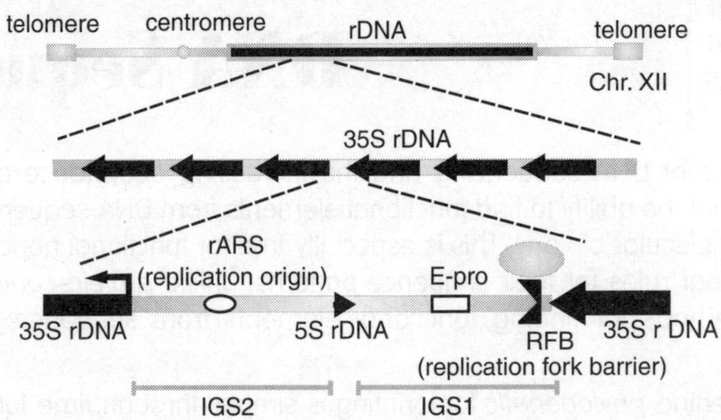

Figure 3.1: Structure of the ribosomal RNA gene repeats (rDNA) in Saccharomyces cerevisiae. S. cerevisiae *has approx 150 tandem rDNA units on chromosome XII. Each unit is 9.1 kb, and includes two rRNA genes (35S and 5S) and two intergenic spacers (IGS 1 and 2). The IGS region is thought to be a hotspot for noncoding functional elements, and some are shown here: a replication origin (MRS) in IGS1, and a replication fork blocking (RFB) site and a noncoding promoter (E-pro) in IGS2. The RFB works as a recombinational hotspot to maintain rDNA copy number by amplification. This recombination is mediated by the Fob1 protein that associates with the RFB, and is regulated by E-pro.*

Each repeat unit consists of the 35S and 5S ribosomal RNA genes (the 5S rRNA gene is only variably present), and these are separated by *spacer* elements. Here, we analyze the *intergenic* spacer (IGS), located between adjacent 35S rRNA genes. In the *Saccharomyces* yeasts, this spacer is divided into two *regions* (IGS1 and IGS2) by the presence of the 5S rRNA gen.

The IGS contains a variety of functional elements, such as promoters, *terminators*, an origin of replication, RFB sites, and other protein binding sites. *Phylogenetic* footprinting of the IGS *identifies* these previously identified elements as well as several other *uncharacterized functional* elements.

REQUIREMENTS

1. The first consideration for a phylogenetic footprinting study is the species to use. To aid in species selection, a phylogeny showing the *relationship* of species related to the species of interest is useful. We used the phylogeny of the *Saccharomyces* yeasts presented in Kurtzman and Robnett to select a number of species for analysis. The relationships of the

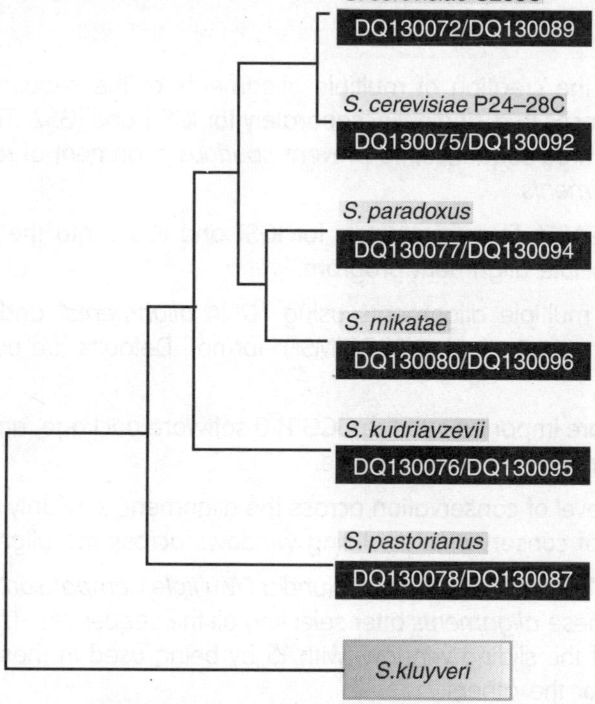

Figure 3.2: Phylogeny of the Saccharomyces *yeast species used in this study. The species/strain names are indicated, as are the accession numbers of the rDNA IGS1/IGS2 sequences (in boxes). For reference, an outgroup species* (Saccharomyces kluyveri) *is also shown in gray.*

species we used here are shown in Figure elsewhere in this chapter

2. DNA sequences for the region of interest are obtained either through sequencing or from DNA databases or genome projects if available. It is recommended that a known *conserved* feature (e.g., part of a gene) be included in the region *analyzed* to act as a *"landmark."* Make sure all sequences are in the same *complement*.

   We previously sequenced the IGS in two parts (IGS 1 and IGS2, divided by the 5S rRNA gene) for our chosen species, and the Genbank *accession* numbers of the sequences used here are shown in Figure 3.2. All sequence *manipulations* were *performed* separately for the IGS1 and IGS2 sequences.

3. Software used in this analysis *(see* Note elsewhere in this chapter for alternative software available):
   (a) DNA sequence alignment: ClustalW (http://align.genome.jp/).
   (b) Multiple alignment editing: GCG 11.0 Package (Accelrys, San Diego, CA). (c) Similarity plots: SimilarityPlot from the GCG 11.0 Package (Accelrys).
   (d) Conserved peak sequence analyses: TESS (http://www.cbil.upenn.edu/ cgi-bin/tess/ tess?RQ=WELCOME).

PROCEDURES

1. The first step is the creation of multiple alignments of the sequences. We used FASTA *formatted* sequences in a single file, separately for IGS 1 and IGS2. The alignments should contain at least three sequences to prevent *spurious* alignment of regions that can *occur* in *pairwise alignments.*
   (a) Enter these FASTA files (separately for IGS1 and IGS2) into the web interface of the ClustalW multiple alignment program.
   (b) Execute the multiple alignments using "*DNA alignments*" and the "*slow/accurate*" options. The output is set to "GCG(MSF)" format. Defaults are used for the remaining parameters.
2. The alignments are imported into the GCG 11.0 software package, and corrected manually for alignment error in the "Editor" mode.
3. To visualize the level of conservation across the alignment, similarity plots are created that graph the level of conservation in sliding windows across the alignment.
   (a) Perform the "*PlotSimilarity*" function (under "*Multiple Comparison*") in the GCG package on each of these *alignments* after selecting all the sequences. The relevant parameter is the size of the sliding window, with 15 by being used in these examples. Defaults were used for the others.
   (b) A graphical representation is then automatically created.
4. Once the similarity plots have been produced, known features can then superimposed onto these graphs. This produces visual *representations* of the alignments *referenced* to known features from one or more of the sequences from which to base *subsequent* analyses, as shown in Figure 3.3.
5. Obviously, peaks of conservation are the interesting points. Reference to Figure 3.3 reveals several relevant points:
   (a) *Promoter elements, terminator* elements, and particularly the gene *coding* regions themselves form conserved peaks.
   (b) Most NOC elements also show remarkable *coincidence* with peaks of conservation, and the example of the RFB site is *highlighted* in Figure 3.4 A.
   (c) Other conserved peaks without known functions are also present.
   (d) Some features that were previously *predicted* are not conserved, *calling* into question their *biological* meaning. These include the Abf1 p binding site in IGS 1 and one of the two Top 1p binding sites *predicted* in the RFB.
   (e) It is important to refer back to the original alignment to look for features that are masked in the *similarity* plots. For instance, the Reb 1p binding site in IGS2 shows only a very small peak of conservation. However, reference to the *alignment* shows that this is a highly *conserved motif* embedded in a very variable region.

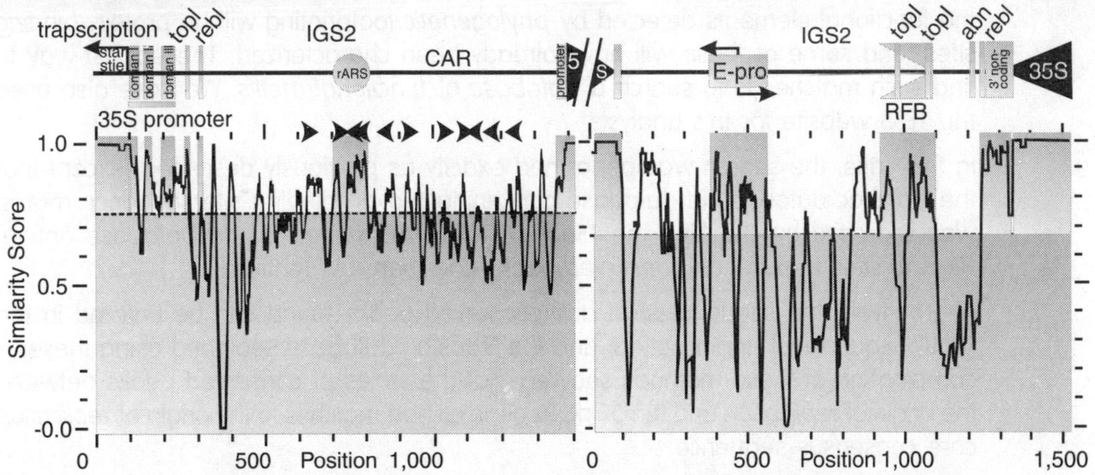

Figure 3.3: Similarity plots of the IGS2 and IGS1 regions. Previously identified features from the IGS are indicated stylistically above the plots. These are the 35S and 5S rRNA genes, 35S promoter and 3' noncoding region, 5S promoter and terminator, origin of replication (rARS; [32]), replication fork barrier site (RFB;) including two Top 1p binding sites, expansion sequence bidirectional promoter (E-pro;), cohesin associating region (CAR;), and two Reb 1 p binding sites and an Abf1p binding site. Conserved peaks that match these elements are boxed in the similarity plots. Sequence matches to the origin of replication core consensus sequence found in conserved peaks are indicated by small arrows above the plot. The dotted line represents the average level of similarity across the alignments. These are different between IGS1 and IGS2 because the average level of similarity is different between the two regions (largely as a result of IGS1 containing more indels). The region of the similarity plots below this line is shaded to emphasize the conserved peaks.

(f) An important question is "how high should a peak of conservation be to imply biological relevance?" However, there is no way to know this a *priori*. We have used the average level of similarity as a proxy for this, but in *reality* this is *arbitrary*. Other evidence, including *ultimately* experimental evidence, is required to show *biological* function.

6. The real object of these studies is to determine what roles *biologically* functional sequence elements play.

To help determine whether or not conserved peaks are biologically functional and to gain clues as to their potential function, it is useful to examine the sequences of the conserved peaks in detail. Examples of some of the *in silico* examinations that can be performed include:

(a) Comparison of the conserved peaks with each other. The sequences of any conserved peaks can be used to search across the whole sequence to identify multiple copies of *conserved* peaks. We have performed this using the TESS website. This has the *advantage* of searching both directions and both strands, and can accept *degenerate* sequences.

Using the "String-based Search Query" page, enter the sequence of the whole region in the "*DNA Sequence*(s)" Input option, and enter the conserved peak sequence(s) in the "Search My Site Strings" String Database option. The search criteria can be modified in the "*String Scoring*" option.

(b) Matches of conserved peaks with previously identified functional motifs. In many cases the functional elements detected by *phylogenetic* footprinting will be protein *binding* sites, and some of these will have already been characterized. The *easiest* way to find such matches is to search a *database* of *functional motifs*. We have also used the TESS website for this analysis.

In this case, the search was performed exactly as previously described, except that the Transfac database (a database containing protein binding site sequence motifs) was searched by checking the "*Search TRANSFAC Strings*" *String* Database option. *Binding* sites that fall on *conserved* peaks can then be identified.

Alternatively, the sequence(s) of any conserved peaks found can be entered in the "DNA Sequence(s)" Input option, and the Transfac database searched using these. A combination of these methods showed that the series of conserved peaks between the origin of *replication* and the 5S rRNA gene all had matches to the origin of *replication* core consensus sequence.

(c) Other things to look for in the conserved peak sequences include the presence of repeats, *palindromic* sequences, and regions of *unusual* sequence *composition*.

7. Our analyses of the rDNA IGS region from these *Saccharomyces* yeast species reveals

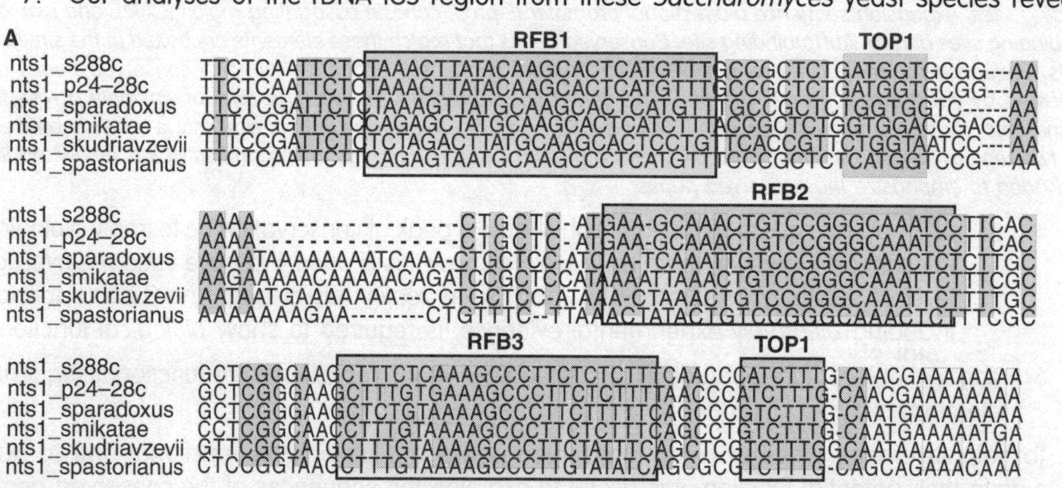

Figure 3.4: Features of the plots. The actual alignments of two regions of interest are shown. Sites completely conserved in all six isolates are boxed in gray. (A) Conservation of the RFB site. The individual RFB elements and predicted Top1p binding sites are boxed. The three RFB elements all overlap with the regions of highest conservation. As can be seen, only one of the two predicted Top1p binding sites is conserved, calling into question the biological function of the other site. (B) Region surrounding the IGS2 Reb1p binding site. The Reb1p binding site is boxed. Although the peak in Figure 3.3 is small, the alignment shows this Reb1p binding site is highly conserved, but is embedded in a background of variable sequence.

that *phylogenetic* footprinting is a simple, yet extremely powerful method for *detecting* a variety of noncoding *functional* elements. Most previously characterized elements in the IGS were obvious as conserved peaks, and for some of those that do not form conserved peaks we have independent reasons to doubt their *functionality*.

Furthermore, there are a variety of conserved peaks for which functions have not been previously described, and we have confirmed that some of these are indeed functional. Indeed, all the conserved peaks identified in the IGS by this *method* that we have tested to date are *functional*, attesting to the great power of phylogenetic *footprinting*. These IGS conserved peaks also represent elements with a variety of functions, therefore this technique is likely to be *applicable* to studies *investigating* many different kinds of *functional* elements.

Finally, the method seems to be very specific: both the origin of replication and RFB sites are believed to contain three *subelements*, and in each case each of these subelements can be identified as its own conserved peak. Single protein binding sites also appear as conserved peaks. Given the simplicity of performing the *analyses* (once the sequences are obtained), phylogenetic footprinting promises to be an important component of *molecular biology* studies that look to relate *noncoding* DNA to biological action.

POINTS TO REMEMBER

1. Choosing species with the appropriate level of *evolutionary* relatedness is the most critical part of phylogenetic footprinting. However, this is easier said than done; the most appropriate species can even change depending on what region is being examined. We recommend choosing species encompassing a range of relatedness, including very closely related species, and more distantly related species. A *phylogeny* is, therefore, very useful for this *species* selection.

   The total phylogenetic distance *separating* all species is likely to be the important *parameter*. In practice this means choosing a good phylogenetic range, rather than a lot of species; three species with appropriate levels of relatedness are likely to give better results than six species from a more *restricted* range. However, if the species are too *distantly* related, some or all of the conserved peaks will disappear. It may be necessary to return and chose a species with an *intermediate* level of *relatedness* if the results from the first pass are not *satisfactory*.

2. In most cases, there are alternatives available to the software we have used.
   (a) There are a number of multiple alignment *programs* available, both as web-based services and as installed programs for personal *computers*. Any of these should be suitable for making the multiple *alignments*.
   (b) Again, there are a number of alignment editor software programs available. Many labs will already have software in use. Features that interactively indicate when bases are matched or color-code the different bases are useful for editing alignments.
   (c) We have also used two freely available software programs to produce similarity plots. These are SynPlot (http://www.sanger.ac.uk/ Users/jgrg/SynPlot/) for Unix systems

and SWAAP (http:// www.bacteriamuseum.org/SWAAP/SwaapPage.htm) for Windows systems

3. We recommend making several alignments using various subsets of the sequences. This allows the user to determine what level of phylogenetic relatedness is the most appropriate for the sequence of interest. As the level of *relatedness* between sequences decreases, the conserved peaks stand out more from the *background* noise. However, when the level of *relatedness* drops off too far, the conserved peaks will *disappear*.

   Furthermore, this also allows the user to detect different types of elements. Functional elements, for a number of reasons, are expected to vary in their rate of evolution. Therefore, some will only be seen when closely related species are used, whereas others will persist longer through *evolutionary* time.

4. The output format selected will depend on the software used to make the similarity plots. Be sure to check what format the multiple alignments should be in for making the similarity plots. The sreformat program from the HMMER package (http://hmmer.wustl.edu/), available as a web-based service (http://bioweb.pasteur.fr/seganal/interfaces/sreformat.htm), is useful for alignment format *interconversions*.

   Furthermore, be careful when switching between different computer systems (e.g., Mac/Unix/Windows) as line breaks and other characters are often coded differently. *Opening* and *resaving* text files in the *native system* is sometimes required.

5. Usually the default *parameters* produce reasonable alignments. However, if there are many gaps the alignment can be poor, and adjusting the parameters (e.g., decreasing the *gap penalty* options) can help. In such cases, other alignment programs that use different *algorithms* (e.g., DIALIGN and multi-LAGAN) may also be useful.

6. The alignment is the basis for phylogenetic *footprinting*; *subsequent* operations are just a way to visualize more easily the information from the alignment. Therefore, it is important to check the alignment for accuracy. However, there is no need to take this too far. Editing the alignment to produce a 1-bp match in a region of low similarity will not affect the overall *results*.

   It is more major errors that should be checked for. In our experience, most relevant alignment errors occur at sites where there are large insertions/deletions in some of the sequences.

7. Here are some basic instructions for using freely available software for making similarity plots:

   (a) SynPlot 0.5.3. This is a command-line program. The input file should be in aligned FASTA format (a2m format). Make sure the format is correct; this is not a simple FASTA format, but instead has the *alignment* gaps included in the FASTA sequences. The *sreformat* program can convert to this format.

   Some of the SynPlot defaults will probably need changing. Change the *sliding* window size (-window) to 15 and the slide length (-increment) to 1 to give the same results as the GCG example. The -nuc width *setting* may also be changed for viewing convenience. Specify the output file name (-out). The user can also include *annotations* for the sequences in GFF format if they exist.

(b) SWAAP 1.0.2. This Windows-based program can accept various alignment formats, although we have used the aligned FASTA format (as in SynPlot) most easily. The alignment should be opened (and reading frame arbitrarily selected even though this plays no role). At this stage, change the defaults under the "*Parameters*" menu if necessary, e.g., Window Size to 15 and Step Size to 1 to give the same results as the GCG example.

To make the output similar to that made by GCG or SynPlot, the "Indel" option should be changed at least to "Treat Pairwise Indels as 5th *Nucleotide*," or even "Remove All Indels" which removes all positions in the alignment that contain an indel. This is because indels seem to be treated as *homologous* characters in this software, thus giving somewhat different outputs.

Then, create the similarity plot by implementing "Calculate Percent Identity Over Sliding Window" under the "*Analyses*" menu, using of course the "*Nucleotide*" option. This creates a table of the values from the sliding windows, which can be plotted directly in the program, or *transferred* to other software (e.g., Microsoft Excel) for plotting.

8. As the similarity plots are simple to produce at this stage, it is worth experimenting with the sliding window size. Remember, these plots are merely graphical representations of the underlying alignments for ease of *visualization*, so the window size can be varied for maximum utility. *Smaller* window sizes will better reveal small conserved motifs, however the *differentiation* from the *background* is less.

9. It will probably be necessary to save the output in a format that can be used by subsequent programs for making figures. We have found that EPS format gives the best results in our situation. To save the file in this format, press "*Print*" once PlotSimilarity produces the image, and select the EPSF option from the "Output Device" item. Change the name in the "Port or File" item to the output name you want.

10. Often, only the region of the plots above a certain threshold (e.g., average identity across the alignment) is shown in figures, highlighting the conserved peaks. This is for ease of reference, and also because the depth of the troughs are determined mainly by the number of indels, rather than the *absolute* level of variation per se, and therefore are not necessarily biologically *meaningful*. We have shaded the lower part of the plots in Figure 3.3.

11. In most cases these features will come from just one of the species in the alignment (in our example the features all come from studies on *Saccharomyces cerevisiae)*. However, if information from several species in the alignment is available, obviously this should be used. Remember to use the *alignment*, not the original reference sequence, as the reference template to plot features, as it is likely gaps will have been introduced into the alignment.

12. The presence of large "*valleys*" in the similarity plot is usually the result of gaps in the alignment, and they are not necessarily the least-conserved regions, per se. The effect of gaps on the similarity plots can be checked by removing the gapped regions from the alignment, e.g., by using the Gapstrip tool at the Los Alamos National Laboratory HIV

website (http://www.hiv.lanl.gov/content/hiv-db/GAPSTRIP/gapstrip.html).

13. Two methods can be used. In one, *degenerate* sequences of the conserved peaks can be made from the alignment. In the other, the sequences of each conserved peak from one reference species can be used as queries to the entire sequence from the same species. A useful way to represent degenerate sequences of *conserved* peaks is as sequence logos, and a tool for making these is available online.

14. When searching for matches to the conserved peaks, the searches should allow for some *mismatch*, as conserved motifs are unlikely to be 100 % conserved.

Chapter 4

Determining DNA Variants

Analysis of DNA evariation, either *somatic* or *inherited* mutation, is important in all fields of *biology*. As all life forms, from virus to humans, are products of their *genetic code*, any minor *alteration* in this code may induce severe or *life-threatening* changes in the organism. Alternatively, *adaptation* and survival advantage in a given *environment* may ensue. Thus, analysis of DNA variation has become a major field of research in biology.

In 1953, Watson and Crick *postulated* the structure of DNA, which may be regarded as the beginning of the *genomic era*. This was followed by the description of PCR in 1971 by Kleppe et al., demonstration of Sanger et al. sequencing reactions with *termination nucleotides* in 1977, and the separation of base substitutions by *denaturant* gradient gel *electrophoresis* (DGGE) in 1983. The familiar PCR amplification, using *thermostable* polymerases, was first demonstrated in 1985.

This discovery accelerated analysis of DNA and development of methods able to analyse DNA. One type of method widely used is the melting gel *technique*, which has been reviewed extensively by Bjorheim and Ekstrom.

The method is based on a theoretical calculation of DNA melting by statistical mechanics calculation and *separation* of variant DNA during *electrophoresis*, where the *migration* of less stable fragments (i.e., *heteroduplexes*) are retarded as compared to the wild-type sequence.

The method, in theory, is applicable to about 95% of the human genome, where the melting temperature is below the stability of the artificial high melting domain, known as the GC-clamp. The method, which *initially* was developed in "*slab*" gel form, has been transferred successfully to six different commercial DNA *capillary* sequencing *instruments*.

DCE requires three steps, design of the appropriate fragment (*software simulation*), PCR *amplification*, and allele separation by *electrophoresis*.

Under the conditions for which DNA fragments are appropriately designed for DCE, PCR *amplifiable*, and given appropriate denaturing conditions (temperature and chemical), heteroduplexes are separated from the wild-type fragments. Therefore, DCE is an *excellent* method for scanning DNA for unknown variants.

REQUIREMENTS

Mathematical Calculation of DNA Melting

Online Internet services, http://www.biophys.uni-duesseldorf.de/local/POLAND/ poland.html.

PCR

1. 10 x buffer IV (ABgene, Epsom, UK).
2. 25 mM $MgCl_2$
3. 4 × 2.5 mM dNTPs.
4. Taq polymerase (ABgene).
5. Primers.
6. DNA template.

Allele separation

Commercial capillary DNA sequencing instrument.

PROCEDURES

Scanning DNA for variants by DCE requires three steps: fragment design, PCR amplification of the target sequence, and allele separation by capillary *electrophoresis*. *Evaluation* of the DNA sequence of interest by use of *statistical* mechanics calculates the melting properties of the DNA and provides information as to whether the fragments can be analyzed by DCE and where the high *melting* temperature tail, the GC-clamp, should be attached. Furthermore, the simulation of allele melting *temperatures* yields information on their expected *separation* by DCE.

The purpose of PCR is to produce enough copies of the target sequence, incorporating the label and attaching the high temperature melting domain. DCE for allele separation alleles was first performed with in-lab assembled instruments, where optical parts like mirror, lenses, laser, and detectors were *mounted* onto an optical bench. Thus, all *handling*, gel *replacement*, sample loading, and recording of *electrophoresis* were done *manually*.

However, the manual handling and technical skill of the users resulted in detection of mutants fractions down to 10^{-6}. This detection limit is superior to most *methods*. With the introduction of *automated* capillary DNA sequencing instruments, more samples could be processed without intervention from the operator], while upholding a limit of detection of 10^{-3}.

Target Sequence of DNA

1. Find DNA sequence of interest in a public database.
2. Select primers for target sequence. The length of the target sequence should be in the range of 80-210bp. Primers can be selected using the following website: http://frodo.wi.mit.edu/primer3/input.htm.
3. Cut and paste the sequence to be amplified into a DNA melting simulation bioinformatics

tool. Observe the theoretical melting profile of the target sequence; simulate GC-clamp and known variants. Figure 4.1 demonstrates the 50% melting *probability* of the Kras gene exon 1, simulated by four computer programs. Please note the difference in melting temperatures; this is a result of *algorithm* approximations and differences between salt *concentration* and dissociation constant settings in each program.

PCR

1. Optimize the PCR reaction, by use of concentration combinations of Mg^{2+}, primers, and dNTPs. This should be performed in combination with different annealing temperatures.
2. Amplify samples by PCR, using the conditions identified by optimization. Table 4.1 provides an example of the composition of a "*standard*" PCR reaction mix. Perform PCR as follows: 95°C, 5 min; 35 cycles of 95°C, 30 s, annealing temperature, 30s, 72'C, 60s (the *annealing temperature* varies between target sequences and is dependent on the *primer* design).
3. Create *heteroduplexes* by denaturing the PCR product at 95°C for 2 min, followed by *reannealing* at 65°C for 30 min (this step can be *programmed* at the end of the amplification *thermocycling* program).

Capillary Electrophoresis

1. Set the running temperature of the capillary electrophoresis *instrument* according to the

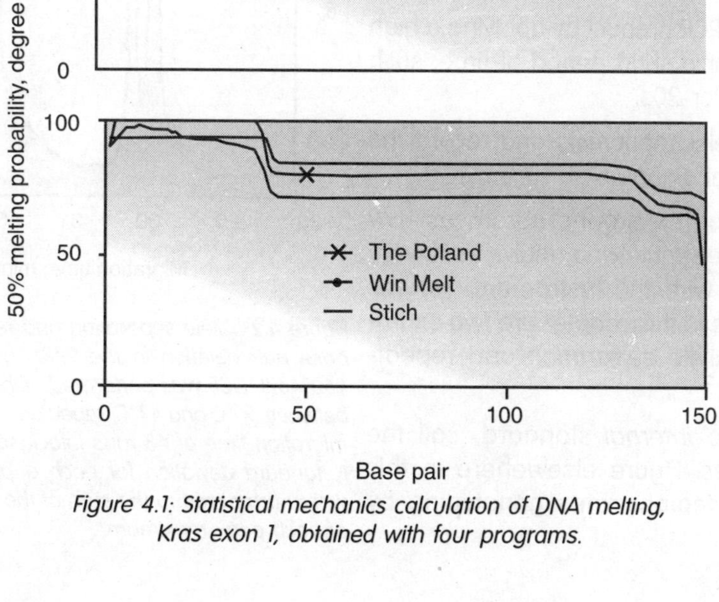

Figure 4.1: Statistical mechanics calculation of DNA melting, Kras exon 1, obtained with four programs.

Table 4.1: 20-μL PCR mix, volume, and concentrations of reagents.

	Concentration	Volume (μL)	Desired concentration
H_2O		12.2	
10 × buffer	10x	2	1 ×
$MgCl_2$	25 mM	2	2.5 mM
Primer 1	5μM	1.2	0.3μM
Primer 2 (6-tam-gc)	5μM	0.6	0.15 μM
DNTP (2.5 mM of each)	10 mm	0.8	400μM
DNA	50 ng/μL	1	2.5 ng/μL
Taq DNA polymerase	5 U/μL	0.2	0.05 U/μL

theoretical melting temperature of the low *melting* domain of the *target* sequence.

For instruments with inappropriate temperature control, temperature cycling (*cycling* temperature capillary *electrophoresis*) can be used.

2. Prepare the instrument for electrophoresis by replacing the polymer using a *syringe* pump or by applying high-pressure nitrogen (instrument dependent).
3. Inject the PCR product by applying a high voltage for a short period of time, such as 10 kV for 20s.
4. Perform electrophoresis and record the fluorescent signal.
5. Inspect the electropherogram as raw data in the sequencing analysis software supplied with the instrument. Figures elsewhere in this chapter are two examples of allele *separation* and repeatability.
6. Using the *internal* standard, call the *genotypes*. Figure elsewehere in this chapter demonstrates genotyping by DCE.

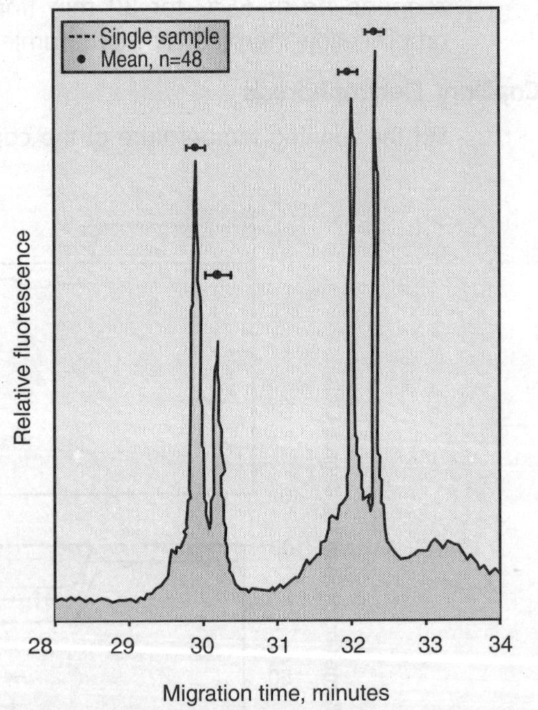

Figure 4.2: Allele separation and repeatability of a one base pair deletion in the PPRA gamma gene (OMIM #601487). DCE was performed with temperature cycling between 50°C and 47°C, repeated 20 times. The average migration time of 48 runs (black square) is shown with I standard deviation for each duplex. Migration times were normalized to the sum of the four-migration times of each peak maximum.

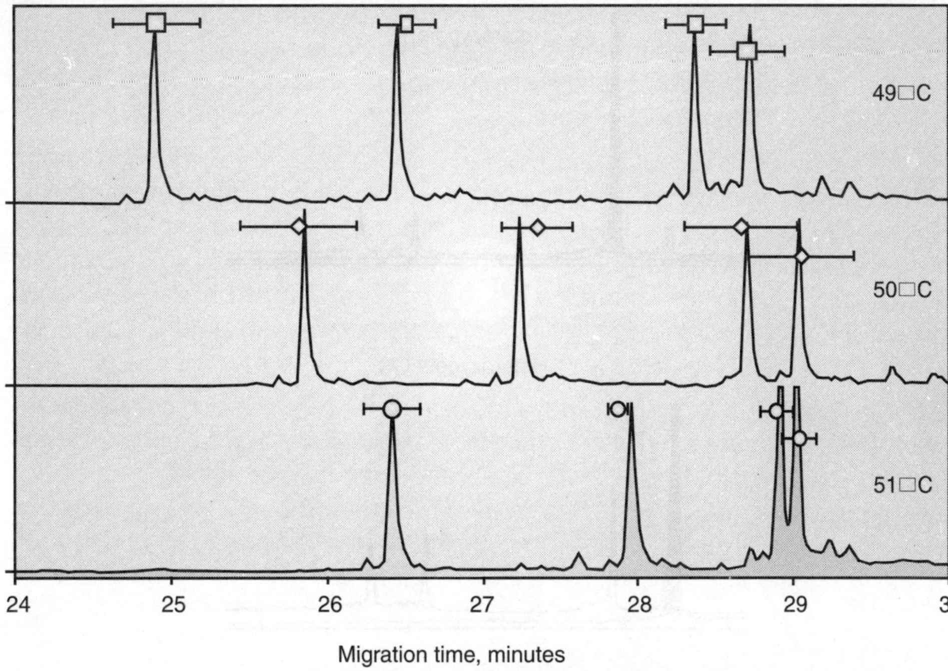

Figure 4.3: Separation of homoduplexes and heteroduplexes (NCBI ref SNP ID: rs10887), using the CEQ 8000 instrument (Beckman Coulter) at different temperature settings. The CEQ8000 instrument hardware was used without modification, and standard CEQ capillary arrays (33 cm), polymer, and separation buffer were used. The sample injection was standardized to 2.5 kV for 15 s while the separation was performed under the standard voltage setting of 4,000 V. Replicates (n = 32) of the samples were analyzed at three temperature settings. The average migration time of 32 runs is shown with 1 standard deviation for each peak (homoduplex and heteroduplex). Please note the baseline separation between the homoduplexes at all temperatures.

POINTS TO REMEMBER

1. Figure elsewhere in this chapter depicts the distribution of 50% melting probabilities in 58440bp from chromosome 1. Here, *95%* of the melting temperature is below 77'C, which is below the stability of the GC-clamp. Please note that the resolution on the x-axis is 0.1 °C.

2. Standard GC-clamp with 6-Fam label:
   5'-6-FAM-CGCCCGCCGCGCCCCGCGCCCGTCCCGCCGCCCCCGC CCG-3'

   If this sequence is used as it is on the "*forward*" or "*reverse*" primer, a standard 20-mer primer with a different label can be used to make the fragment-specific internal standard. For large-scale studies *featuring* many *fragments*, primer cost reduction can be *achieved* by in-PCR attachment of a GC-clamp and labeling. This is illustrated in Figure elsewhere in this chapter.

   The purpose of the GC-clamp is to prevent strand dissociation during electrophoresis. When the low *melting* domain is subjected to the appropriate *denaturing* condition

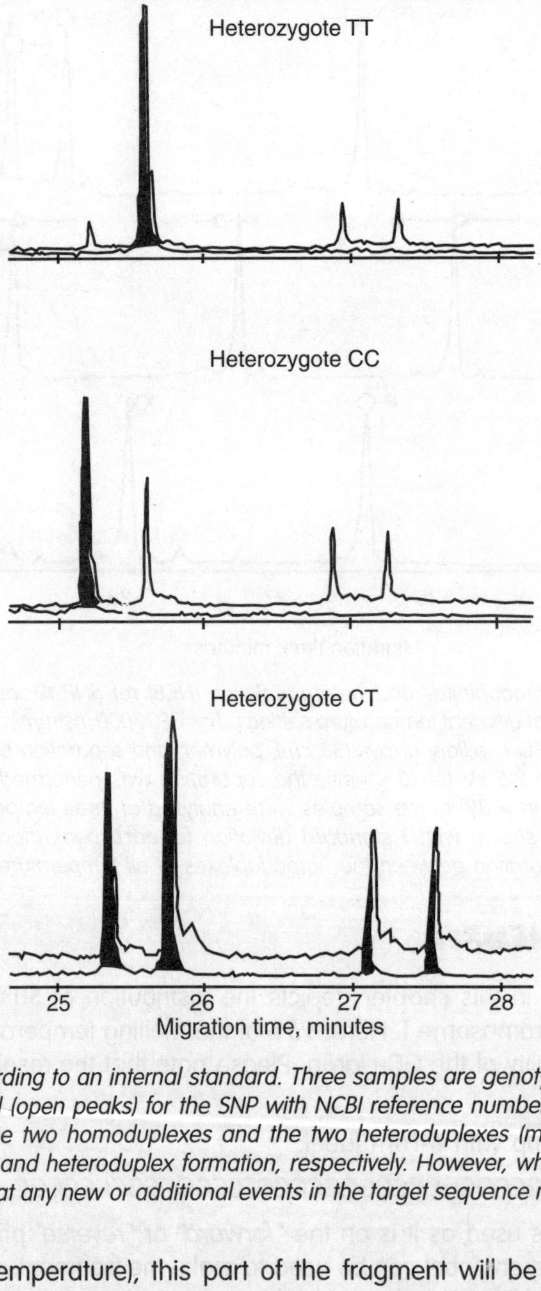

Figure 4.4: Genotyping according to an internal standard. Three samples are genotyped (solid peaks) and scored against an internal standard (open peaks) for the SNP with NCBI reference number rs375384. The four peaks in the internal standard are the two homoduplexes and the two heteroduplexes (mismatch of Watson and Crick strands) formed during PCR and heteroduplex formation, respectively. However, when using DCE for genotyping, the user should be aware that any new or additional events in the target sequence many alter the peaks positions in the electropherogram

(chemical and temperature), this part of the fragment will be in a state of *equilibrium* between dsDNA and ssDNA. Because of the branching, the DNA fragment will be retarded during electrophoresis.

Therefore, fragments with minor differences (substitution, deletion, or insertion) generally have a different average *velocity* through the *polymer* from the *consensus sequence*, which results in the observed *separation* of the *alleles*.

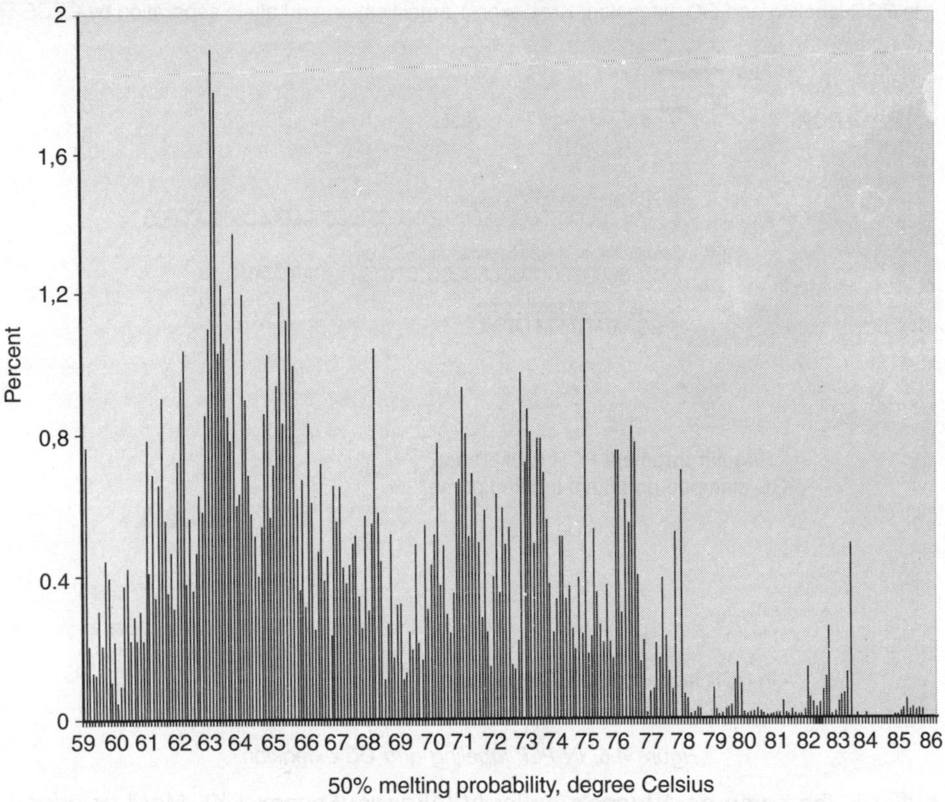

Figure 4.5: Melting probabilities of 54440bp from chromosome I

In PCRlabeling and GC extension, Kras mmn1amplification and allele separation by CTC

3. Calculation of the melting properties of the DNA of interest can be performed by various software packages or online Internet programs. Information about *commercial* software or online *bioinformatics* tools can be found via the following *Internet* links:

   http://bioweb.pasteur.fr/seganal/interfaces/dan.html. http://www.changbioscience.com/primo/primomel.html. http://stitchprofiles.uio.no/ctemp_prof.htm.

   http://bioinformatics.org/meltsim/wiki/Main/HomePage. http://www.biophys.uni-duesseldorf.de/local/POLAND/poland.html. http://www.medprobe.com/uk/sequencher-demo.html.

   We have been successfully using WinMelt (Medprobe, Oslo, Norway) to simulate changes in *thermodynamics* and used the theoretical melting temperature as a guide for the temperature setting during *electrophoresis*. We found a linear relationship between *stability* changes and separation achieved by DCE.

4. The buffer and $_{\text{MgCl2}}$ are prepared in the laboratory to keep the PCR cost as low as possible. As a 10 × PCR buffer, we regularly use sterile-filtered 750 mM Tris-HC1 (pH 8.8 at 25^{0}C), 200 mM $(NH_4)_2SO_4$, with 0.1 % (v/v) Tween 20 added to the buffer after sterile filtration. This

In PCR labeling and GC extension, Kras exon 1 amplification and allele separation by CTCE

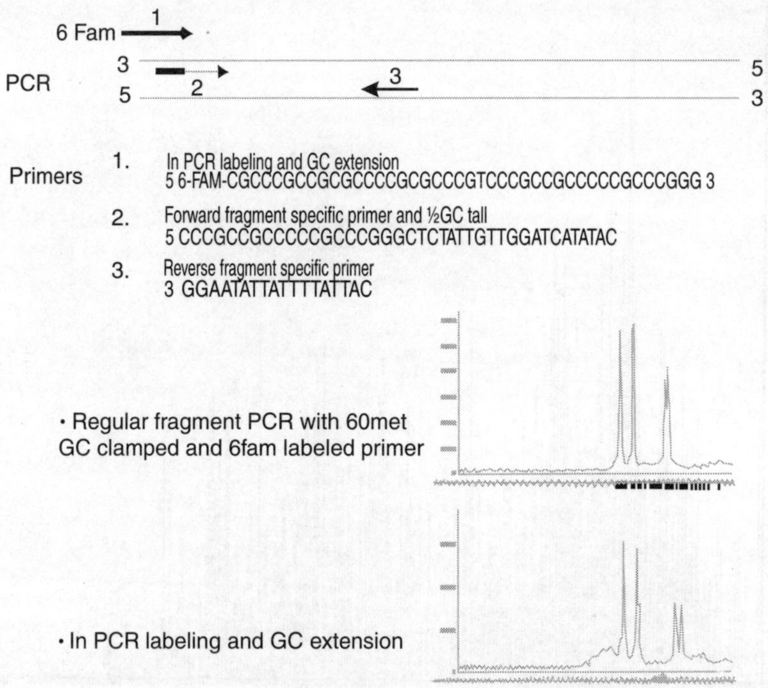

Figure 4.6: IN-PCR labeling and GC extension

buffer is the same as ABgene's buffer IV (ABgene, Epsom, UK). MgCl, is added into the reaction mixture from a separate vial, because this facilitates PCR optimization with respect to Mg^{2+} *concentration*.

5. The selection of enzyme should be based on the detection limit needed for the study. Using Taq *polymerase* gives an error rate in the range of 1×10^{1} to 1×10^{-5} *misincorporations* perbp per cycle, while PFU (Stratagene, La Jolla, CA) has been shown the produce tenfold fewer errors.

   Therefore, studying low mutant fractions requires the use of PFU polymerase. However, we *successfully combined* the two *enzymes* (ratio 10:1, Taq:PFU), preserving the *processivity* of TAQ and the fidelity of PFU.

6. We tested six capillary DNA sequencing instruments, from different *manufacturers*, and found that alleles can be *sufficiently* separated by DCE with ABI 310 and ABI 3100 (Applied Biosystems, Foster City, CA), CEQ 8000 Genetic Analysis (Beckman Coulter, Fullerton, CA), MegaBACE 1000 and MegaBACE 4000 (GE Healthcare Bio-Sciences Corp., Piscataway, NJ), and SCE 2410 (Spectrumedix, State College, PA). Standard *capillaries* and *sieving* matrix for each instrument were used, making this method readily available to any user with access to such instruments.

7. An example of a bioinformatics tool that can be used to access a *nucleic acid* sequence is found at http://snpper.chip.org/. This "database" presents gene search, dbSNP data,

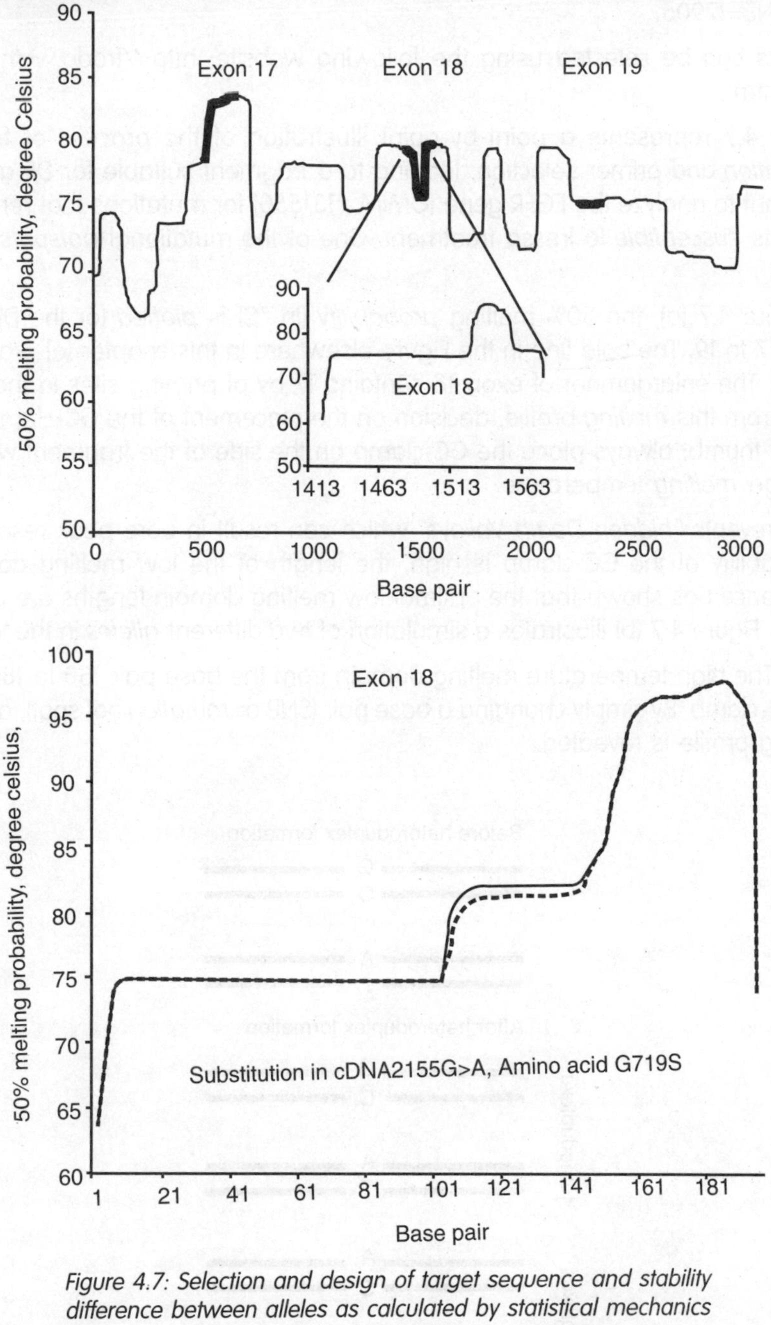

Figure 4.7: Selection and design of target sequence and stability difference between alleles as calculated by statistical mechanics

sequence search options, and *annotated* sequences (with intron, exon, coding, and SNP) in a very *straightforward* format.

For illustration purposes, annotated "epidermal growth factor receptor isoforma" canbe-foundat the following link: http://snpper.chip.org/bio/show-sequence/?TYPE=

U&GENE=17905.

8. Primers can be selected using the following website: http://frodo.wi.mit.edu/ primer3/ input.htm
9. Figure 4.7 represents a point-by-point illustration of the process of fragment design *simulation* and primer selection, leading to a fragment suitable for DE analysis. Assume we want to analyze the EGFR gene (OMIM #131550) for mutations that render lung cancer patients *susceptible* to Iressa treatment. One of the mutational *hotspots* is sited in exon 18.

   In Figure 4.7 (a), the 50% melting *probability* (in °C) is *plotted* for the DNA sequence of exon 17 to 19. The bold line in the Figure elsewhere in this chapter (a) graph indicates the exons. The enlargement of exon 18 contains 17 by of priming sites in the intron on each side. From this *melting* profile, decision on the placement of the GC-clamp is made. As a rule of thumb, always place the GC-clamp on the side of the fragment with the "highest" average *melting* temperature.

   This prevents "*hidden Death Valleys*" which can result in pore peak resolution. Although the stability of the GC-clamp is high, the length of the low melting domain is limited. Experience has shown that the optimal low melting domain lengths are between 80 and 210bp. Figure 4.7 (b) illustrates a simulation of two different *alleles* in the target sequence.

   Note: The high temperature melting domain from the base pair 150 to 180 is the effect of the CG-clamp. By simply changing a base pair (SNP or *mutation* hot spot), a slightly different melting profile is revealed.

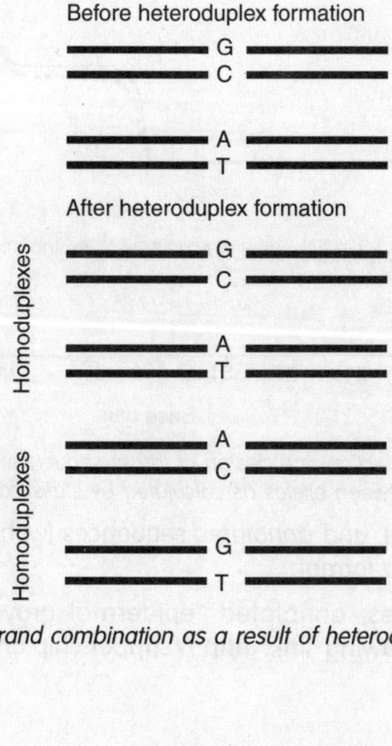

Figure 4.8: Strand combination as a result of heteroduplex formation

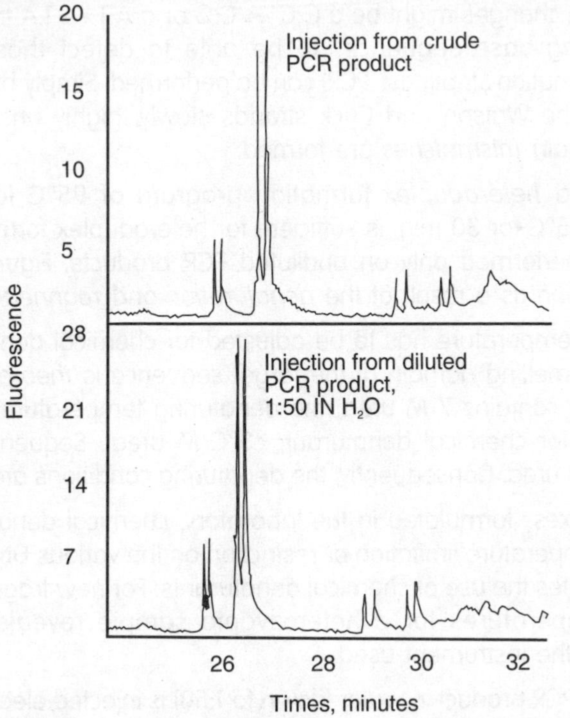

Figure 4.9: Effect on electrochemical injects by diluting PCR product in H_2O.

This melting difference is sufficient to separate *homoduplex alleles.*

10. While optimizing the chemical conditions of the PCR is important for most assays involving amplification of DNA, the number of thermocycles also can affect the *specificity* of the *amplification*. The number of final copies (with a maximum of 10^{11}copies/.µl) is a function of the starting copy number (No), amplification efficiency (eff) and the number of cycles (C). The following *formula* can be used to calculate the *final* copy number after PCR: $N = No\ (eff)^{c}$

    Because one of the primers is labeled with 6-fam, we generally use twice the amount of the unlabeled primer than the labeled primer. This is to prevent *over-amplification* of the labeled strand compared to the *unlabeled.* Table elsewhere in this chapter is an example of a standard 20-µL PCR mix.

11. Troubleshooting a PCR reaction is beyond the scope of this chapter. The objective for primer design is to *design primers* specific for the target sequence, which anneal to the correct sites and facilitate *amplification* of the desired sequence. Once this is achieved, the labeled PCR products are ready for allele *separation* by DCE. However, a good *laboratory practice* is always to optimize each *fragment* with regard to Mg^{2+} *concentration* and *annealing temperature.*

12. Although many minor alterations (base substitution, deletion, and insertion) in DNA results in *thermodynamic* changes, some changes do not alter the thermodynamics of the target

sequence. These changes might be a G:C → C:G or a A:T → T:A transition, *depending* on the *neighbouring* base sequence. To be able to detect those changes, a simple *heteroduplex* formation step (post-PCR) can be performed. Simply by *denaturing* the dsDNA and *annealing* the Watson and Crick strands slowly, highly unstable (as compared to matched base pair) *mismatches* are formed.

A regularly used *heteroduplex* formation program of 95°C for 2 min, followed by reannealing at 65°C for 30 min, is sufficient for heteroduplex formation. Please note that this should be performed only on undiluted PCR products. Figure 4.8 depicts possible strand *combination* as a result of the *denaturation* and *reannealing.*

13. The denaturing temperature has to be adjusted for chemical denaturant in the polymer. Therefore, if the melting domain of the target sequence is *theoretically* melting at 75 °C and the *polymer* contains 7 M urea, the denaturing temperature is set to 54°C (75°C – 21°C, correction for chemical denaturant, ¢3°C/M urea). Sequencing polymer regularly contains 6 or 7 M urea. Consequently, the denaturing conditions are corrected accordingly.

    For sieving matrixes formulated in the laboratory, chemical denaturant can be omitted. However, the temperature limitation or restriction on the various DNA capillary sequencing instruments dictates the use of chemical denaturants. For new fragments, a serial analysis at different temperatures for a heterozygote sample reveals the best separation temperature for the instrument used.

14. Crude or diluted PCR product in water (down to 1:50) is injected electrochemically, following the standard injection procedure recommended by the instrument manufacturer. Figure 4.9 demonstrates the relative small differences in injection efficiency introduced when PCR products are diluted in water. The PCR product was injected by applying 10kV for 20s and alleles were separated by CTCE in a MegaBACE 1000.

    The peaks are two homoduplexes and two heteroduplexes, respectively. Please note the nontemplate addition of adenosine, which is the stutter peak to the right of each duplex. We observed degradation of DNA diluted in water stored in 4°C.

    Therefore, the diluted product should be stored at –20°C or below when not in use. The dilution of PCR products in "commercial loading solutions" or *formamide* is not recommended, as this may denature the dsDNA.

15. DCE has been used for the genotyping of DNA variants by separating alleles and comparing the pattern with an internal standard. The internal standard is made from a heterozygote sample by *reamplification* with a 20-mer primer specific for the GC-clamp. This primer has a different *fluorophore* attached to its 5¢ end from the primer used to amplify the samples.

    As the GC-clamp is standardized, the same 20-mer labeling *technique* can be used for different fragments. Amplification of the internal standard with differential labeling is performed on a 1:1,000, or 1:10,000, diluted *heterozygote* PCR product in H_2O. Due to the high template copy number, fewer *thermocycles* are needed (less than 25). Primers:

    Forward: 5'-Alternative label-CGCCCGCCGCGCCCCGCGCC3' Reverse: Unlabeled primer

    In Figure 4.10, the bold lines represent the GC-clamp and the *short* lines with arrows

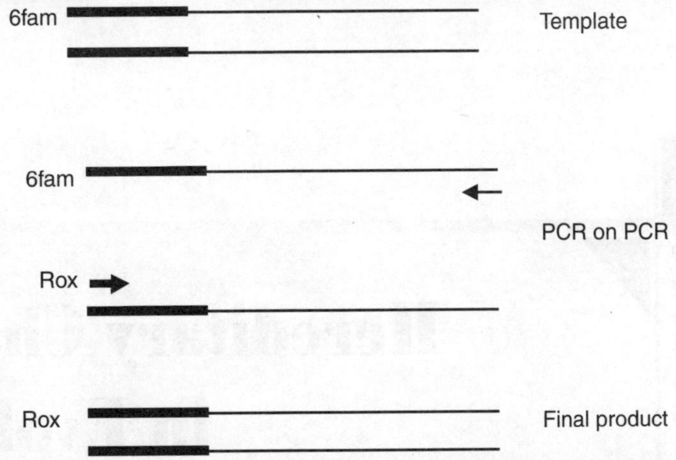

Figure 4.10: Label "switchning" used to construct an internal standard.

indicate the forward and *reverse* primers. The *alternative* label is instrument *dependent* but could be, for example, carboxy-X-rhodamine (ROX).

16. We noted that labeled ssDNA molecules have a "tendency" to migrate into the *heteroduplex* region during DCE. Hence, using unbalanced primer *concentrations* reduces the risk of *amplifying* the labeled ssDNA butm allows the amplification of unlabeled ssDNA, which has no effect on the analysis.
17. The electrophoresis cost per sample (using standard conditions, as recommended by the instrument manufacturer) is about $1 (MegaBACE 1000). However, we optimized the DCE method using the MegaBACE 1000 with respect to cost per analysis.

    By formulating the sieving matrix and running buffer in the *laboratory*, we got more runs out of the capillary arrays and thereby reduced the polymer cost. Hence, the cost to us of running the *electrophoresis* with the MegaBACE 1000 is about 5 cents per sample.

Chapter 5

Hereditary Changes in Evolution

The concept of population is intimately connected with evolution. Evolution may be *parsimoniously* defined as the process of change over time of the allele *frequencies* within populations. In this chapter, we consider the *parameters* and statistics required to describe the genetic properties and genetic changes in populations of *diploid sexual* organisms. The *criteria* used to characterize a population must take into account the *dynamic* properties of that population. Populations are undergoing constant change because the individuals that comprise the population are themselves changing from birth to mate selection, reproduction, and *eventually* death. In addition, individuals may migrate into or away from a particular population.

When a portion of a population relocates, the members of that subpopulation introduce a sample of the alleles from the original population into the new locale. The subpopulation may then grow to become different from the original population, possibly leading to a *speciation* event. When individuals migrate into a population from another distinct population, they may introduce new alleles. Added to these processes of change is the *inexorable* accumulation of point mutations, which adds new *alleles* to any population (other types of mutations, such as deletion or *inversion*, usually *inactivate* genes rather than create new alleles).

It is evident that populations and the forces acting upon them can be complex, thus requiring correspondingly complex statistical models. There are a number of reasons why we wish to study populations. Arguably the most important is that *variation* within populations is the basis for natural selection. Another *reason* is to infer population history. For example, we can measure allele frequencies in human populations and use them to infer *migration* patterns that affected recent human evolution. In addition, we might look for alleles associated with genes implicated in particular genetic diseases. A special case of this is *analyzing* genes associated with diseases for which particular human *subpopulations* have elevated risk (e.g., adult diabetes in some native American populations in the southwestern United States).

BASIC PRINCIPLES

Every diploid organism has two copies of each genetic locus carried on pairs of autosomes (chromosomes other than sex chromosomes). A *locus* is an identifiable region on a chromosome, and it may correspond to a gene or to a physical marker such as a *sequence-tagged* site (STS). We discuss genes for simplicity. Gene *nomenclature* can vary from organism to organism, and we use conventions like those used in human genetics. *Genes* are represented as uppercase italicized letters (up to four for humans), and *phenotypes* are represented by the corresponding Roman letters: phenotype *A* corresponds to gene *A*.

We are concerned with *diploid* organisms, and the alleles residing on different members of a corresponding pair of *chromosomes* are separated by a forward stroke: A_1/A_2 means that allele A_1 of gene *A* resides on one chromosome and allele A_2 resides on the other. The two gene copies corresponding to a particular locus in an organism may or may not be *exactly* identical. For example, in human populations from equatorial Africa, the β-globin gene *HBB* may appear in at least two forms: type A, which is associated with the normal *phenotype*, and type S, which is associated with sickle-cell disease.

These alternative forms of the same gene are examples of alleles. During meiosis, alleles corresponding to a particular locus *segregate*, which means that one copy of any locus appears in any given gamete. In contrast, two different genes on the same chromosome do not segregate unless *recombination* has occurred.

If two alleles at a given locus are identical in an individual, then that *individual* is said to be *homozygous* for the genes at that locus. If the two alleles are different, then the individual is *heterozygous* with respect to the genes at that locus.

A special case of homozygosity is *autozygosity,* which means that two alleles are identical because they have descended from the same common ancestral allele. Suppose that phenotype A is associated with allele A_1, and that phenotype A is associated with allele A_2. If an A_1/A_2 heterozygote has phenotype A, then A_1 is said to be the *dominant* allele and A_2 is said to be the *recessive* allele. In humans, *cystic fibrosis* is associated with a mutant form of the *CFTR* gene (located on *chromosome* 7).

Individuals who have two copies of the normal gene, or one copy of the normal allele and one copy of the defective allele, are not affected. but persons who have two defective alleles are affected. The wild-type allele is thus *dominant,* and the disease is said to be autosomal recessive. Sometimes the phenotype of the heterozygote is intermediate between those of the parents, each of whom is *homozygous* for different *alleles.* This *circumstance* is called incomplete dominance.

Sometimes the phenotype associated with a gene fails to appear because of the particular *constellation* of other genes in that *individual* or particular environmental *circumstances.* The probability that a gene confers the *phenotype* associated with it is called its penetrance. It is very important to note that a large proportion of *phenotypes* in complex organisms depend upon the *cumulative* effects of several genes.

Such phenotypes are called *polygenic traits.* For example, height and skin color in humans do not depend on single genes but rather on the collaboration of a number of genes. Skin color in humans is thought to be *controlled* by three to six loci. Teasing out the contributions of *multiple* genes, each of which might contribute in a *quantitative* way (so-called *quantitative trait loci*), can be a complicated exercise in genetics.

HEREDITARY CHANGES IN MAN

In this section, we use populations drawn from different regions of the world to show how *genetic* variation is distributed across human populations. Because only a small proportion of alleles are unique to a *particular population*, knowing what population or region an individual belongs to does not help much in *predicting* which alleles the individual will have.

On the other hand, by studying information across different loci, we can often predict which population or region an individual comes from by knowing which *combination* of alleles is present in the individual.

The data are taken from 52 populations in seven geographic regions of the world measured for genotypes at 377 loci. An example of a region is sub-Saharan Africa, and examples of populations within this region are *Bantu, Mandenka, Yoruba, San, Mbuti Pygmy*, and *Biaka Pygmy*.

The loci examined were *microsatellites,* which are tandem repeats of short *k*-words. Corresponding to each locus is a characteristic number of different alleles. An example of the allele frequencies for two loci measured for seven different *geographical* regions is shown in Figure 5.1. For the particular locus D12S2070, there are eight alleles. Clearly, the allele frequencies differ for different regions.

For example, one allele represents 85%, of the total in native American populations, but that same allele represents only 7% of the total in populations from the Middle East. *Allele* frequencies for other loci may not vary much *across* regions, as *illustrated* by D6S474.

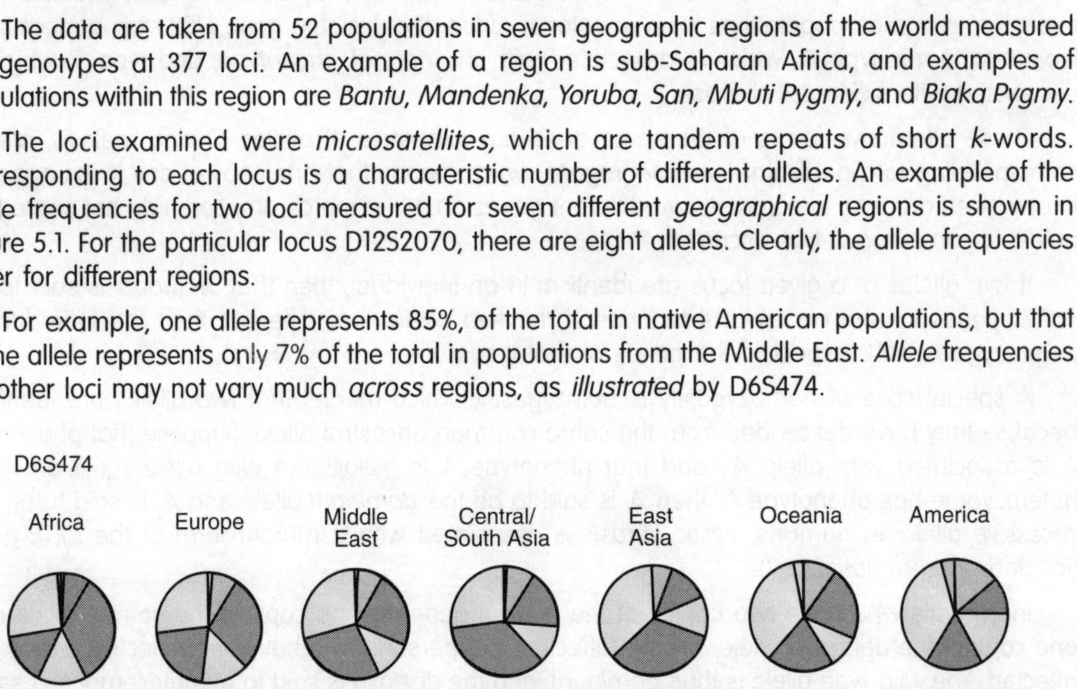

Figure 5.1: [This figure also appears in the color insert.] Allele frequencies associated with two microsatellite markers in human populations taken from seven different regions. Each allele has a different color code, and the size of the sector in the pie chart indicates allele frequency.

Biostatistics Calculation

We now introduce some statistics for describing population variation. Let K be the number of alleles for any given locus, and let p_i be the relative frequency of allele j. The probability F that two randomly chosen genes at a locus are identical is

$$F = \sum_{i=1}^{K} p_i^2. \tag{5.1}$$

We note that the population frequencies p_i are unknown parameters of our population and must be estimated from a sample of individuals from that population. F can then be calculated from the estimated allele frequencies. The *heterozygosity* H for any locus is defined as $H = 1 - F$:

$$H = 1 - \sum_{i=1}^{K} p_i^2. \tag{5.2}$$

Computational Example 5.1: Calculating heterozygosities

Heterozygosity provides a measure of the amount of variation within a population. The results illustrate how H behaves for different numbers of alleles and allele probability distributions. Observe that for a fixed number of alleles, the *heterozygosity* is greater if the distribution of the allele *probabilities* is uniform than if some alleles *predominate.*

Also, for uniform allele distributions, the heterozygosity is greater if the number of alleles is greater. Heterozygosities can be calculated within populations, within regions, or globally. Table 5.1 *exemplifies* the *typical* structure of the underlying data for a single locus having three alleles.

There would be similar tables for every other locus, each with its characteristic number of alleles. The *heterozygosity* reported for each population measured over 100 loci would be the *heterozygosity* averaged over all 100 loci.

Examples of Calculated Heterozygosities

For the data in Table elsewhere in this chapter, we can calculate a variety of heterozygosities.

Population 1, region 1:

$$H_{Pop1-1} = 1 - \left[(3/10)^2 \_ (5/10)^2 + (2/10)^2\right] = 0.620.$$

Region 1 as a whole:

$$H_{R_1} = 1 - \left[(15/30)^2 + (10/30)^2 + (3/30)^2\right] = 0.611.$$

Average of populations in region 1:

$$H_{R_1,P} = (0.62 + 0.62 + 0.46)/3 = 0.567.$$

World as a whole:

$$H_W = 1 - \left[(31/120)^2 + (44/120)^2 + (45/120)^2\right] = 0.658.$$

Table 5.1: Example of data structure and heterozygosities for a single gene having three alleles examined in four world regions, each containing three populations. The three rightmost columns show the number of instances of the alleles *a, b, c.*

		Allele		
		a	*b*	*c*
Region 1	Pop 1-1	3	5	2
	Pop 1-2	5	3	2
	Pop 1-3	7	2	1
	Total for region	15	10	5
Region 2	Pop 2-1	3	3	4
	Pop 2-2	4	3	3
	Pop 2-3	3	4	3
	Total for region	10	10	10
Region 3	Pop 3-1	0	1	9
	Pop 3-2	1	1	8
	Pop 3-3	0	2	8
	Total for region	1	4	25
Region 4	Pop 4-1	2	7	1
	Pop 4-2	1	7	2
	Pop 4-3	2	6	2
	Total for region	5	20	5
World		31	44	45

In addition to the direct measures of heterozygosity, it is useful to consider particular averages. If there are B populations in a region with heterozygosities H_1, H_2, ... , H_B, then the average *heterozygosity* for that region is

$$H_{R,avg} = \frac{1}{B}\sum_{i=1}^{B} H_i. \tag{5.3}$$

We can also calculate the average regional heterozygosity over R regions,

$$H_R = \frac{1}{R}\sum_{i=1}^{R} H_{R_i}, \tag{5.4}$$

where the H_{Ri} correspond to the *heterozygosities* of each region as a whole, obtained by pooling data from all *populations*(i.e., not calculated as in [5.3])

Now we are prepared to take a look at the actual population heterozygosities estimated for different human populations. The first column of heterozygosities are the regional data calculated directly from the allele *frequencies* of the *pooled populations*-not calculated as averages. The bottom line, first column, is computed from a pool of all the data, taking the world as a single region.

Data above the last entry in the second column ($H_{R,avg}$) were calculated as averages of population heterozygosities using (5.3). The average regional heterozygosity, H_R, calculated according to (5.4), is 0.733. The average *subpopulation* heterozygosity, H_R, can be calculated from the entries above it as a weighted average using the numbers of each population as *weighting* factors.

Table 5.2: Heterozygosities for 377 human microsatellite alleles by population and region. H_R: region as whole; $H_{R,avg}$: average of populations in region. The number of populations in each region is given in parentheses. Source: Rosenberg NA et al. *(2002)* www.sciencemag.org/ cgi/content /full/298/5602/2381/DC1.

Region	HR	HR,avg
sub-Saharan Africa	0.792	0.774 (6)
Europe	0.753	0.751 (8)
Middle East	0.761	0.756 (4)
Central/S. Asia	0.759	0.752 (9)
East Asia	0.730	0.723 (18)
Oceania	0.695	0.683 (2)
America	0.644	0.599 (5)
World	0.771	0.727

Remember that the heterozygosities provide measures of population variation. With the data in Table elsewhere in this chapter and the population averages described above, we can estimate the fractions of human variation that occur within populations, among populations within regions, and among regions of the world. Let H_T be the heterozygosity for the total population of the world. Since we can write

$$H_T = H_T - H_R + H_R - H_S + H_S,$$

we see that

$$1 = \frac{(H_T - H_R)}{H_T} + \frac{(H_R - H_S)}{H_T} + \frac{H_S}{H_T}. \tag{5.5}$$

Table 5.3: Example of allele frequencies for two populations measured with respect to three genes having two alleles. Both alleles of all genes are present in both popula-tions. While the minor allele of each gene is somewhat common in each population, particular combinations of alleles may be quite rare (e.g., *A2/A2, B2/B2, C2/C2 in* population I).

Gene	Allele Frequencies			
	Population I		Population II	
	1	2	1	2
A	0.8	0.2	0.2	0.8
B	0.8	0.2	0.2	0.8
C	0.8	0.2	0.2	0.8

The first term on the right-hand side of (5.5), the fraction of variation that occurs among regions of the world, is (0.771 – 0.733) /0.771 = 0.049. The second term, the fraction of variation that occurs among populations within a region, is (0.733 – 0.727)/0.771 = 0.008. The last term, the fraction of variation within populations, is 0.727/0.771 = 0.943. This means that 94% of the human *variation* in the world occurs at the level of local *populations*, and only about 5% of the variation occurs among *regions* of the world.

On average, for each locus, there is more variation between two African people than there is between the average African and the average Chinese person. The last *statement* does not mean that we can't distinguish East Asians from Africans. A simple example that shows why can be illustrated from the allele *frequency* data in Table elsewhere in this chapter.

Suppose that we have two populations, I and II, and estimated allele frequencies at three unlinked loci, each of which has two alleles. We score a *particular* individual (Joe) for his *genotype* with respect to these three genes and find him to be A_1/A_1, B_1/B_1, and C_1/C_1 that is, *homozygous* with respect to the first alleles of genes *A, B,* and C.

The joint probability of this occurrence for an individual belonging to population I is $(0.8)^2 \times (0.8)^2 \times (0.8)^2 = 0.26$. The joint *probability* of this genotype for an individual belonging to population II is $(0.2)^2 \times (0.2)^{2\times} (0.2)^2 = 0.00006$. Clearly, Joe's genotype is far more common in population I than in population II, and we would conclude that Joe is more likely to be "I-ese" than "II-ian."

Structure of Human Population

The world's human population displays *population structure:* there are clear subpopulations indicated by the data. The populations are hierarchical in the sense that each local population can be grouped into *successively* larger inclusive groupings (local population * regional population_* world population). Because the local populations are genetically distinct, the total population for the world as a whole is said to be *stratified.*

For such populations, there are particular relationships between heterozygosities calculated as averages of subpopulation data and heterozygosities calculated from pooled data. Note in Table 5.2 that the heterozygosity computed from population values, $H_{R,avg}$, is always less than H_R, the *heterozygosity* computed by pooling all individuals in a region (the region as an whole). This type of *behaviour* is a general *occurrence* when a population is broken down into subpopulations (the individual populations in the example above). This can be shown as follows.

Suppose we wish to compare the heterozygosities for a total population, which is broken down into B subpopulations, to the average heterozygosity of the subpopulations. We perform measurements on one gene having *K* alleles.

We let p_i be the average fraction of allele *i* in the population as a whole and p_{ib} be the fraction of allele *i* in subpopulation *b*. The total population homozygosity, F_T, is

$$F_T = \sum_{i=1}^{K} p_i^2,$$

and the homozygosity of subpopulation *b is*

$$F_b = \sum_{i=1}^{K} p_{ib}^2,$$

The average homozygosity for the subpopulations, *Fs,* is

$$F_S = \frac{1}{B}\sum_{b=1}^{B} F_b$$

(analogous to (5.3)). We are interested in the difference between F_S calculated by averaging the *Fb* for the subpopulations, and F_T calculated from pooled data:

$$F_S - F_T = \frac{1}{B}\sum_{b=1}^{B}\sum_{i=1}^{K} p_{ib}^2 - \sum_{i=1}^{K} p_i^2. \tag{5.6}$$

In Exercise 3 we show that

$$F_S - F_T = \sum_{i=1}^{K}\left[\frac{1}{B}\sum_{b=1}^{B}(p_{ib} - p_i)^2\right]. \tag{5.7}$$

The terms in the brackets are seen to be the variances of the frequencies (across subpopulations) of each allele, measured with respect to the population average. In terms of variances σ_i^2,

$$F_S - F_T = \sum_{i=1}^{K} \sigma_i^2. \tag{5.8}$$

In terms of heterozygosities, $H_T = 1 - F_T$ and $HS = 1 - F_S$, so we finally obtain

$$H_T - H_S = \sum_{i=1}^{K} \sigma_i^2. \tag{5.9}$$

Since the variances σ_i^2 are always positive, the heterozygosity for a pooled population is always larger than the average of the heterozygosities of the subpopulations of which it is *constituted.*

The quantities we have been discussing are closely related to the F-statistics defined by Wright (1951). These statistics were called fixation indices, and they describe the heterozygosities of structured populations at different levels,

$$F_{ST} = \frac{H_T - H_S}{H_T},$$

$$F_{SR} = \frac{H_R - H_S}{H_R},$$

$$F_{RT} = \frac{H_T - H_R}{H_T},$$

where subscripts *S, R,* and *T* stand for subpopulation, region, and total in the same sense that we have been using them above. These subscripted quantities should not be confused with homozygosities, *F.* Note that the terms in (5.5) are related to the *fixation indices.*

ALLELIC RECOMBINATION

The discussion up to now has not included *recombination,* which permits alleles on one chromosome to *substitute* for alleles on other copies of that chromosome. When describing a population in terms of individual loci, it *suffices* to report allele frequencies corresponding to each locus. Sometimes we are interested in several different loci that reside on the same *chromosome.* The specification of the alleles for loci on the same chromosome is called the *haplotype.*

The distinction between genotype and haplotype is shown in Figure 5.2. Without recombination, new haplotypes would be generated only by new mutations. However, new haplotypes are produced by recombination because chromosome pairs usually recombine during gamete formation. Given a group of individuals affected by a particular genetic disease phenotype, how do we identify the gene or set of genes that confer that *phenotype*?

One way is to identify linkage of the locus associated with the disease with mapped genetic markers. Such *linkage analysis* may be performed by using families containing affected individuals. One *limitation* of such an approach is that the numbers of individuals in *families* available for genetic study are usually rather small, which means that the number of *meioses* represented in the pedigree (family tree) is correspondingly small.

As we shall the next section, low numbers of meioses mean that there will be a low number of recombination events, which means that any linkage detected will probably be to genetic markers that are comparatively far away from the locus causing the disease.

Another limitation is that the interpretation of linkage analysis data is tied to a model of inheritance. An alternative to linkage analysis is *association analysis,* which seeks statistical relationships between alleles within a larger population and the disease *phenotype.* An example of *association* analysis is the *case-control* study.

The population can now be much larger than one or a few families, so many meioses usually will have occurred since the appearance of the disease-causin gallelein the population. If a particular allele of gene *X* is usually co-inherited with the disease state, then the disease locus is probably

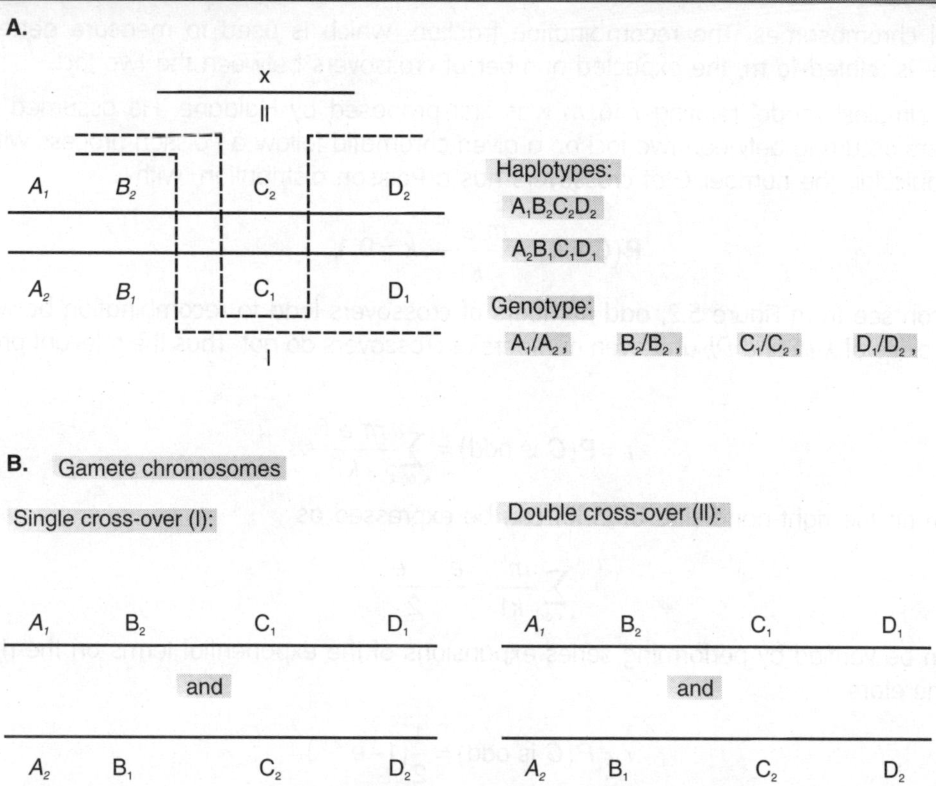

Figure 5.2: Distinctions between haplotypes and genotypes, and the effects of recombination. Panel A: A portion of two chromosomes is shown containing four genes, each of which has two alleles. Two different recombination events (broken lines) are diagrammed. Recombination I involves a single crossover, while recombination II involves a double crossover. The products from these events are illustrated in Panel B.

close to gene X. To apply this logic, we need to know the sizes of genomic regions for which *allelic* combinations have not been disrupted by recombination. Such regions are also called haplotype blocks.

Recombination and Distance Between Genes

We first develop an expression that relates the probability of recombination to distance. Two crossover events are diagrammed in Figure elsewhere in this chapter. Crossover I involves a single genetic exchange in interval x and joins the left segment of the top chromosome to the right segment of the bottom *chromosome* and vice versa.

An odd number of crossover events anywhere in region x will produce products that are recombinant for markers to the left and right of region x, in this case alleles of genes A and D. Crossover II involves two genetic exchanges in region x. Even numbers of exchanges may generate products recombinant for genes that lie to the left (or right) of x and genes lying within x, but they do not generate products recombinant for genes that *flank* region x.

We are interested in how frequently haplotypes are shuffled by genetic exchange. The *recombination fraction* r is the probability that alleles at two loci on a chromatid come from *different* parental chromosomes. The recombination fraction, which is used to measure genetic map distance, is related to m, the expected number of crossovers between the two loci.

The simplest model relating *r* to *m* was first proposed by Haldane. He assumed that the crossovers occurring between two loci on a given chromatid follow a Poisson process with mean m. In particular, the number C of crossovers has a Poisson distribution, with

$$\mathbf{P}(C=k)=\frac{m^k e^{-m}}{k!}, k=0,1,\ldots.$$

As we can see from Figure 5.2, odd numbers of crossovers lead to recombination between loci on both sides of *x (A* and *D),* and even numbers of crossovers do not. Thus the relevant probability is

$$r=\mathbf{P}(C \text{ is odd})=\sum_{k \text{ odd}}\frac{m^k e^{-m}}{k!}. \tag{5.10}$$

The sum on the right-hand side of (5.10) can be expressed as

$$\sum_{k \text{ odd}}\frac{m^k}{k!}=\frac{e^m-e^{-m}}{2}.$$

(This can be verified by performing series expansions of the exponential terms on the righthand side.) Therefore.

$$r=P(C \text{ is odd})=\frac{1}{2}\left(1-e^{-2m}\right). \tag{5.11}$$

This is the Haldane mapping function, which relates recombination frequency (probability) with map distance, *m.* When m is small, we can neglect terms of order m^2 and higher in a series expansion of the exponential term in (5.11) to see that for short distances $r \approx m$. Note that as *m* gets very large in (5.11), the probability of recombination approaches 0.5. This means that relative to a chosen locus on one chromosome, there is a 50:50 chance that an allele at a distant locus has become associated with the partner chromosome.

In such a case, the loci appear to be *unlinked.* (Note that because of random assortment of chromosomes during meiosis, the probability that genes appearing on *different* chromosomes will appear in the same gamete is also 0.5.) Loci that are close together are therefore less likely to be separated by recombination than are loci that are far apart, and for *sufficiently* short distances, there is an *approximately* linear *relationship* between recombination frequency and distance.

This is the rationale for measuring genetic distances in terms of recombination frequencies, and the map units used for eukaryotes are *centimorgans* (named after Thomas Hunt Morgan). One *centimorgan* distance between two loci corresponds to having (on average) one recombination event *separating* these loci every 100 meioses. Because meiosis is different in male and female mammals (*oocytes* are arrested in meiosis I in females), the relationship between map distances in human males and females is different.

In males, it is estimated that there are 0.92 cM per Mb, but for human females, there are 1.7cM/Mb. The sex-average relationship between recombination frequency and distance is thus very approximately 1 cM/Mb. These numbers are only averages: recombination rates vary from chromosome to chromosome and at different positions in any particular chromosome. Local recombination rates measured at genetic map resolution range from 0.1 cM/Mb to 3 cM/Mb for Chromosome 3.

In some regions rates are even larger, as described in elsewhere in this chapter. The total genetic map length of an *organism* can be obtained by multiplying the number of cM/Mb by the physical length, or more *accurately* by concatenating map distances *separating* all loci. Using the latter approach, the genetic length of the human female autosomal genome is estimated to be 4821 cM, and the corresponding length for the human *male autosomal* genome is about 2590cM.

Measurement and Recombination

Appropriate genetic markers that can easily be assayed are required for genetic mapping or measurement of *recombination*. If recombination is to be detected between two loci, there must be more than one allele for each locus. For example, the recombination events I and II shown in Figure5.2 (bottom) produced new *haplotypes* that are different from the *parental haplotypes*.

Given suitable assays, these haplotypes could be detected. However, suppose that the genotype for the rightmost marker was D_1/D_1; if we were scoring recombination using loci *A* and *D,* we would not be able to detect recombination because the state of the allele at position *D* (D_1 /D_1) would remain unchanged even if recombination occurred.

What is required are *polymorphic markers*-genes or loci that have two or more alleles with sufficiently high frequency in the population. We say that a locus is polymorphic if it has at least two alleles with frequency greater than 1/100. (This cutoff is arbitrary, serving only to remove alleles with extremely small frequencies.)

There are three major considerations affecting the utility of any type of genetic marker: (a) the *density of the marker* along the genome, (b) the *type of assay* of marker alleles, and *(c),amenability to high-throughput screening*. Genes were the earliest markers employed in genetics.

With estimates of 25,000 genes in the human genome, the marker density is roughly 110^5 bp. Traditionally, genes were scored by different methods, depending upon the phenotype that they confer. For example, different blood group alleles might be distinguished by by *immunological* assays. Genes *contributing* to complex *traits* might have no corresponding simple *phenotypic* assay.

If different assay methods are needed for scoring different loci, prospects for high-throughput screens based on gene products are poor. Recombinant DNA *technologies* provided assays for new types of genetic markers that could be scored using the same experimental platform.

One approach is gel electrophoresis of DNA digested by a particular restriction endonuclease followed by *Southern blotting* (i.e., transfer of *resolved* DNA fragments to membranes for *hybridization* to a labeled probe).

One type of marker that can be detected in this way is the *restriction fragment length polymorphism (RFLP),* which arises when a particular restriction endonuclease cleavage site is inactivated (i.e., mutated) in a subpopulation. Another type of polymorphism is the presence or absence of a *transposable element,* often Alu elements for human DNA. There are about 10^6 Alu sequences in the human genome, but different human populations may display characteristic Alu insertions or deletions, which alter sizes of restriction fragments from particular chromosomal regions. Thus, Alu elements are another source of RFLPs. Microsatellite and minisatellite *polymorphisms, variable number of tandem repeat (VNTR)* polymorphisms, can sometimes be detected by gel electrophoresis and Southern blotting. These consist of tandem repetitions of DNA segments occurring at numerous locations throughout the human genome. The number of repeated bases may differ for different *individuals* or populations. One type (*minisatellite* DNA) tends to have extensive repetition of elements ranging in size from 14 to 500 bp.

These are often found in telomeric regions of chromosomes. Another type (microsatellite DNA) comprises shorter blocks of sequence with repeat sizes of 1 to 13 bp. Di-, tri-, and tetra-nucleotide repeat units are common. *Microsatellite* DNA is distributed throughout the genome. Examples of di- and tri-nucleotide repeats are $(CA)_n$ and $(AAT)_n$, with the polymorphism arising from differing values of n among members of the population.

The advent of the polymerase chain reaction (PCR) revolutionized genome analysis because of its sensitivity (as little as one molecule can be detected) and specificity (particular regions within a vast "sea" of genomic DNA can be selectively *amplified*). PCR is an *automatable* method that can be used to assay for VNTR and restriction site *polymorphisms,* obviating the need for Southern *blotting* (which is hard to automate and employs *radioactive* probes). A disadvantage of PCR is the required design and synthesis of specific primer pairs for each region to be amplified.

A single-nucleotide polymorphism (SNP) is variation in the identity of the base appearing at a particular position in the genome. There are an estimated 1.5×10^6 -10^7 polymorphic SNP sites in the human genome (Cargill et al., 1999; *International* SNP Map Working Group, 2001; Venter et al., 2001), corresponding to *resolution* at a level of hundreds of base pairs. Detection can employ either DNA sequencing or *oligonucleotide* array *technologies,* which means that SNP analysis can be readily automated and multiplexed.

If detection is by hybridization, the underlying principle is that mismatched bases reduce the melting temperature of DNA-DNA hybrids. In the simplest example, four different oligonucleotide features, each having one of the four different bases at the polymorphic site, can be arrayed on a solid substrate.

The state of the base at the polymorphic site is indicated by which of the four features hybridizes with the DNA being tested. An alternative detection method is primer extension. The underlying principle in this case is the ability of DNA polymerase to add the correct base to a growing chain during synthesis along a DNA template. The *oligonucleotide* probe sequences arrayed on a solid substrate act as primers, so designed that the next base to be added *corresponds* to the *polymorphic* site.

The polymorphism can be detected by supplying dXTP molecules bearing distinctive fluorescent tags and detecting the wavelength of *fluorescence emission* after DNA synthesis. When the *primers* are not *immobilized,* other detection methods such as standard DNA sequencing gels or MALDI-TOF mass spectrometry may be employed.

NON RANDOM ALLELIC ASSOCIATION

Linkage disequilibrium (hereafter abbreviated LD) refers to the nonrandom association of alleles in haplotypes. It is measured by comparing the proportion of an observed *haplotype* with the proportion that would be *predicted* based upon the population frequencies of the alleles at each *locus*. In this section, we discuss a number of properties and *consequences* of LD.

Quantitative Description of LD

Linkage disequilibrium may be quantified as the difference between the observed and predicted frequencies for allele combinations at two or more loci. The predicted frequencies are computed using the population frequencies of the alleles. For a system in which each locus has two alleles, let $P_{A_1B_1}$ and $P_{A_2B_2}$ be the *probabilities* in the population of *haplotypes* A_1B_1 and A_2B_2, respectively.

The allele frequencies of the genes contributing to these haplotypes are P_{A_1}, P_{B_1}, P_{A_2} and P_{B_2}. Note that the allele frequencies are population quantities, which are *unaffected* by recombination. We introduce a quantity D that measures the difference between these observed and *predicted* frequencies:

$$D = p_{A_2B_2} - p_{A_2}p_{B_2}. \tag{5.12}$$

If the A_1B_1 and A_2B_2 haplotypes are in excess of their predicted frequencies, as indicated above, then there must be corresponding deficits in the haplotypes A_1B_2 and A_2 B_1 (having probabilities $P_{A_1B_2}$ and $P_{A_2B_1}$, respectively):

$$\begin{aligned} p_{A_2B_1} - p_{A_2}p_{B_1} &= -D. \\ p_{A_1B_2} - p_{A_1}p_{B_2} &= -D. \end{aligned} \tag{5.13}$$

Some straightforward algebra shows that

$$D = p_{A_1B_1}p_{A_2B_2} - p_{A_1B_2}p_{A_2B_1} \tag{5.14}.$$

(taking into account the fact that $P_{A_1}P_{B_1} + P_{A_2}P_{B_2} + P_{A_1}P_{B_2} + P_{A_2}P_{B_1} = (P_{A_1} + P_{A_2})(P_{B_1} + p_{B_2}) = 1$). The value of D can be either positive or negative, and the sign depends on the arbitrary labeling of the alleles. If $D = 0$, the loci are said to be in *linkage equilibrium*.

The magnitude of D depends upon the probabilities of the individual alleles. For example, in the special case for which $P_{A_1} = P_{A_2} = P_{B_1} = P_{B_2} = 0.5$, the maximum value of D occurs when $P_{A_1B_1} = 0.5$ (which requires that $P_{A_2B_2} = 0.5$ and that $P_{A_1B_2} = P_{A_2B_1} = 0.0$). The value of D in this case is 0.5^2, or 0.25. Alternatively, with the probabilities of all alleles again set at 0.5 and $P_{A_1B_1} = P_{A_2B_2} = 0$, then $P_{A_1B_2} = P_{A_2B_1} = 0.5$, and from (5.14), $D = -(0.5^2) = -0.25$.

In practice, it is convenient to express the data in terms of $|D'|$, where $D' = D/D_{max}$ and D_{max} is the maximum value of D. We show in Exercise 5 that

$$\begin{aligned} D_{\max} &= \min\{p_{A_2}p_{B}, p_{A_1}p_{B_2}\}, \text{ if } D > 0, \\ D_{\max} &= \min\{p_{A_1}p_{B_1}, p_{A_2}p_{B_2}\}, \text{ if } D < 0. \end{aligned} \tag{5.15}$$

$|D'|$ has the convenient property that $0 \le |D'| \le 1$, with $|D'| = 1$ corresponding to complete LD.

Another convenient measure of LD, r^2, is obtained by dividing D^2 by $P_{A_1}P_{A_2}P_{B_1}P_{B_2}$

$$r^2 = \frac{D^2}{p_{A_1}p_{A_2}p_{B_1}p_{B_2}}. \tag{5.16}$$

The quantity r^2 is the square of the *Pearson* product-moment correlation coefficient. which we encountered in elsewhere in this section. The *equivalence* between (5.16) and the definitional form for r^2 can be seen by noting that for binary alleles of gene X (X_1 and X_2), the variance is $\sigma^2 X = P_{X_1}P_{X_2}$. This means that the *denominator* of (5.16) is $\sigma^2_A\sigma^2_B$.

Another way of expressing r^2 with the use of (5.12) is

$$r^2 = \left(\frac{p_{A_1B_1}}{p_{A_1}p_{B_1}} - 1\right)\left(\frac{p_{A_2B_2}}{p_{A_2}p_{B_2}} - 1\right) \tag{5.17}$$

The quantity r^2 measures the *correlation* between alleles at the two sites. For markers that have complete disequilibrium and that have the same allele frequency, it takes the maximal value 1.0 (= [(0.5/0.25 −1)]²). If there is complete equilibrium, the numerators and denominators of the ratios in (5.17) are equal and $r^2 = 0$.

Decaying Linkage Disequilibrium

LD decays over time. It is also possible to obtain an expression for D at any generation in a population in terms of the recombination parameter (recombination fraction) r between twc markers. Suppose that we want to calculate the change in frequency of a haplotype in going from one generation to the next.

Consider a particular A_1B_1 haplotype. This could have arisen as a nonrecombinant copy of an A_1 B_1 haplotype in the previous generation (frequency $(1 - r)p_{A_1B_1}$) or as a result of recombination in the previous generation between two haplotypes that were A_1* and *B_1 (the * denoting that the allele at the other locus is immaterial). The frequency of these latter events is justn $r\,p'_{A_1}p_{B_1}$. Combining the possibilities, we see that the fraction $p'_{A_1B_1}$ of haplotype A_1 B_1 *is*

$$p'_{A_1B_1} = (1-r)\,p_{A_1B_1} + rp_{A_1}p_{B_1}. \tag{5.18}$$

If we subtract $p_{A_1}p_{B_1}$ from both sides of (5.18), we get

$$p'_{A_1B_1} - p_{A_1}p_{B_1} = (1-r)\,p_{A_1B_1} - (1-r)\,p_{A_1}p_{B_1}.$$

which, with the definition of D in (5.12) and the fact that the allele frequencies do not change with time for a simple recombination model, becomes

$$D' = (1-r)D. \tag{5.19}$$

This recursion formula can be applied over t generations to relate the LD D_t at generation t to the initial linkage disequilibrium D_0:

$$D_t = (1-r)^t D_0. \tag{5.20}$$

Equation (5.20) indicates how linkage disequilibrium changes over time. LD will decrease to 0 after a large number of generations because $0 < r \leq 0.5$. We can estimate the number of generations necessary to reduce LD to half of its present value. We take $r = 0.01$ (corresponding to 1 cM-conveniently close linkage for association studies). We solve the equation

$$\frac{D_t}{D_0} = \frac{1}{2} = (1-r)^t$$

for t, with r = 0.01. The result is $t \sim 69$. Taking the human generation time to be 20 years, we calculate that the human LD half-life is about 1400 years for markers separated by 1 cM.

Factors Affecting LD

In the previous section, we modeled LD using a population model that allowed recombination to occur. Other forces acting on populations can lead to association of alleles. For example, smaller *populations* will experience greater sampling *variation*, leading to *genetic drift*. Effects of genetic drift are balanced by recombination (which breaks up haplotypes) and *mutation* (which introduces new *haplotypes*).

The magnitude of LD between two polymorphic loci in a stable population is expected to depend on population size, recombination fraction, and mutation rate. Moreover, LD can also reflect the precise mutational history of a population, *even if no recombination occurs.* Consider, for example, a haplotype A_1 B_1 C_1 in an ancestor, with distances $A - B$ and $B - C$ being identical. Now assume that a particular set of *mutations* occurs in the *pedigree* shown in Figure 5.3.

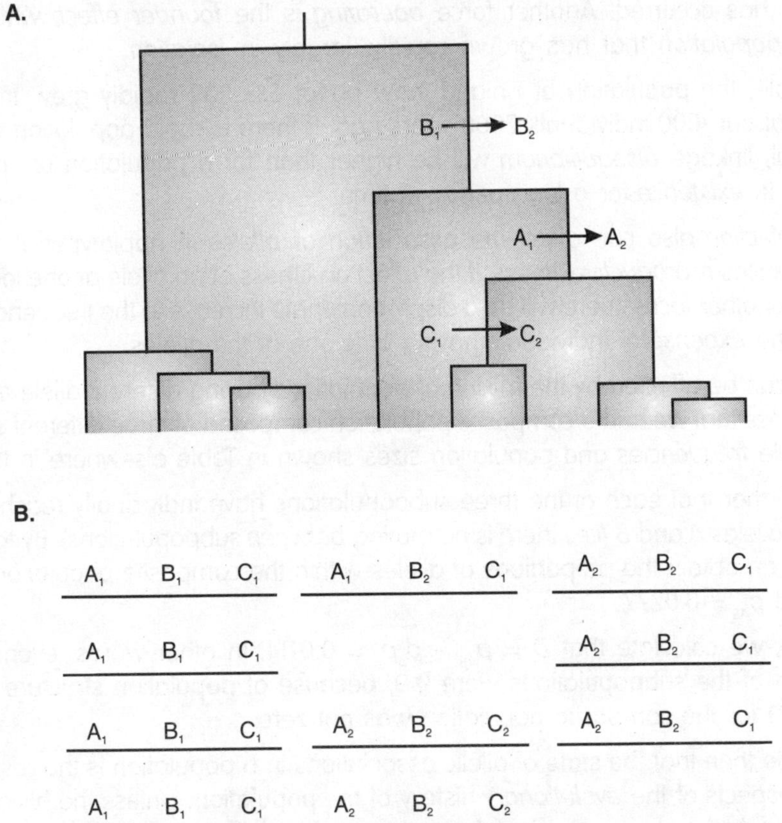

Figure .5.3: Measures of linkage disequilibrium are dependent upon the genealogical history of mutations, independent of any recombination. A particular genealogical tree for ten haplotypes is shown in panel A. The haplotypes corresponding to the leaves of the tree immediately above each column are diagrammed in panel B.

Table 5.4: Example of effects of population stratification on estimated linkage disequilibrium. Probabilities of allele *A* at locus A and allele *B* at locus B are reported. Each individual subpopulation is at linkage equilibrium.

Population	Population size *N*	P_A	P_B	P_{AB}	*D*
I	1,000	0.3	0.5	0.15	0
II	2,000	0.2	0.4	0.08	0
III	10,000	0.05	0.1	0.005	0
I+II+III	13,000	0.0923	0.177	0.0277	0.114

The computed values of LD for loci *AB* and *BC* are not the same, even though no recombination has occurred. Another force *operating* is the *founder effect,* which is observed with a small *population* that has grown rapidly, largely in *isolation*.

For example, the population of Finland (now about 5×10^6) rapidly grew from a founding population of about 1000 individuals 2000 years ago. If there is rapid population expansion over a short interval, linkage *disequilibrium* will be higher than for a population of *comparable* size that has been in *existence* for a long period of time.

Natural selection also can affect the association of *alleles* in haplotypes if the individual's genotype influences *reproductive fitness.* If the effect on fitness of an allele at one locus is changed by an allele at another *locus,* there will be a disproportionate increase in the frequency of *individuals* with both, at the *expense* of individuals having only one of the alleles.

Finally, LD can be affected by the mixing of populations having different allele *frequencies*. For example, suppose that we had a composite population composed of three different *subpopulations* having the allele *frequencies* and population sizes shown in Table elsewhere in this chapter.

Suppose further that each of the three subpopulations have individually reached equilibrium with respect to alleles *A* and *B (i.e.,* there is no mating *between* subpopulations). By taking weighted averages, we can obtain the proportions of alleles within the composite population: $p_A = 0.0923$, $p_B = 0.177$, and $p_{AB} = 0.0277$.

From (5.12), we calculate that $D = p_{AB} - p_A p_B = 0.0114$. In other words, even though the *D* values for each of the subpopulations were 0.0, because of population structure (stratification), the computed D for the composite population was not zero.

We conclude then that the state of allelic associations in a population is the result of *interplay* among many aspects of the *evolutionary* history of the population. Unless the history is "known," it is hard to model LD between pairs of *loci.*

This means that in general, data for the present state of LD in populations are insufficient for making inferences about *parameters* such as recombination fraction, *population size,* or *mutation* rate.

HUMAN GENOME AND LD

There are two major reasons for interest in the pattern of *linkage* disequilibrium in the human population. First, there are medical reasons, as indicated under the head allelic reombination. Recall that identification of genes associated with genetic diseases depends upon *relating* the disease *phenotype* with readily storable genetic markers, either by using *pedigree* analysis or through statistical association.

If there are few genetic markers on the map, disease states associated with any one of them may not be physically close, meaning that the hunt for the disease gene must cover an extensive area. Using markers such as microsatellite repeats has produced *genetic* maps with a resolution of about 0.5 cM. Association of such markers with a disease gene would (on average) allow us to restrict the search to a 500,000 by region of the genome.

The numerous single-nucleotide polymorphisms in the human genome provide an opportunity for an even finer scale of mapping (hundreds to thousands of by map resolution). However, if there is substantial LD, a much smaller number of "tag SNPs" may serve to mark regions in linkage disequilibrium, which in turn may contain the *disease-causing* alleles. The second major reason is that patterns of LD reflect the evolutionary history of human populations over the last 250,000 years (e.g., effects of *selection, migration, population expansion,* and *admixture*).

Different populations might have different patterns of LD, which would reflect the population history. The ability to solve these problems has been augmented by DNA sequencing technologies and the consequent ability to detect single-nucleotide *polymorphisms* (SNPs). Remember that SNPs are the character states at *individual nucleotide* positions in a genomic sequence. In most cases examined so far, SNPs are *biallelic:* there are just two alleles, a major and a *minor allele.*

This means that we can employ the equations developed in elsewhere in this chapter to describe LD in the human genome. When we inquire about LD patterns in a genome, we are really asking about the frequency distribution of chromosomal lengths having ID¢l or r^2 values exceeding some arbitrary threshold value. These patterns could *equivalently* be described in terms of recombination frequencies within intervals *separating* the genes.

It is helpful to think about LD patterns in terms of some simple models. For example, suppose that the recombination parameter *r* is approximately constant throughout the length of a chromosome. In that case, we might model the locations of *recombinational crossovers* as a *Poisson process,* which would produce an exponential distribution of lengths for segments in *linkage* disequilibrium if all *recombination* events were to occur in a *single generation.*

Such a model applied over several generations will produce regions with high LD: loci in close proximity are less likely to be separated by *recombination* than distant loci, and for close loci, there simply may not have been enough *meioses* during the history of the population to produce shortrange *equilibrium.* This is illustrated in Figure 5.4. *Alternatively,* we might try to imagine some simple patterns to which the observed data might *conform.*

For example, we might imagine that the genome is divided into *haplotype blocks,* which consist of consecutive chromosomal loci showing high levels of local LD, separated by short regions within which many recombination events have occurred. This is equivalent to a *punctate recombination* model (where the recombination fraction r is not constant). Whether or not these or more complicated *models* are appropriate, we want statistics to *describe* the observed patterns. For the haplotype block model, these might be (a) mean block size, (b) numbers of blocks per Mb in a *region*, and (c) the *fraction* of the region contained in (or covered by) *haplotype* blocks.

Descriptions of LD in terms of haplotype blocks are influenced by a number of different factors:

— The history of the sampled populations: Populations sampled after a bottleneck will display greater amounts of LD, and populations that have experienced exponential growth will have lesser amounts of LD, than populations that have continued at stable population sizes for a long period of time. As noted above, pooled samples may produce LD by admixture.

— Sample size: The estimated haplotype frequencies are only a sample from a larger population, and estimates of the population frequencies from the sample frequencies are less reliable when the number of genomes sampled is small.

— Methods for determining *haplotypes*: *Haplotypes* may be determined directly, but often they have been inferred by *statistical* methods based upon allele frequencies in the sample (i.e., they are estimated from *diploid* genotypes). Such estimates may be less reliable than direct measures, because of required assumptions in a population model.

— Spacing between SNPs: If the marker density is high (SNPs have close spacing), short segments in LD can be detected, but if marker densities are low, only larger *haplotype* blocks will be observed.

— Cutoffs imposed to exclude rare alleles: To simplify *interpretation* of patterns, some investigators employ only more common alleles to define haplotypes. But rare alleles are found in the "*younger*" haplotypes, and common alleles are *biased* toward older haplotypes. Older haplotypes have had more time for recombination to occur, so that the length distribution of haplotype blocks may be *skewed* toward lower values than in younger haplotypes.

The observations above imply that any statistic describing LD in a particular case is meaningful only within the context of the *methods* and sampling *protocols* employed for making the estimate. Computational Example 5.2 illustrates how SNPs can be used to identify genomic regions that display LD. In this illustration, there are five SNP loci, and the state of each is represented by one of only two alleles.

Invariant loci are indicated by dashes in this illustration, but in the literature, SNP characters are usually listed in positional order with no representation of the positions between the SNPs. The measure of LD employed is *lD'l,* calculated for all possible *pairings* of alleles [$n(n-1)/2$ pairings of *n* alleles]. The result is displayed at the end of the box in a triangular grid, with each element shaded to represent the magnitude of *lD¢l* for the corresponding pair of alleles.

If the order of *SNPs* representing rows and columns is as shown, then the elements representing LD between SNP *j* + 1 and SNP *j* appear in order along the diagonal set of elements (reading from bottom left to top right). Haplotype blocks would appear as clusters of shaded elements near the diagonal. (Only one block, comprised of SNPs 3 and 4, is illustrated in this simplified example.) Figure 5.5 presents the results for real data.

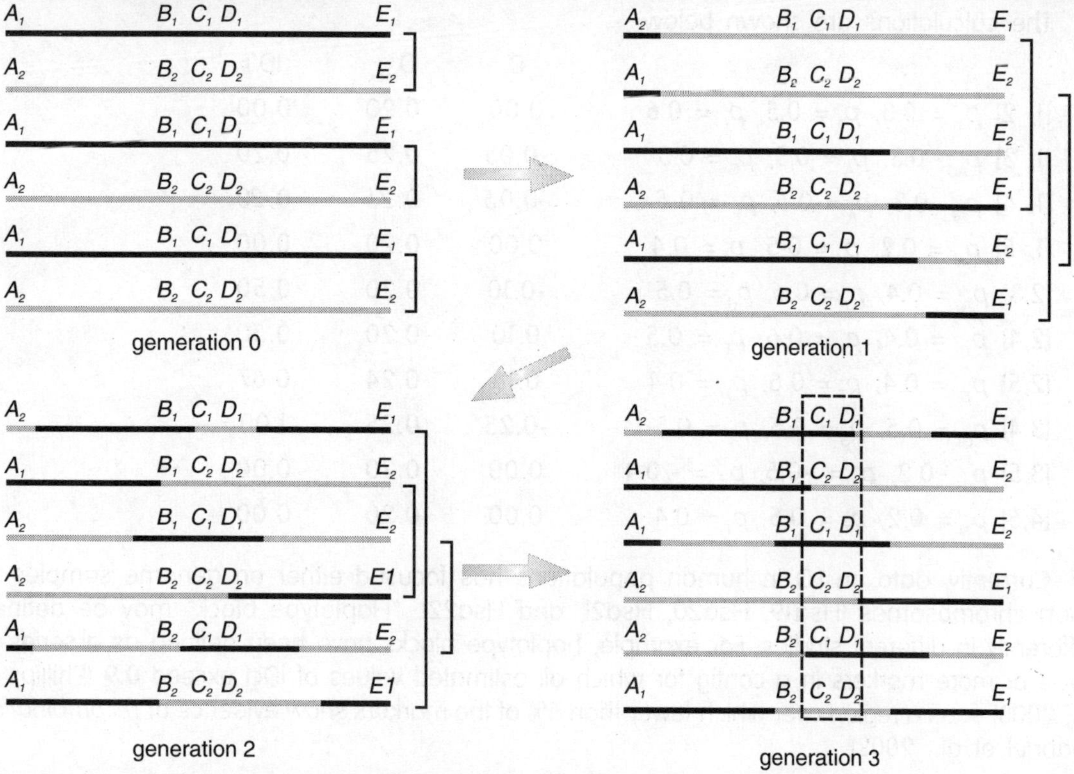

Figure 5.4. Relationship between recombination and haplotype blocks for a population represented by six chromosomes. Initial chromosomes (generation 0) having alleles A_1, B_1,..., E_1 are indicated by black lines, and initial chromosomes having alleles A_2, B_2, ... E_2 are shown by grey lines. Chromosomes for subsequent generations are produced by recombination between indicated pairs of chromosomes (brackets). In this simplified model, the probabilities of recombination within all subintervals of the same length are identical.

Computational Example 5.2: Calculation of pairwise linkage dis-equilibrium statistic |D|

Below are the SNP data for our example.

```
                         SNP number
  1                           2  3  4                    5
--A---------------------------T--G--T--------------------T---
--T---------------------------T--G--T--------------------T---
--A---------------------------A--G--T--------------------A---
--T---------------------------T--G--T--------------------A---
--A---------------------------T--G--T--------------------A---
--T---------------------------T--C--T--------------------T---
--A---------------------------A--C--A--------------------A---
--T---------------------------T--C--A--------------------A---
--A---------------------------T--C--A--------------------T---
--T---------------------------T--C--A--------------------A---
```

The calculations are shown below.

	D	D_{max}	|D'|
[1, 2] $p_{AT} = 0.3$, $p_A = 0.5$, $p_T = 0.6$	0.00	0.20	0.00
[1, 3] $p_{AG} - 0.3$, $p_A = 0.5$, $p_G = 0.5$	-0.05	0.25	0.20
[1, 4] $p_{AT} - 0.3$, $p_A = 0.5$, $p_T = 0.5$	-0.05	0.25	0.20
[1, 5] $p_{AT} = 0.2$, $p_A = 0.5$, $p_T = 0.4$	0.00	0.20	0.00
[2,3] $p_{TG} = 0.4$, $p_T = 0.6$, $p_G = 0.5$	-0.10	0.20	0.50
[2,4] $p_{TT} = 0.4$, $p_A = 0.6$, $p_T = 0.5$	0.10	0.20	0.50
[2,5] $p_{TT} = 0.4$, $p_T = 0.6$, $p_T = 0.4$	0.16	0.24	0.67
[3,4] $p_{GT} = 0.5$, $p_G = 0.5$, $p_T = 0.5$	-0.25	0.25	1.00
[3,5] $p_{GT} - 0.2$, $p_G = -0.5$, $p_T = -0.4$	0.00	0.20	0.00
[4,5] $p_{TT} = 0.2$, $p_T = 0.5$, $p_T = 0.4$	0.00	0.20	0.00

Currently, data on LD in human populations has focused either on genome samples or short chromosomes (Hsa19, Hsa20, Hsa2l, and Hsa22). "Haplotype block" may be defined differently in different studies. For example, haplotype blocks have been defined as a series of three or more markers in a contig for which all estimated values of lD¢l exceed 0.9 (Phillips et al., 2003) or as a region over which fewer than 5% of the markers show evidence of *recombination* (Gabriel et al., 2002).

The study of Hsa19 (Phillips et al., 2003) indicated that with a median 5.5 kb spacing between SNPs, about 32% of the chromosome was contained in haplotype blocks. The observed distribution of block lengths resembled an *exponential* distribution, and simulations indicated that similar distributions could be produced without invoking selection, special,*population* histories, or recombination *hotspots*.

The median block length was approximately 20 kb. In the case of Hsa21, individual chromosomes were *separated* by using somatic cell hybrids, so that haplotypes were measured directly (Patil et al., 2001). The mean SNP *spacing* was 1300 bp, and an optimization method was employed to define blocks such that the entire genome was covered. About *80%* of the chromosome was *contained* in blocks defined by three or more SNPs, and the mean size of these blocks was about 16 kb.

The mean size of all blocks (including smaller ones) was 7.8 kb. The dependence of inferred haplotype block sizes on marker *spacing* is shown in Table 5.5. This emphasizes the assertion that marker *spacings* that are too large will fail to detect smaller blocks of *disequilibrium*, in effect *sampling* only the tail of the LD block length distribution function.

The large number of SNPs allows estimation of recombination rate variation across the human genome. Although Innan et al. (2003) found that the observed LD for chromosomes Hsal9 and Hsa21 could be modeled by using uniform but lower-than-average recombination rates, other experiments indicated that recombination "hotspots" are present in the Class II region of the major histocompatability complex (MHC) in humans (Jeffreys et al., 2001; Kaupi et al., 2003).

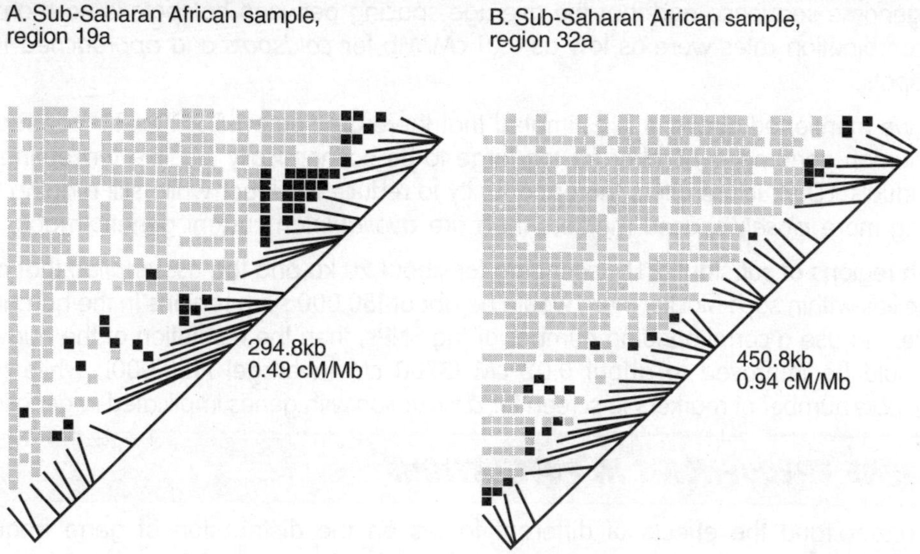

Figure 5.5: Plots of pairwise |D'| for two different human genomic regions drawn from a sample of subSaharan African populations. Plotting is analogous to the diagram in Computational Example 5.2. Dark grey squares indicate strong LD, light grey represents intermediate LD, and medium grey indicates no LD. Markers and marker spacings are indicated by lines below and to the right of the triangular grids.

Substantial LD and apparent haplotype blocks were noted based upon genetic data alone, and recombination events were detected experimentally by assaying SNPs in DNA extracted from single sperm cells. Five recombination hotspots were identified, each extending for 1-2 kb and lying between the observed haplotype blocks. Whereas the average genomic recombination rate is about 1.1 cM/Mb, the rates at these hotspots range from about 3 cM/Mb up to 130 cM/Mb-100- to 1000-fold higher than rates within the haplotype blocks themselves.

A more *comprehensive* statistical study *employing* SNP data for Hsa20 indicated that punctate recombination (i.e., presence of hotspots and coldspots) is a general feature of the human genome (McVean et al., 2004). This study found that half of all *recombination* events occur within just 10%

Table 5.5: Effects of marker spacing on predicted haplotype block length.

Distance between markers (kb)	Predicted mean block length (kb)
0.1	0.5
1.0	4.0
5.0	16.8
10.0	32.4
50.0	148.6

of the genome sequence and that the average spacing between hotspots is no more than 200 kb. Recombination rates were as low as 0.01 cM/Mb for *coldspots* and *approached* 100 cM/Mb for hotspots.

As we mentioned above, it is estimated that there are some $1.5 \times 10^6 - 10^7$ common SNP loci in the human genome, a number far too large to be economically feasible for routine screening of individuals. LD structure offers the opportunity to reduce the labor while still reaping the benefit of having more *closely* spaced markers than are *available* on current genetic maps.

With regions of substantial LD extending for about 20 kb and the *assumption* that most of the genome lies within such blocks, there would be about 150,000 such regions in the human genome. If we were to use a corresponding number of tag SNPs, then the resolution of the human genetic map would be improved to about 0.02 cM (3700 cM (average) /150,000), while retaining a *manageable* number of markers to screen for *association* with genes implicated in genetic diseases.

GENE FREQUENCIES IN POPULATIONS

To understand the effects of different forces on the distribution of gene frequencies in populations, it is helpful to use stochastic or deterministic models. We saw an example of the latter type when we discussed the rate of decay of LD under heading decaying linkage disequilibrium. In this section, we introduce some Markov chain models to describe gene frequencies in finite populations.

We begin with the simplest model, which was introduced by Fisher (1930) and Wright (1931). The assumptions we make to build this model are:

— The population size N is constant from generation to generation.

— Organisms are diploid (so there are 2N copies of each gene).

— All members of each generation reproduce *simultaneously*: *generations* donot overlap.

— Mating among individuals is random.

— Allele frequencies are not perturbed by mutation, migration, or selection.

The term "random mating" deserves further explanation. To form an offspring, choose an individual at random from the population and then choose one if its gametes at random. Return the chosen individual to the population, and repeat the experiment. This results in two gametes that form an individual in the next generation. This procedure is repeated N times to form the next generation.

Our model, diagrammed in Figure 5.6, describes a population in terms of a *gene pool* from which genes are randomly drawn to constitute the next generation. To describe a population containing N individuals, it is not necessary to keep track of each individual. Instead, we follow the fat(of the 2N copies of each gene. Suppose that each gene is either allele A of allele B. If at generation n there are i copies of allele A, then the number of A alleles at generation $n + 1$ has a binomial distribution with 2N trials anc success probability $P = i/2N$.

Because of the random components in this model, allele frequencies change over time, even though the model does not allow for mutation or selection This process is called genetic drift. The random number of offspring per individual implies that an initial population allowed to evolve over

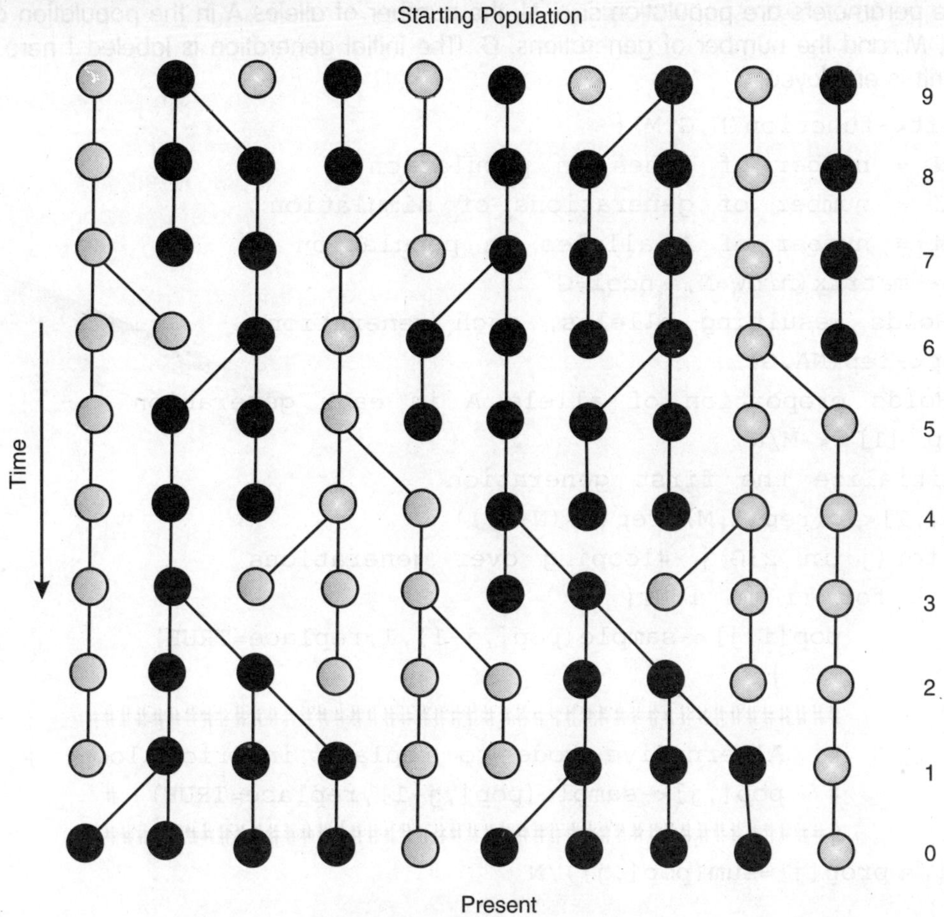

Figure 5.6: Schematic illustration of Wright-Fisher model for a population of size N = 5, with two alleles (A = white circles, B = black circles) at equal starting frequencies. Each horizontal line of 2N = 10 circles corresponds to a different generation. The total population size does not change from generation to generation. Copies of each gene to be reproduced for the next generation are chosen randomly, so that some may be copied more than once and others not at all. This leads to genetic drift, *illustrated here by the excess of B alleles in the present population (0 generations ago). Note that three generations ago there was an excess of W alleles.*

t generation: will produce different outcomes for different "trials." (In the real world, w(observe populations that have experienced only one trial during the course of evolution.)

The simulation below illustrates genetic drift in a population foi a gene having two different alleles. Notice that the allele frequencies evolve differently for different trials. At some point, one allele or the other "wins" (*fixation* of that allele) and the other allele is lost (it is "unlucky"). The surviving allele has become fixed, not as a result of selection based upon any benefit it may confer but merely as a result of the random nature of the process.

Computational Example 5.3: Simulation of genetic drift

We can see the effects of genetic drift for a two-allele system by employing the R function below. The parameters are population size, N; the number of alleles A in the population at the beginning, M; and the number of generations, G. (The initial generation is labeled 1 here.) The function drift is employed:

```
drift<-function(N,G,M){
# N = number of genes in population
# G = number of generations of simulation
# M = number of A alleles in population
pop<-matrix(nrow=N, ncol=G)
# Holds resulting alleles, each generation
prop<-rep(NA,G)
# Holds proportion of allele A in each generation
prop [1] <-M/N
#Initialize the first generation
pop[,1]<-c(rep(1,M),rep(0,(N-M)))
   for(j in 2:G){ #looping over generations
      for (i in 1:N){
         pop[i,j]<-sample(pop[,j-1],1,replace=TRUE)
            }
         ##################################################
         #  Alternative code to replace interior loop: #
         #  pop[,j]<-sample(pop[,j-1],replace=TRUE)  #
         ##################################################
      prop[j]=sum(pop[,j])/N
   }
return (prop)
}
```

We perform four different runs for 100 generations with an initial population of 60 genes, with M = 30:

```
> tmp<-matrix(nrow=4,ncol=100)
> for(i in 1:4){tmp[i,]<-drift(N,G,M)}
```

Plots for each run are shown in Figure 5.7.

```
> par(mfrow=c(2,2))
> plot(1:G,tmp[1,],type="s",lty=1,ylim=c(0,1.0))
# Repeat for tmp [2, ] , ... , tmp [4,]
```

Observe that in one of the runs (panel A) the A allele has drifted to a relative frequency of 1.0, and in two of the cases (panels B and D) the A allele has become extinct. In one run (panel C), the A allele has been enriched but has not been lost or fixed in 100 generations. Notice that, in this case, the A allele almost became fixed by generation 94, after which the other allele began

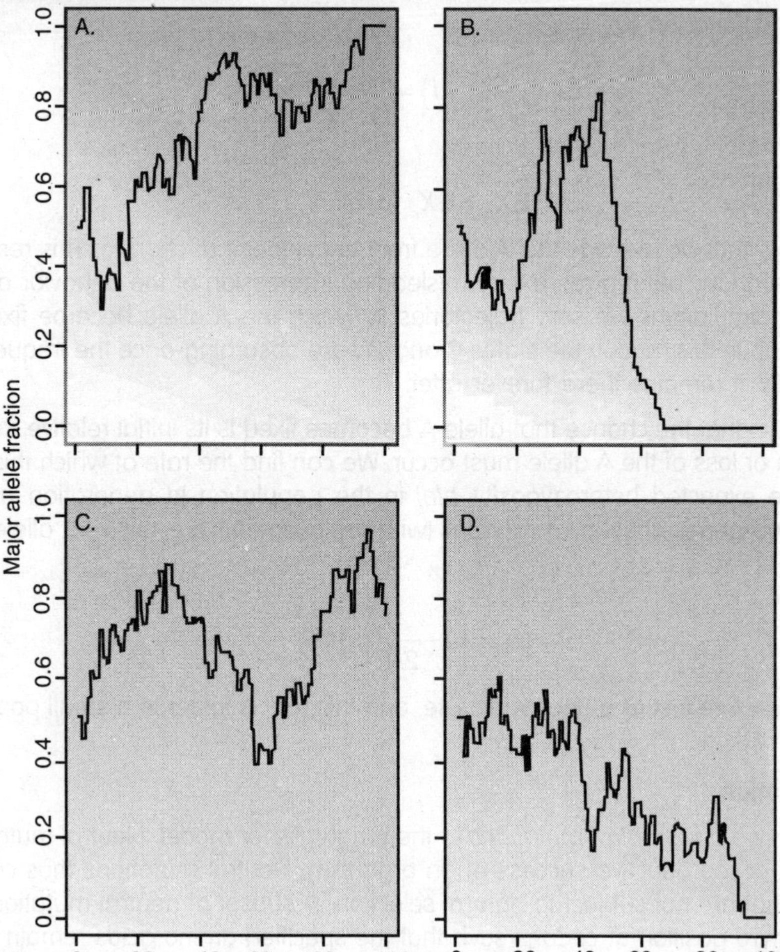

Figure 5.7: Simulation of genetic drift after 100 generations for 60 biallelic genes, with both alleles initially at the same frequency. Four independent realizations of this process are illustrated. One allele or the other becomes fixed in simulations A, B, and D. For simulation C, neither allele has been fixed after 100 generations. The plots were produced using:

to experience a run of "good luck."

The Wright-Fisher Model as a Markov Chain

If we denote the number of *A* alleles in the population in generation *n* by X_n, then we recognize that the sequence $X_0, X_1, \ldots$ is a Markov chain, the set of possible outcomes being {0, 1, 2,..., 2*N*}. The transition matrix of the chain is given by

$$p_{ij} = \binom{2N}{j}\left(\frac{i}{2N}\right)^j\left(1-\frac{i}{2N}\right)^{2N-j}, \quad i, j = 0, 1, \ldots, 2N.$$

We note that, conditional on Xn, = i,

$$\mathbf{E}\left(X_{n+1}\middle|X_n = i\right) = 2N \times \frac{i}{2N} = i,$$

so that

$$\mathbf{E}X_n = \mathbf{E}X_o \text{ for all } n.$$

This result says that *on average* the *A* allele frequency does not change. This result is akin to the Hardy-Weinberg law, but it gives a very misleading impression of the behavior of the model. Recall that in our simulations we saw trajectories in which the A allele became fixed! Because there is no mutation in this model, the states 0 and *2N* are absorbing-once the frequency of allele A reaches 0 or *2N,* it remains there forever after.

It can be shown that the chance that allele A becomes fixed is its initial relative frequency. We know that fixation or loss of the A allele must occur. We can find the rate at which this loss occurs by calculating the expected heterozygosity *h(n)* in the population in generation *n.* This is the probability that two genes chosen at random (with replacement) are different alleles. It can be shown that

$$h(n) = \left(1 - \frac{1}{2N}\right)^n h(0). \tag{5.21}$$

Variation is therefore lost at a geometric rate, and this rate is faster in a small population than in a large one.

Mutation in Evolution

In this section, we add *neutral mutations* to the Wright-Fisher model. Neutral mutations do not affect the survival or reproductive success of an organism. Neutral mutations thus can lead to a variety of alleles that are not subject to natural selection. A subset of neutral mutations are those that occur in the third position of codons such that the specified amino acids remain unchanged.

We employ the *infinitely many alleles model,* which states that each mutation produces a novel allele. This implies that any *homozygous* individuals are also autozygous because, under this mutation model, if two alleles are the same, they must have descended from the same ancestor. This also implies that alleles cannot back-mutate. Of course, the number of alleles cannot actually be infinite, but this is a good approximation, provided the actual number of possible alleles is sufficiently large.

For example., a look at a table of the genetic code shows that, given any specification of first and second positions in a codon, there are on average 2.7 third-position variants that will specify the same amino acid. For a polypeptide chain containing about 400 amino acid residues (a reasonable average for eukaryotes), there are about 2.7^{400} possible alleles involving third positions of codons and producing the same amino acid sequence.

This number does not take into account amino acid composition, which would affect the average number of third position variants. This works out to approximately 3.5×10^{172} alleles that would code for the same polypeptide chain. This number is not infinite but still is very large. We now compute some population statistics in terms of this model.

As before, we track $2N$ copies of each gene at each generation. Let μ be the rate of mutation of each gene; this rate is assumed to be constant for all genes and for all generations. We obtain an expression for the probability F_t that two genes are identical by descent in generation t. This event can occur in two ways: either both genes are copies of a single gene from generation $t-1$ (probability $1/2N$) and neither copy is mutant, or the two genes are nonmutant copies of different genes in the previous generation and those two genes are identical by descent. Thus,

$$F_t = \frac{1}{2N}(1- \ )^2 + \left(1-\frac{1}{2N}\right)(1- \ )^2 F_{t-1}. \tag{5.22}$$

If enough generations pass, genetic drift and mutation lead to a steady-state condition in which the probability of homozygosity does not change. After this point, $F_{t-1} = F_t = F$, a constant. Because μ in 5.22 is a very small number (around 10^{-6} amino acid substitutions per coding region per year), we can ignore terms in μ^2. Also, since N is ordinarily a large number, we can also ignore terms in μ/N. Expanding the terms $(1-\mu)^2$ in (5.22), applying the approximations, and rearranging gives the result

$$F = \frac{1}{1+4N}. \tag{5.23}$$

The heterozygosity H is just $1-F$ (by definition), so

$$H = \frac{4N}{1+4N}. \tag{5.24}$$

Equation (5.24), unlike (5.2), was derived in terms of a model, without explicit knowledge of the individual allele frequencies. As we see later, the product $4N\mu$ appears repeatedly in other contexts, so for convenience we write

$$\theta = 4N\ . \tag{5.25}$$

We will discover in anywhere else in this chapter that 0* is the expected number of differences in two sequences drawn at random from a population of size N. The parameters p and N are usually confounded-they appear together in θ as a product and from population data from a single generation alone they cannot be estimated independently. The quantity θ is called the *mutation parameter.*

A STOCHASTIC PROCESS

We note that (5.22) was found by asking how the relevant statistic in generation *t*, F_t, could have been obtained, given its value in the prior generation. In other words, rather than considering how the statistic changes as we go forward in time, we found it practical to look backward. In a sense, this is a very natural thing to do when considering the genetics of populations.

We are given a population as it exists now, and we may want to make inferences about how it reached its current state. (Predicting the future is beyond the realm of observational science unless one can afford to wait a suitably long time to check predictions.) The *coalescent* (Kingman, *1982) is* a very useful stochastic process that allows us to model the ancestry of genes in the population. This section describes the nature of the coalescent.

Coalescence Model

We illustrate how the coalescent works by discussing two aspects: the time to the most recent common ancestor (T_{MRCA}) for two gene sequences from the sampled population and the expected number of pairwise differences between a pair of sequences. We use the Wright-Fisher model. The population contains 2*N* copies of each autosome or autosomal single-copy gene at each generation. (Don't worry about mutation yet-that is dealt with at the end of the argument.)

We discuss the model in germs of an autosomal gene. We count time backward from the present, so at the present $g = 0$, and successive generations back in time are at times $g = 1, 2,$ .... (We use *g* to denote generation in this section, as a reminder that time is running back into the past.) Each gene in generation $g - 1$ had an ancestor in generation *g*. (Note the direction for the generation scale!) The coalescent model is implemented by allowing each gene in generation *g* - 1 to "choose" its ancestor from among the 2*N* gene copies that existed in generation *g*.

Clearly, some of the gene copies in generation *g* may be chosen multiple times, and others may be chosen not at all. The process is repeated, going back from generation *g* to generation $g + 1$. Because some gene copies are not chosen in each generation going back, the number of ancestors becomes smaller and smaller until the lineages coalesce to a single ancestor some number of generations ago.

Note that the process is stochastic: this means that different "runs" of this process yield different values of the $T_{MRCA.}$ We seek the expectation of this quantity for two genes. Pick one gene out of the 2*N* copies in generation $g - 1$. It came from one of 2*N* ancestors in generation *g*. The probability that a second gene in *generation* $g - 1$ came from the same parent is $1/2N$. The probability that the second gene came from a different parent is therefore $1 - 1/2N$. The *probability* that two genes have not coalesced to a common ancestor within *g* generations is the product of the probabilities that they have not coalesced at *generations* 1,2,...,g:

$$\mathbf{P}(\text{First coalescence} > \text{g generations}) = \left(1 - \frac{1}{2N}\right)^{g}. \tag{5.26}$$

We see that T_{MRCA} for a pair of genes has a geometric distribution with

$$\mathbf{P}(\text{First coalescence} > \text{g generations ago}) = \frac{1}{2N}\left(1 - \frac{1}{2N}\right)^{g-1}, \; g = 1, 2, \ldots.$$

From previous studies, it follows that

$$\mathbf{E}T_{MRCA} = 2N.$$

When the population size *N* is large, we can approximate the distribution of the time to the most recent common ancestor by an exponential distribution with mean 1. To do this, we measure time in units of 2*N* generations. Then, for $t = g/2N$,

$$\mathbf{P}(\text{First coalescence} > t \text{ units ago}) = \left(1 - \frac{1}{2N}\right)^{2Nt} \approx e^{-2}.$$

Thus, in a large population with time measured in units of 2N generations, the time to the most recent common ancestor of a pair of genes has probability density function f_2 (t) given by

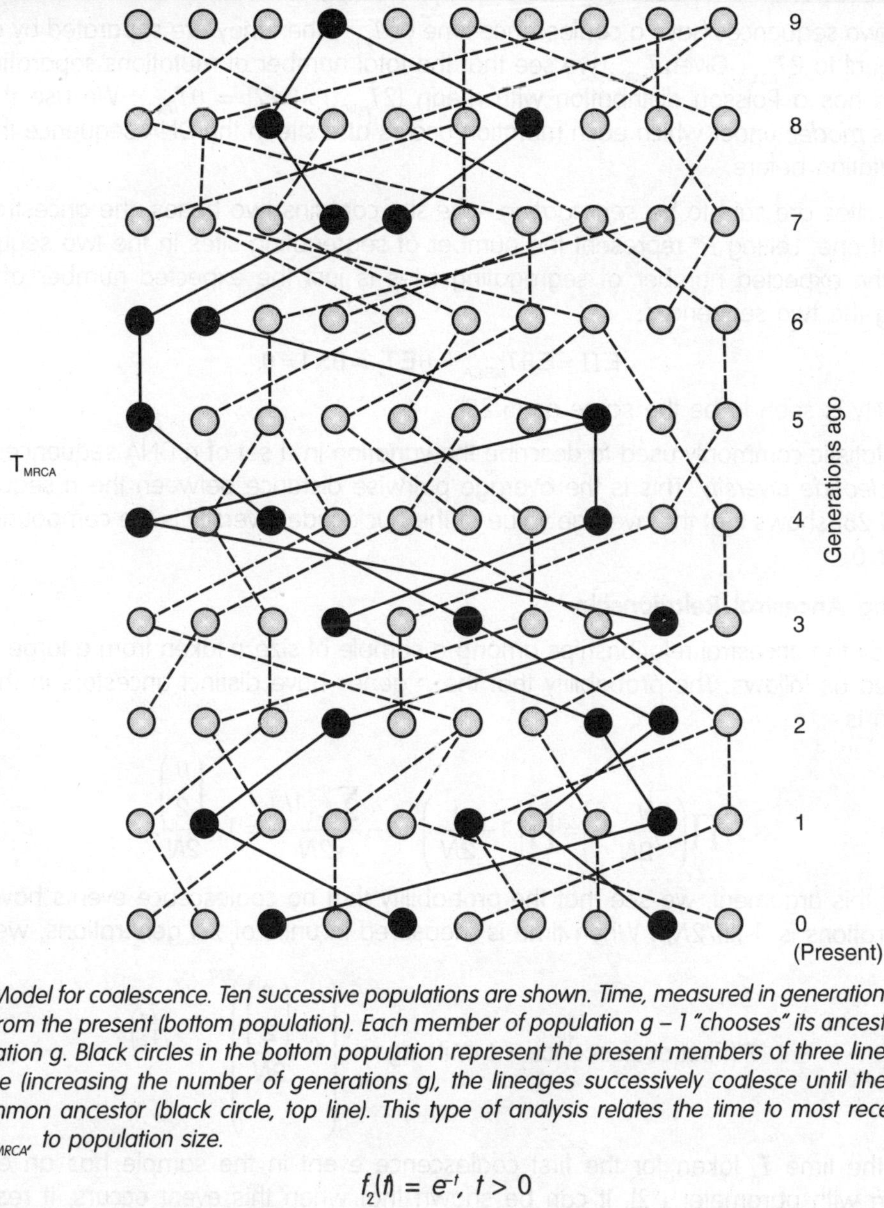

Figure 5.8: Model for coalescence. Ten successive populations are shown. Time, measured in generations, is counted backward from the present (bottom population). Each member of population g – 1 "chooses" its ancestor randomly from population g. Black circles in the bottom population represent the present members of three lineages. Going back in time (increasing the number of generations g), the lineages successively coalesce until they unite at a shared common ancestor (black circle, top line). This type of analysis relates the time to most recent common ancestor, T_{MRCA}, to population size.

$$f_2(t) = e^{-t},\ t > 0 \tag{5.27}$$

Variation in Two DNA Sequences

We can use the previous results to obtain the expected number of differences between any two DNA sequences sampled in the present generation. The number of mutations that occur along a single lineage of length g generations has a binomial distribution with parameters g and *t*. With $g = 2Nt$ and $\mu = \theta/4N$ and *N* large, we can use the Poisson approximation to the binomial distribution to see that on our new timescale the number of mutations along a lineage of length *t* time units has a Poisson distribution with mean $\theta t/2$.

If the two sequences have a coalescence time of $T_{MRCA,}$ then they are separated by an amount of time equal to $2T_{MRCA.}$ Given $T_{MRCA,}$ we see that the total number of mutations separating the two sequences has a Poisson distribution with mean $(2T_{MRCA}) \times (\theta/2) = \theta T_{MRCA}$. We use the *infinitely many sites model,* under which each mutation occurs at a site in the DNA sequence that has not had a mutation before.

These sites are said to be *segregating—the* site contains two bases, the ancestral one and the mutant one. Letting P^* represent the number of segregating sites in the two sequences, we see that the expected number of segregating sites is just the expected number of mutations separating the two sequences:

$$\mathbf{E}\Pi = \mathbf{E}\theta T_{MRCA} = \theta \mathbf{E} T_2 = \theta \times 1 = \theta. \tag{5.28}$$

This quantity is seen to be the same as (5.25).

One statistic commonly used to describe the variation in a set of n DNA sequences is the so-called *nucleotide diversity.* This is the average pairwise distance between the *n* sequences; the result in (5.28) shows that the average value of the nucleotide diversity is the compound mutation parameter θ.

Establishing Ancestral Relationship

To study the ancestral relationships among a sample of size *n* taken from a large population we proceed as follows. The probability that the *n* genes have distinct ancestors in the previous generation is

$$\prod_{j=1}^{n-1}\left(\frac{2N-j}{2N}\right) = \prod_{j=1}^{n-1}\left(1-\frac{j}{2N}\right) \approx 1-\frac{\sum_{j=1}^{n-1} 1/j}{2N} = 1-\frac{\binom{n}{2}}{2N}. \tag{5.29}$$

Repeating this argument, we see that the probability that no coalescence events have occurred in *g* generations is $(1-(2)/2N)^g$. When time is measured in units of $2N$ generations, we obtain

$$\mathbf{P}(\text{First coalescence} > t \text{ units ago}) = \left(1-\frac{\binom{n}{2}}{2N}\right)^{2Nt} \approx e^{-\binom{n}{2}}.$$

Thus the time T_n taken for the first coalescence event in the sample has an exponential distribution with parameter (*2). It can be shown that when this event occurs, it results in the coalescence of precisely two randomly chosen members of the sample; the possibility of three or more members of the sample coalescing simultaneously can be ignored. At that time, the sample has $n - 1$ distinct ancestors.

We can repeat the previous argument to see that, with time measured in units of *2N* generations, we wait a further amount of time T_{n-1} having an exponential distribution with parameter (n*21). and then choose two of those $n - 1$ ancestors to coalesce. This process of randomly joining pairs of ancestors continues back to the common ancestor of the sample.

In summary, the waiting times $T_n, \ldots, T_3, T_2$ for coalescence events are independent of each other and have exponential distributions with

$$\mathrm{E}T_j = \frac{2}{j(j-1)}\, j = n, n-1, \ldots, 2. \tag{5.30}$$

At each coalescence event a randomly chosen pair of ancestors is chosen to coalesce.

We can find the expected time to the most recent common ancestor T_{MRCA} of the sample of n genes by noting that

$$T_{MRCA} = T_n + T_{n-1} + \ldots + T_2,$$

so that

$$\begin{aligned}
\mathrm{E}T_{MRCA} &= \mathrm{E}(T_n + T_{n-1} + \cdots + T_2) \\
&= \mathrm{E}T_n + \mathrm{E}T_{n-1} + \cdots + \mathrm{E}T_2 \\
&= \frac{2}{n(n-1)} + \frac{2}{(n-1)(n-2)} + \cdots + \frac{2}{2\times 1} \\
&= 2\left(\frac{1}{n-1} - \frac{1}{n}\right) + 2\left(\frac{1}{n-2} - \frac{1}{n-1}\right) + \cdots + 2\left(\frac{1}{1} - \frac{1}{2}\right) \\
&= 2\left(1 - \frac{1}{n}\right).
\end{aligned}$$

Thus, the expected height of the ancestral tree linking n genes is about 2 coalescent time units. The variance of the height can be computed in a similar way. When measured on the original time scale, the expected height of the coalescent tree is $2N \times 2(1 - 1/n)$ generations, or about $4N$ generations in a large sample. The coalescent can be thought of as a random bifurcating tree whose properties can be studied by simulation.

In Figure elsewhere in this chapter, some random coalescent trees for samples of size 5 are shown. Notice the large variability in the height of the trees, and observe that most of this variability is due to the deep coalescence events, those near the top of the tree. On average, over half of the height of the coalescent tree comes from the deepest coalescence.

Estimating Recombination

Experimental estimation of recombination and mutation rates is very difficult, primarily because these rates are so small. In this section, we outline two model-based approaches for estimating mutation rates.

Under the coalescent model, mutations are placed on the tree according to independent Poisson processes of rate $\theta/2$ down each branch of the tree. Since a coalescent tree has j branches of length T_j, the length of the coalescent tree of a sample of size n is

$$L_n = \sum_{j=2}^{n} jT_j, \tag{5.31}$$

and the mean length, from Exercise 11, is

$$\mathbf{E}L_n = 2\sum_{j=1}^{n-1} \frac{1}{j}. \tag{5.32}$$

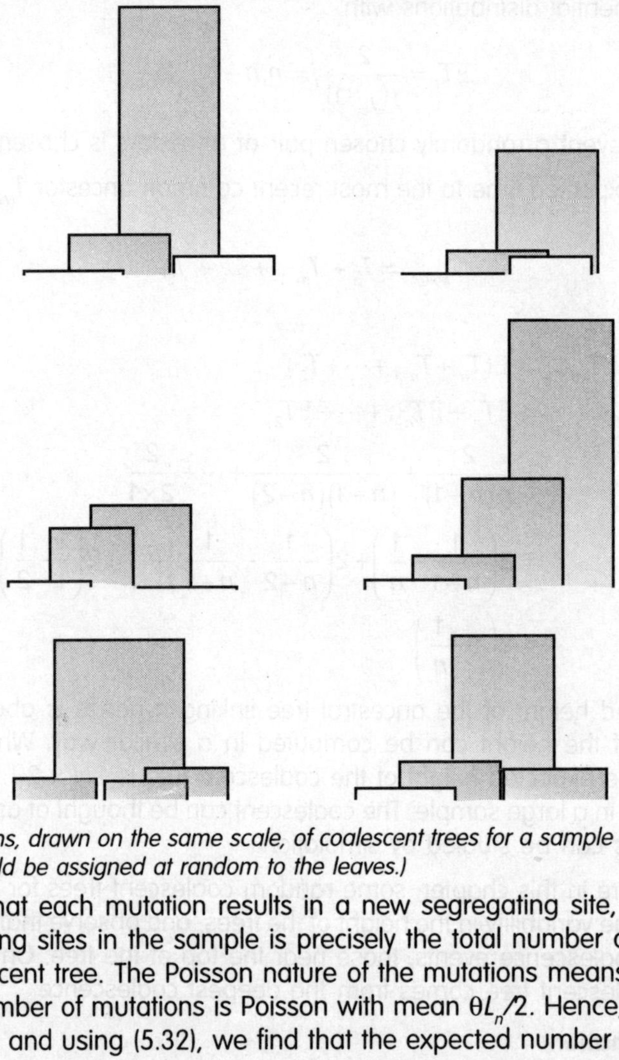

Figure 5.9: Six realizations, drawn on the same scale, of coalescent trees for a sample of n = 5. *(In each tree, the labels 1, 2, 3, 4, 5 should be assigned at random to the leaves.)*

If we suppose that each mutation results in a new segregating site, then we see that the number of segregating sites in the sample is precisely the total number of mutations that have arisen on the coalescent tree. The Poisson nature of the mutations means that given the length L_n of the tree, the number of mutations is Poisson with mean $\theta L_n/2$. Hence, by averaging over all possible tree lengths and using (5.32), we find that the expected number of segregating sites in our sample is

$$\mathbf{E}(\text{Number of segregating sites}) = \theta\sum_{j=1}^{n-1}\frac{1}{j}. \tag{5.33}$$

We can use the result in (5.33) to provide an estimator of the parameter 0*. If we observe S segregating sij es in the sample of *n* (aligned) sequences, then we can use the Watterson estimator

$$\theta_W = \frac{S}{\sum_{j=1}^{n-1}\frac{1}{j}}. \tag{5.34}$$

This estimator is, by design, unbiased (that is, $\mathrm{E}\theta_w = \theta$). We can compute its variance, from which we find a rather surprising result: the variance of the estimator decays at a rate proportional to 1/log *n*. (as opposed to a rate proportional to 1/*n* that would be expected for estimates of a population parameter based on an independent sample). It is precisely the dependence caused by the relatedness of genes in the sample that reduces the amount of information contained in that sample.

This phenomenon is typical of estimators based on coalescent models. In Exercise, the properties of θ_w are studied by simulation. An alternative approach to inference for population parameters is a Bayesian one. In this approach, we find the *posterior* distribution of θ given the number of segregating sites S observed in the sample. We start by specifying a prior distribution π(θ) for the parameter θ, and then compute the posterior distribution $f(\theta|S)$ via Bayes' Theorem in the form

$$f(\theta \mid S=k)=\frac{\mathbf{P}(S=k \mid \theta)\pi(\theta)}{\mathbf{P}(S=k)} \tag{5.35}$$

Thus, we see that the normalizing constant is

$$\mathbf{P}(S=k)=\int \mathbf{P}(S=k \mid \theta)\pi(\theta)d\theta.$$

All our inferences about θ are contained in the posterior density rather than point estimates such as θ_W. Finding the form of the posterior density usually requires a computational approach, one of which is outlined below.

We solve an apparently harder problem first. We simulate observations from the posterior distribution of θ and the coalescence times *T2 ... ,* T_n, given that we observe $S = k$ segregating sites in the sample. Write T = $(T_2 ...., T_n)$, and note that the posterior distribution of (θ, *T*) given $S = k$ is given by

$$f(\theta,T \mid S=k) \propto \mathbf{P}(S=k \mid \theta,T)\pi(\theta)f(T), \tag{5.36}$$

where $f(T)$ is the prior probability density for T obtained from the coalescent model. The likelihood P $(S = k|\theta, T)$ can be found as follows. Writing $L = 2T_2 + ... + nT_n$, we see that

$$\begin{aligned}\mathbf{P}(S=k \mid \theta,T) &= \mathbf{P}(\text{Poisson with mean } \theta L/2=k) \\ &= e^{-\theta L/2}(\theta L/2)^k/k!,\end{aligned} \tag{5.37}$$

the last coming from the form of the Poisson distribution in (3.2). A rejection algorithm for simulating observations from (5.36) is as follows:

1. Simulate an observation θ having density π(θ), and simulate *T* having independent exponential distributions with means given in (5.30).
2. Calculate $L = 2T_2 + .. . + nT_n$ and $h = e^{-\theta L/2}(\theta L/2)^k/k!$.
3. Simulate *U* from a uniform density on (0, 1). If $U \leq h$, accept the observation (θ, *T*) *if* $U < h$, and ignore it otherwise. Return to step 1.

Because the probability density of accepted observations is proportional to the product of P$(S = k \mid \theta, T)$ and π(θ), the algorithm does indeed generate observations from the posterior density in (5.36). The θ values then have the distribution (5.35) that we wanted. Implementation in *R* and further details are discussed in Exercise enywhere in this chapter.

CONCLUDING REMARKS

Our introduction to stochastic models of gene frequencies and estimation of population parameters such as the mutation rate 0 has necessarily been very brief. In particular, we have, not made explicit reference to the effects of re-combination or population expansion in this setting. There is an extensive body of literature about coalescents with recombination.

It provides a useful theoretical tool for interpreting patterns of LD in natural populations, and for inference about recombination rates. For a taste of this, see Nordborg and Tavare (2002) for example.

Chapter 6

Obtaining Genomic Sequence

This tutorial will guide the user through the process of *aligning* multiple whole genome sequences with *Mercator* and MAVID. Both programs are *freely* available and allow *researchers* to align moderately sized genomes on a single laptop.

The combination of Mercator and MAVID is an example of a hierarchical strategy for *aligning* genomes. First, Mercator is used to *construct* an *orthology* map between the input genomes, which is a high-level one-to-one mapping between *genomic segments*.

The second step is to run MAVID, a global multiple alignment program, on the sets of orthologous (and colinear) segments specified by the orthology map. The result is a set of multiple alignments with the property that every nucleotide is part of at most one multiple alignment.

For the tutorial, we will align the genomes of three fruit fly species: *Drosophila melanogaster*, *Drosophila yakuba*, and *Drosophila ananassae*. The genome sequences of the first two species are organized into chromosomes, whereas that of the third is currently comprised of more than 10,000 unmapped scaffolds.

The tutorial will begin with the downloading of the raw genome sequences. The author will then describe how to prepare the genome sequences and create an orthology map between them using Mercator.

Procedures for comparatively *scaffolding* genomes and discovering rearrangement *breakpoints* with Mercator will also be described. The tutorial will conclude with the *gener-ation* of nucleotide-level alignments using MAVID, and the *extraction* of a specific interval from the *resulting* whole-genome alignment.

REQUIREMENTS

For the purposes of this tutorial, it is assumed that a UNIX-like computing environment (e.g., Linux, Mac OS X, or Cygwin on Microsoft Windows) is being used. All software distributions listed in Table elsewhere in this chapter should be *downloaded* and compiled. Compiled binaries should all be made available through the PATH environment variable.

PROCEDURES

This tutorial will specify every command, in order, for the processing and alignment of three fruit fly genomes. Commands to be run will be specified by lines beginning with $. The output for some commands will be shown and *truncated* output will be *indicated* by eliipses (...). Approximate running times for selected commands will be specified as comments. *Running* times are for an Apple PowerBook with a 1.25 GHz PowerPC G4 processor and 1 GB of RAM.

Table: 6.1: Websites of Programs for the Alignment of Multiple Whole Genomes.

Mercator	http://bio.math.berkeley.edu/mercator/
MAVID	http://bio.math.berkeley.edu/mavid/
RepeatMasker	http://www.repeatmasker.org/
WU-BLAST	http://blast.wustl.edu/
SNAP	http://homepage.mac.com/iankorf/
BLAT	http://www.cse.ucsc.edu/-kent/

We will begin by starting in an empty directory and *creating subdirectories* for the input and output files of the alignment process.

```
$ mkdir input
$ mkdir output
```

Detecting Sequence of Genome

Genome sequences can be obtained from many sources on the internet. Most sources are either genome *sequencing* centers or databases that collect from *many primary* sources. We will download the *D. melanogaster* release 4 and *D. yakuba* release 2 assemblies from a database site, the University of California Santa Cruz (UCSC) Genome Browser (http://genome.ucsc.edu). The D. ananassae CAF1 assembly will be *downloaded* from the AAA Drosophila website.

```
$ cd input
$ # Define a variable for the UCSC URL
$ GOLDENPATH=http://hgdownload.cse.ucsc.edu/goldenPath/
$ # Download the DroMel genome from UCSC (39 MB)
$ wget $GOLDENPATH/dm2/bigZips/chromFa.zip
$ my chromFa.zip DroMel.zip
$ # Download the DroYak genome from UCSC (49 MB)
$ wget $GOLDENPATH/droYak2/bigZips/chromFa.tar. gz
$ my chromFa.tar.gz DroYak.tar.gz
$ # Download the DroAna genome from AAA (317 MB)
$ wget http://rana.lbl.gov/drosophila/caf1/dana_caf1.tar. gz
$ my dana_caf1.tar. gzDroAna.tar. gz
```

In the case that any of these assemblies are no longer found at the URLs previously cited, the author has placed copies of them at http://bio.math. berkeley.edu/mercator/tutorial/. To check that the downloaded assemblies are valid and to get some *basic statistics* about them, we will use the faCount, faLen, and stats *utilities* (*Mercator distribution*).

The faCount utility *calculates nucleotide frequencies* within each input FASTA record (*chromosomes* or contigs in our case) and the faLen utility simply outputs the length of each sequence. Combining faLen with stats, which *calculates* some basic *descriptive* statistics of a set of numbers, allows us to calculate useful statistics for the draft assembly of *D. ananassae.*

```
$ unzip -p DroMel. zip faCount
#seq len     A       C      G      T       N       cpg
chr4 1281640 415025  225495 224520 416500  100     40533
chrM 19517   8152    2003   1479   7883    0       132
chrU 8724946 1494654 978040 986285 1522538 3743429 186802
...
$ tar zxOf DroAna.tar.gz dana/scaffolds. bases faLen stats
             N = 13749
           SUM = 230993012
           MIN = 55
  1ST-QUARTILE = 1191
        MEDIAN = 1517
  3RD-QUARTILE = 3575
           MAX = 23697760
          MEAN = 16800.7136519
           N50 = 4599533
```

From the output of the last command, we see that half of the bases in the *D. ananassae* assembly are in scaffolds of length 4,599,533 or greater (this is the N50 statistic for a genome assembly).

Laboratory Preparation

Unfortunately, it is often the case that two whole genome sequences downloaded from the Internet are in different formats, so some work must be done to prepare the sequences for *alignment.*

Masking Genome Sequence

For the best genome annotations and alignments, the genome sequences must be *"masked"* for repeats. *See* Note elsewhere in this chapter for details on the different ways in which a sequence can be masked. Fortunately for us, the sequences obtained from the UCSC Genome Browser website are already softmasked with RepeatMasker and Tandem Repeats Finder. For the *D. ananassae* sequence, we will need to do the masking ourselves.

We will use the RepeatMasker program, as well as the nmerge (WU-BLAST distribution, often required by RepeatMasker) and faSoft Mask (Mercator distribution) utilities.

```
$ # Extract sequence for DroAna (1 min)
$ tar zxOf DroAna.tar.gz dana/scaffolds. bases > DroAna.fa.unmsk
$ # Mask interspersed repeats (19 hours)
$ In -s DroAna.fa.unmsk DroAna.fa.int
$ RepeatMasker -no_is -nolow -species drosophila DroAna.fa.int
RepeatMasker version open-3.1.5
Search engine: WUBlast

analyzing file DroAna.fa.int
identifying matches to drosophila genus sequences in batch 1 of
6036
$ # Mask low complexity repeats (13 hours)
$ In -s DroAna.fa.unmsk DroAna. f a. low
$ RepeatMasker -no-is -noint -species drosophila DroAna.fa.low
RepeatMasker version open-3.1.5
Search engine: WUBlast

analyzing file DroAna. f a. low
identifying simple repeats in batch 1 of 6036
identifying more simple repeats in batch 1 of 6036
identifying low complexity regions in batch 1 of 6036
...
$ # Merge masking into one hardmasked file (1 min)
$ nmerge DroAna.fa.int.msk DroAna. f a. low.msk > DroAna.fa.msk
$ # Create softmasked file (2 min)
$faSoftMask DroAna.fa.unmsk DroAna.fa.msk > DroAna.fa
```

Preparation of Database Files

For efficiency purposes, we need to put our FASTA-formatted sequences into another format. The author has developed a file format, the Sequence Database format (SDB), that allows for fast random access to multiple sequences stored in a single file. We will use the fa2sdb utility to put our softmasked genomes in to SDB format.

```
$unzip -p DroMel.zip|fa2sdb -c DroMel.sdb
$tar zxOf DroYak.tar.gz I fa2sdb -c DroYak.sdb
$cat DroAna. fa|ffa2sdb-c DroAna.sdb
```

To get a listing of the *D. melanogaster* chromosomes and their lengths, we can use the sdbList utility.

```
$ sdbList -l DroMel.sdb
chr2L 22407834
chr2R 20766785
chr2h 1694122
chr3L 23771897
```

To get the sequence from a specific genomic interval, we can use the sdbExport utility.

```
$ # Get sequence of 2nd coding exon of gene "dachshund"
$ sdbExport -r DroMel.sdb chr2L 16477453 16477480 -
>chr2L:16477453-16477480-
ATGCCTATCGATCAAGCCACCAGAAAG
```

Use of Mercator

The simplest way to use Mercator for orthology map creation is to use coding exons as map anchors. Therefore, we need to obtain gene annotations for each of our genomes. For the *D. melanogaster* and *D. yakuba* genomes, we will simply download annotations. For the *D. ananassae* genome, we will have to produce our own annotations through the use of gene prediction software.

First, download annotations for *D. melanogaster* and *D. yakuba* from the UCSC Genome Browser and. convert them to GFF using the utility program ucsc2gtf (Mercator distribution).

```
$ # Obtain annotations for DroMel
$ wget $GOLDENPATH/dm2/database/flyBaseGene.txt.gz
...
$ zcat f lyBaseGene. txt. gz I ucsc2gtf flybase > DroMel.gff
$ # Obtain annotations for DroYak
$ wget $GOLDENPATH/droYak2/database /genscan.txt.gz
...
$ wget $GOLDENPATH/droYak2/database /xenoRefGene. txt.gz
...
$ zcat genscan.txt.gz|ucsc2gtf genscan>DroYak.gff
$ zcat xenoRef Gene. txt.gz|I ucsc2gtf xenoRefSeq >> DroYak.gff
```

Notice that we have combined two independent annotations *of D. yakuba* into one GFF file. Users can use, as many annotation sets as they like and, in fact, the more the better (sensitivity is all that matters).

Now we will generate an annotation *of* the *D. ananassae* genome using the SNAP gene prediction program (wrapped by the runSnap script, Mercator distribution). The program zff2gtf (Mercator distribution) is used to convert from SNAP's ZFF format to GFF.

```
$ # Run SNAP with D.melanogaster parameters (2 hours)
$ runSnap/usr/local/snap/HMM/fly \
< DroAna.fa.int.msk>DroAna.zff
$ cat DroAna.zff|zff2gtf--source=SNAP>DroAna.zff
```

Creating Input Files for Mercator

With SDB and GFF files for each genome in hand, we are now ready to generate the input files for Mercator. The easiest way to do this is with the makeMercatorInput script (Mercator distribution). We simply supply the names of the assemblies as arguments to this script. The makeMercatorInput script will look in the current directory for each genome's SDB and GFF file.

```
$ # Create input files for Mercator (15 min)
$ makeMercatorInput DroMel DroYak DroAna
Making chromosome file for DroMel...done
Making anchors for DroMel...done
Extracting protein sequences for anchors ... done
Making chromosome file for DroYak...done
Making anchors for DroYak...done
Extracting protein sequences for anchors ... done
Making chromosome file for DroAna...done
Making anchors for DroAna...done
Extracting protein sequences for anchors ... done
BLATing anchors pairwise...
DroMel-DroYak
Loaded 10029188 letters in 98948 sequences
Searched 7247210 bases in 53254 sequences
...
```

This script performs the following tasks:

1. Creates a file for each genome specifying the names and lengths of the sequences that make up that genome.
2. Creates a set of nonoverlapping anchor intervals for each genome from the CDS records of the GFF files.
3. Creates a file for each genome of the protein sequences coded for by each of the anchor intervals.
4. Compares the protein sequences of each genome pairwise using the BLAT (8) program to create "hit" files.

Also required by some components of Mercator and by MAVID is a phylogenetic tree relating the input species. The branch lengths of the tree should be the expected number of substitutions per site along each branch.

The tree must be in Newick format (http://evolution.genetics. washington.edu/ phylip/ newicktree.html). We will put our tree in the file treefile.

```
$ echo "((DroMel:0.1,DroYak:0.1):0.4,DroAna:0.6);">treefile
```

Running Mercator

Running Mercator is simple and fast once all of the input files have been generated. Because the *D. ananassae* assembly is still in scaffolds, we will tell Mercator that it should be treated as a draft genome by using the -d flag.

```
$cd..
$ mercator -i input -o output DroMel DroYak -d DroAna
Loading input files ...
Loading chromosome files ...
DroMel 13 chromosomes
DroYak 21 chromosomes
DroAna 13749 contigs Loading anchor files ...
DroMel 53254 anchors
DroYak 98948 anchors
DroAna 89541 anchors Loading hit files ...
DroMel-DroYak 75082 hits (2380 filtered)
DroAna-DroMel 75397 hits (4355 filtered)
DroAna-DroYak 110120 hits (3324 filtered)
Sorting edges...
Time spent loading files: 16 seconds
Making map...
...
Assembling draft genomes...
Number of runs: 1177 (using 46614 cliques)
Checking cliques...
Map-making completed
Number of runs: 1177
Number of cliques: 46614
Mean run length: 39.6041
Median run length: 19
Max run length: 513
Min run length: 1
Coverage of DroMel anchors: 98.4133 %(52409/53254)
Coverage of DroYak anchors: 81.5964 % (80738/98948)
Coverage of DroAna anchors: 81.738 % (73189/89541)
Writing coverage files ...
Coverage of DroMel: 82.3921 %
Coverage of DroYak: 69.2232 %
Coverage of DroAna: 58.1449 %
...
Run time: 38 seconds
```

```
$ # Mercator has finished, let us look at the output files $
cd output
$ is

DroAna.agp          DroMel.coverage     genomes
DroAna.anchors      DroMel.mgr          map
DroAna.coverage     DroYak.agp          pairwisehits
DroAna.mgr          DroYak.anchors      runs
DroMel.agp          DroYak.coverage     pre.map
DroMel.anchors      DroYak.mgr
```

After running the main Mercator program, we now have an orthology map where the orthologous intervals are defined by the boundaries of the landmarks in the file "pre.map" and a map with the breakpoint regions cut in half in the file "map."

Genome's Component Sequence

When a genome is specified as "draft" to Mercator (using the -d option), the program will attempt to comparatively scaffold that genome's component sequences. That is, it uses information from the other genomes to orient and join the draft genome's contigs or scaffolds. Mercator specifies the comparative scaffolding of a draft genome in the form of an AGP file (http://www.ncbi.nlm.nih.gov/Genbank/WGS.agpformat.html).

Later steps in the alignment process will not be aware of comparative scaffolding, so we must provide updated SDB files for each genome. In our alignment, *D. ananassae* has been comparatively scaffolded by Mercator, so we must "assemble" its component sequences into a new SOB file using the sdbAssemble program (Mercator distribution). For the other genomes, we will simply make a link to original SDB files.

```
$ sdbAssemble ../input/DroAna.sdb DroAna.sdb<DroAna.agp
$ In -s../input/DroMel. sdb
$ In -s../input/DroYak.sdb
```

Landmarks for Orthologons Segments

Because Mercator has only used exons as landmarks for determining orthol-ogous segments, the exact boundaries of the orthologous segments are not yet determined. If we wish to refine the boundaries of the identified orthologous segments, we can use the breakpoint finding program included in the Mercator distribution. This program attempts to find the best position within each "break-point region" (intervals in between segments identified in the "pre.map") at which to break and add the left and right intervals to the flanking segments.

This procedure may be skipped if the exact boundaries of the segments are not required. Locating breakpoints involves a number of steps. Note that SDB files for each genome must be present in the current directory (output), as set up in the last section.

```
$ # The breakpoint finding algorithm requires the tre
$ In -s ../input/treefile
```

```
$ # Convert the orthology map into a more general homology map
$ omap2hmap genomes < pre.map > rough. homology. map
$ # Create the graph relating the breakpoint regions
$ makeBreakpointGraph rough. homology. map treefile
$ # Make pairwise alignments for breakpoint regions (2 hours)
$ mkdir bp_alignments
$ makeBreakpointAlignmentInput —out-dir=bp_alignments
$ mavidAlignDirs —init-dir=bp_alignments
$ # Find a good configuration of breakpoints (8 min)
$ f indBreakpoints rough.homology.map treefile edges\
bp_alignments>breakpoints
$ # Refine the map by splitting the breakpoint regions
$ breakMap breakpoints<rough.homology.map>better.homology.map
$ # Convert back to the orthology map format
$ hmap2omap genomes < better. homology. map > better. map
```

Orthology Mapping

Now that we have an orthology map, we are ready to run a global multiple alignment program on each orthologous segment set identified by the map. To help in the alignment process, we will give the alignment program a set of "constraints:" short intervals within the orthologous segments that we know should be aligned. These constraints are derived from the sequence similarities identified between the anchors given to Mercator. To make the constraints file, we run the following command:

```
$ # Convert pairwise hits to alignment constraints (2 min)
$ phits2constraints -i .. /input < pairwisehits > constraints
```

The input files for MAVID are then generated by makeAlignment Input.

```
$ # Create directories and files for alignment (3 min)
$ mkdir alignments
$ makeAlignmentInput —map=better.map . alignments
```

Nucleotide-Level Alignment Step

With the input for MAVID generated, all that is left is to run MAVID on the sequences for each orthologous segment set. Each segment set is stored in a separate subdirectory. This is a good step at which to parallelize, but if that is not an option, the mavidAlignDirs script (Mercator distribution) can be used. *See* Note elsewhere in this chapter for details on the nucleotide-level alignment step.

```
$ # Align all sequence files in directory structure (13 hours)
$ mavidAlignDirs —init-dir=alignments
```

We now have a multiple whole-genome alignment of *D. melanogaster, D. yakuba,* and *D. ananassae.*

Slice Alignment Program

We may now extract parts of the whole-genome alignment that are of particular interest using the sliceAlignment program (Mercator distribution). For example, we may wish to get the alignment of the second coding exon of the gene *dachshund.* The sliceAlignment program outputs alignments in multi-FASTA format, so we will use the fa2clustal utility (Mercator distribution) to put the exon alignment into a more readable form.

```
$ sliceAlignment alignments \
DroMel chr2L 16477453 16477480 - > exon.mfa
$ fa2clustal < exon.mfa
CLUSTAL

DroMel  ATGCCTATCGATCAAGCCACCAGAAAG
DroYak  ATGCCTATCGATCAAGCCACCAGAAAG
DroAna  ATGCCTATCGATCAAGCCACCAGAGAG
        ***************************
```

This tutorial has taken the user through the basic steps of creating a multiple whole-genome alignment using Mercator and MAVID.

There are many additional details and options that have been left out of this tutorial at each step. More details are available in the full documentation of each of the programs.

POINTS TO BE NOTED

1. *Obtaining genome sequences.* To download genomes from the UCSC Genome Browser, it is easiest to go through the "Downloads" section of the website. For the assembly of interest, click on the "Full data set" link to access complete genome sequences as compressed FASTA files. Other databases from which to obtain genome assemblies are at the NCBI5m.nih.gov) and Ensembl (http://www.ensembl.org).
2. *Preparing the genome sequences.*
   a. *Masking repeats.* An "unmasked" FASTA formatted file has all characters in uppercase. A masked sequence can either be "hardmasked" or "softmasked." In hardmasked files, characters that are part of repetitive sequence are changed to Ns, whereas in softmasked files they are changed to lowercase.

      Unmasked and softmasked sequence may also have Ns, which are commonly used to indicate assembly gaps. Ideally, we would like our genome sequences to be softmasked, so that we have repeat annotations as well as full sequence information.

      Masking repeats is a bit of an art, and the author will not go into all of the details here. Very briefly, one needs to mask both interspersed and simple (or low complexity) repeats. Masking of these two types of repeats should be done separately because gene finding is best done on sequence hardmasked for interspersed repeats (simple repeats can occur within genes).

b. *Creating sequence database files.* There are four command-line utilities made available in the Mercator distribution for handling SDB files. The Mercator library code may also be used for writing C++ programs that access SDB files directly. The command-line utility f a2 sdb is used to create or append to a SDB file from sequence records in FASTA format. DNA sequences may be compressed (two nucleotides per byte) inside of a SDB file if the -c option is specified.

The sdbExport utility is used for the extraction of specific genomic intervals from a SDB file. It can extract one or more intervals at a time and outputs sequences in FASTA format. The sdbList utility is used to list the names and lengths (with the -1 option) of the records inside of a SDB file. Lastly, the sdb2fa utility is used to convert a SDB file into FASTA format.

3. *Obtaining gene annotations.* Gene annotations for many genomes can be obtained at the same database sites that provide whole genome sequences. For the UCSC Genome Browser site, annotations can be obtained either through the "Table Browser," or directly from the "Downloads" section. If annota-tions are not available online, you can produce them using gene prediction software.

The easiest prediction programs to use in this case are single-genome *ab initio* gene finders (e.g., geneid, GENSCAN [10], and SNAP). Regardless of how the annotations are obtained, they need to be converted to the GFF format (http://www.sanger.ac.uk/Software/formats/GFF/). Three scripts (genscan2gtf, ucsc2gtf, and zff2gtf) in the Mercator distribution are available for converting to GFF from some common formats.

Mercator requires that GFF annotations have CDS records (lines with "CDS" in the feature field) for the coding intervals of each exon. It is critical that the "frame" field be specified for each CDS record in the GFF files. This field allows Mercator to translate each coding exon correctly.

4. *Generating input for Mercator.* For custom jobs (e.g., to parallelize some tasks), you may wish to generate the input for Mercator without using the makeMercatorInput script. In such cases, consult the README file in the Mercator distribution for exact specifications of the various input files that are required. Some routines of makeMercatorInput are customizable via command-line options. Use the - -help option to get full usage information.

5. *Constructing an orthology map with Mercator.* Mercator has a number of user-settable parameters that may be specified as command-line options. The options that affect Mercator's performance are - -min-run-length, —prune-pct, —join-distance,- -max-eval, - -repeat-num, and - -repeat-pct. Consult the Mercator README file for descriptions of these options.

6. *Comparatively scaffolding draft genomes.* When Mercator comparatively scaffolds the components of a "draft" genome, it joins components that it believes should be adjacent to each other into new sequences with names beginning with assembled. For example, in our fruit fly alignment, the scaffold_13770, scaffold-13165, and scaffold_13337 sequences from the *D. ananassae* assembly are joined into a new sequence called assembled6, with a string of Ns separating the component sequences within assembled6.

The number of separating Ns may be specified by Mercator's - -padding command-line option. These Ns are meant to indicate gaps of unknown length between the component sequences.

7. *Refining the map via breakpoint finding.* The breakpoint finding process can be very computationally intensive, depending on the input genomes. If a cluster is available to the user, it is a good idea to parallelize the mavidAlignDirs step. When running the findBreakpoints program, accuracy may be traded for speed via the - -resolution opti on. Breakpoints will be found more accurately with larger "resolution" values.
8. Generating input for MAVID. MAVID requires, at a minimum, three input files. These files are a phylogenetic tree in Newick format, unmasked sequences in a multi-FASTA file, and a hardmasked version of the multi-FASTA file. When Mercator is used, alignment constraints may be given to MAVID via the -c command-line option. In this tutorial, the makeAlignmentInput and mavidAlignDirs take care of generating and passing the correct files to MA VID.
9. *Aligning orthologous segments with MAVID.* Although the focus of this tutorial is on the application of Mercator and MAVID, the hierarchical strategy for whole-genome alignment allows for the components to be substituted with similar programs independently of each other. For example, in cases where the orthol-ogous segments are very small, CLUSTAL W could be used to do the multiple nucleotide alignment instead of MAVID.

   However, there is a signif-icant advantage to using MAVID as the nucleotide-level aligner with Mercator: alignment constraints. By using the alignment constraints output by Mercator, MAVID can more accurately align coding regions and is able to process longer sequences.
10. *Extracting subalignments.* The sliceAlignment program is designed to efficiently extract subalignments from a multiple whole-genome alignment. It extracts alignments based on the coordinates given as input for a specified reference genome A single interval may be given as command-line arguments or multiple intervals can be given on the standard input. With multiple intervals as input, the program will be very efficient if the intervals are sorted by their start coordinates.

Chapter 7

Phylogeny at the Sequence Level

The increasing amount of fully *annotated* genomic sequences permits now to gain essential information on the *molecular dynamics* at the basis of evolution. *Genome* evolution involves variation at the sequence level (base substitutions or *insertion/deletions*), but also changes on a larger scale, like *rearrangements* on the gene *organization* along the genome.

Comparative analysis of gene order in multiple genomes containing homologous genes may be thus of significant help in the investigation of general or lineage-specific *genome plasticity*, as well as in the inference of phylogenetic relationships. In *particular*, conserved gene *contiguity* within a chromosome (sometimes also referred to as conserved gene synteny), even if interrupted by *intervening* genes, suggest potential functional *couplings* of the genes involved.

GeneSyn is a software tool for the automatic identification of conserved gene order across multiple annotated genomes, that detects sets of genes *appearing* in the same order (measure that can be defined in different ways) in any subset of the *genomes*.

REQUIREMENTS

Role of Computer

1. Any Pentium-class (or equivalent) computer with at least 256 Mb of RAM (for typical executions of the program), running a standard distribution of any of these operating systems: Unix (any version), Linux (any), MAC OS (any). The computer has to be connected to the Internet to download the software *package*.
2. A C compiler must be installed on the computer to compile the program. The *program* can be compiled also under Windows with any *Unix-Linux emulator* (like *Cygwin*) installed.

FASTA-Like Format

GeneSyn needs as input at least two sequences of gene *identifiers* (that from now on we will call improperly "*genomes*"). Each genome is defined in FASTA-like format: that is, it must start with a line with ">" followed by the sequence name. Then, on a new line, there must be the sequence of gene identifiers *annotated* in the genome. An example is the following:

```
>cin
cox2 cob pro nad41 his arg gin
>csa
cox2 cob nadl pro nad41 his arg gin
```

Gene identifiers can contain letters, numbers or any other *symbol*, and can be separated by any number of white spaces or newline characters. It is essential that the same genes (i.e., *orthologous genes*) are denoted in the same way in their *respective* genomes. The program is *case-insensitive*, that is, it reads for example Gln and gin as if they were the same gene.

In the previous example, all the genes are defined as *appearing* on the same strand. Double stranded output can be defined by putting a minus sign before the identifier of genes appearing on the complementary DNA strand, without spaces:

```
>cin
cox2 cob pro nad41 his arg gin
>csa
cox2 cob -nadl pro nad41 his arg gin
```

Thus, in the example, gene nadl of the second genome appears on the reverse strand. The assessment of *orthologous* genes, to which the same name is assigned, can be made through *retrieval* of specialized *databases*, like COG, or by using specific algorithms.

Genes do not have to be present in all the input genomes, and, vice versa, the same gene name can appear more than once in a single genome (e.g., for denoting *paralogous* genes). Notice however that some advanced *parameters* are available only on input without duplicated genes in a *single genome*.

PROCEDURES

In the following, we first introduce (1) how to download and install the software on a local computer, then explain (2) how to run the program with standard parameter settings, (3) how to set *additional parameters*, and (4) how to read the output of the *program*.

GeneSyn Package

The GeneSyn package is available free of charge for *noncommercial* use. Because the software is written in the C language, a C *compiler* (e.g., gcc) must be installed on the computer to compile the program. No other special *library* is needed. The program can be compiled also under Windows with any Unix-Linux *emulator* (like *Cygwin*) installed.

1. Connect to http://www.pesolelab.it/, and click on the "Tools" link appearing in the page. Another page with a list of several tools will appear, including GeneSyn.
2. Follow the corresponding link, and you will reach a page that permits to download the program. Click on the appropriate link, and select the "Save" option.
3. The file downloaded will be named GenesynX. X. tar. gz, where X.X are the numbers

denoting the version of the software available. Current version is 1.1, thus, for example, the package is called Genesyn1.1. tar. gz. In the following, replace 1.1 with the digits of the version number you downloaded.

The commands described in the following have to be entered from a command-line prompt. Note that most of the Linux/Unix systems are case-sensitive, and thus commands entered should respect the upper/lowercase characters appearing in file names.

4. To unpack the file, enter gunzip Genesyn1.1 .tar.gz. File Genesyn1.l .tar.gz now has become Genesyn 1.1.tar. Enter tar xvf Genesyn 1.1 .tar. At this point, a directory named Genesyn 1.1 should have appeared.
5. Enter cd Genesyn 1.1 to enter the newly created directory. It contains a file named "*compile,*" and a directory named "src" (containing the *source code*).
6. Edit file compile, and put after CC= the name of the C compiler available on your computer (the compiler is set to gcc, if ti is a available on your computer you do not have to edit the file). To check whether gcc is available on your computer, enter gcc -v. If a message showing the version number of gcc *appears*, then it is available. Otherwise, consult your *sysmtem* administrator to find out whether any C compiler is installed on your system.
7. Enter./complile.
8. If everything went well, two executables named genesyn and grunner will have appeared. Make sure that, wherever you place the *executables*, *genesyn* and *grunner* are located in the same directory.
9. To test if the program works, enter ./test. It will launch the program on a sample input file contained in the *GeneSyn* package (example.tfa), so you can also get familiar with the input and output format.

Default Parameters

The following commands have to be entered from a command-line prompt.

1. To run the program with default settings, enter

   `./genesyn input-file <percent>`

   where

   a. `input-file is the name of the input file,`

   b. <percent> must bean integer number between 1 and 100 denoting the minimum *percentage* of input genomes in which genes in the same order have to appear. In other words, the program will output subsequences of genes common to the given *percentage* of the input *genomes*.

      A subsequence of a sequence is a sequence that can be obtained by deleting any number of items (in this case, genes) from the *original* sequence. Therefore, the subsequences will consist of genes *appearing* in the same order in the *genomes* where the subsequences appear.

2. The two previously described *parameters* are required, exactly in that order. With default parameters, the program will:
   a. Assume that the input genomes are linear.
   b. Look for genes (*subsequences*) appearing in the same order on the same strand.
   c. Look for genes (subsequences) appearing in the same order and *separated* by any number of *interleaving* genes.
   d. Output solutions at least three genes long.

Additional Parameters

After the basic command-line elements:

```
./genesyn input-file percent
```

a number of additional parameters can be *specified*. More than one *parameter* of the ones described in the following can be added after the required ones, in any order.

Circular Genomes

In some cases, it is useful to assume that input genomes are circular, e.g., in the case of *bacterial* or *mitochondrial* genomes. To set this option, add -c to the *command-line*:

```
./genesyn input-file percent -c
```

Thus, given input file

```
>Sequence 1
genel -gene4 gene3 -gene2 gene5
>Sequence 2
-gene4 gene3 -gene2 gene5 genel
```

with percent = 100, the program will report genel -gene4 gene3 -gene2 gene5 as a common subsequence. Without the –c option, it reports only -gene4 gene3 -gene2 gene5. Notice that this option can be set only if the *genomes* do not contain *duplicated* genes.

Both Strands of the Gene

It might happen that the same genes appear in the same order in two or more genomes, but on different strands (as explained before the reverse strand can be *specified* by putting a minus sign ahead of gene names). With default parameters the program assumes that *common subsequences* must appear on the same strand. To set this option, add -r to the command-line:

```
./genesyn input-file percent -r
```

In this way, for example, given as input file

```
>Sequence 1 genel gene4 gene3 gene2 gene5
>Sequence 2
-gene5 gene2 -gene3 gene4 -genel
```

with percent = 100, the program will report genel gene3 gene5 as appearing in both genomes. In Sequence 2, it appears on the reverse strand.

Strand-Independent Subsequences

Even if input genes are annotated on both strands of the genomes, the program can be run to find conserved gene order *regardless* of the *strand* on which single genes are *annotated*. To set this option, add -n to the command-line:

```
./genesyn input-file percent -n
```

In this way, for example, given as input

```
>Sequence 1
gene1 gene4 gene3 gene2 gene5
>Sequence 2
-gene1 gene2 -gene5 gene4 -gene3
```

with percent = 100 the program will report subsequence gene1 gene4 gene3 as appearing in both genomes.

Cap Value for Subsequences

With default parameters, the program reports subsequences common to a number of the *input* genomes, as long as the genes appear in the same order, regardless of the number of *interleaving* genes separating them in their genomes. A *maximum* gap value (expressed as a number of genes) can be set on the command-line:

```
./genesyn input-file percent -g <value>
```

where <value> must be a non-negative *integer* number. For example, by setting "-g 0" the program will report only subsequences where the genes are adjacent in their *respective* genomes (maximum gap is zero). Note that this option can be set only if the genomes do not contain duplicated genes.

Gap Value in Basepairs

A maximum value for gaps separating genes can be defined according to the distance of the genes in the genomes, defined as the number of *basepairs* separating them. In this case, however, also the input file has to be *formatted* in a different way. An example is the following:

>Sequence 1

gene1 s_{11} gene4 s_{12} gene3 s_{13} gene2

>Sequence 2

-gene5 s_{21} gene2 S_{22} -gene3 s_{23} gene4

that is, gene names must be *separated* by an integer number (s_{11}, S_{12}, and so on in the example) denoting the number of basepairs between them (e.g., it can be the distance *separating* their respective TSSs). Clearly, the distance between *nonadjacent* genes will be the sumof the distances *associated* with the genes separating them. In the example, the distance between gene1 and gene3 in Sequence 1 will be given by $s_{11} + s_{12}$.

Display on the Screen

The output of the program appears on the *screen* and it is simultaneously written into a file whose name is displayed on the screen at the end of the *execution*. The output file name has the form gs <number> . res, where <number> is an integer that changes from execution to *execution*.

The output file starts first of all with the command-line that had been used to produce the output, followed by a *brief description* of the *parameter* settings that were used, and finally the list of the subsequences that satisfied the input *constraints*. An example is presented in Figure 7.1 showing the output of the program obtained on the example.tfa file provided with the package and with the parameters set in the test script.

As explained in the preamble, the program looked for *subsequences* appearing in 17 genomes out of 34 (the percentage was set to 50%), at least 9 genes long (-m 9 in the command-line). The -r option was also set, so the' program looked for subsequences *appearing* on either strand of the input *genomes*.

After the heading, the output lists the subsequences *satisfying* the input constraints, followed by the number of genomes *containing* them (17 occurrences), and a schematic *representation* of where the subsequence was found (one if it was found, zefo otherwise) and on which strand. In the example, the first subsequence was found in input genomes 1, 2, from 4 to 10, and from 14 to 20 on their *"forward"* strand.

The second subsequence, instead, was found in genomes from 4 to 10, 12, 14, and from 17 to 20 on the reverse strand; and in genomes from 23 to 27 on the forward strand. Genomes are numbered according to the order they appear in the input file, starting from 1.

```
./genesyn example.tfa 50 -r -m 9
Looking for subsequences appearing in at least 17 genomes out of 34
Subsequences have to be at least 9 elements long
Processing both strands of input genomes


COX1 COX2 K ATP8 ATP6 COX3 NAD3 COB M
17 occurrences
         123456789   0123456789   0123456789   01234
Forward :  110111111  1010111111   1000000000   000001
Backward   :   000000000   0000000000   0000000000   00000

V RRNL L1 NAM NAD4L F -N -NAD3 -COX3
17 occurrences
         123456789   0123456789   0123456789   01234
Forward : 000000000  0000000000   0001111100   000001
Backward   :   000011111   1010100111   1000000000   00000
```

Figure 7.1: A sample output of GeneSyn.

POINTS TO REMEMBER

1. The algorithm GeneSyn is based on finds subsequences of the input genomes by building an indexing structure called directed acyclic word graph (DASG). More in detail, a DASG for a set of sequences is a directed acyclic graph such that:
   a. On vertex (called source) has only *outgoing* edges.
   b. Each edge is *labeled* with an element of the input sequences (in this case, the name of a gene).
   c. On each path from the source node, the concatenation of the labels of the edges *encountered* corresponds to a subsequence of at least one of the input sequences.
   d. Because the graph is acyclic, for any vertex V of the graph, no path starting at V can *return* back to V itself.

      A *q*-DASG can be defined in a similar way, with the only difference that subsequences corresponding to paths in the graph must appear in at least q of the input sequences. Thus, as a first step, the algorithm builds a q-DASG according to the percentage *parameter* provided as input.

      Once the graph is built, finding subsequences common to at least *q* of the input sequences is straightforward, because it involves performing a *traversal* of the graph.

      Although, in general, finding the longest common subsequence of a set of sequences is a NP-Hard problem (that is, no solution can be obtained in *reasonable* time), the problem becomes tractable if an element appears at most once in a sequence.

      Anyway, the *program* accepts as input genomes in which a gene *identifier* can appear any number of times, but in this case only default parameters can be used.
   e. In case an error occurs during the *execution* of the program, usually a *self-explanatory* message appears. The most frequent cause of error is because of the *misplacement* of the *executable* files genesyn and grunner. If the program crashes for any other reason, users are welcome to contact the authors via email: giulio.pavesi@unimi.it.

      If the program reports that no subsequences were found, and instead a result was expected, the most common reason is that different names were provided for the same gene in different genomes.
   f. The input file must be a plain text file. A common source of trouble, often ignored, is the fact that text files have a different format in different *operating systems* (e.g., a text MS-DOS/Windows file is different from a Unix/Linux one).

      Although most of *Unix/Linux* systems usually recognize MS-DOS files and process them correctly, it is often a good idea to process a MS-DOS file transferred to a Unix/Linux computer with the "dos2unix" utility.

      Also, do not run the program on a MS-Word document (that contains all the Word formatting characters, and it is not a text file) or do not prepare the text input file by *cutting* and *pasting* the text from a Word document: most often than not, paste invisible Word *formatting* characters that will cause the input to be misread by the program. Rather, save Word file as a *text file*.

g. The duration of a run of the program *strongly* depends on the input parameters that were provided, especially the percentage and gap values. In any case, the time *seldom* exceeds a few *minutes* on tens of genomes with a few hundred genes with no gap threshold and percentage set to around half of the sequences. If input genomes contain *duplicated* genes, instead, the time and *memory* usage might grow exponentially (on the number of duplicated genes).

Chapter 8

Repositioning of Centromere

The centromere *repositioning* phenomenon consists in the move of the *centromere* along the *chromosome* during evolution. This *phenomenon* is relatively *frequent*, and has been documented in *primates, nonprimate mammals*, and *birds*. It implies the inactivation of the old *centromere* and the *rapid progression* of the newly seeded centromere toward the complex organization that probably *stabilizes* its activity.

Both events have a huge impact on chromosomal *architecture*. The segmental duplicon clusters at 6p22.1 and 15q24-26 are clear examples of remains of *inactivated ancestral centromeres*. These *duplicons* are dispersed in a relatively large area (approx 10 Mb), and contribute to the bulk of *nonpericentromeric* segmental duplications that constitute approx 5% of the human genome. Human *neocentromeres* are rare newly created centromeres formed in a previously noncentromeric location and devoid of *alphoid* sequences.

Two human neocentromeres emerged in the same 15q24-26 chromosomal region where an *ancestral* centromere inactivated, establishing an *intriguing* connection between neocentromeres and *inactivated anccstral centromeres*.

An other human neocentromere was found to be located in the same 3q26.1 region where an *evolutionary* centromere was seeded and fixed in Old World monkey (OWM) *ancestor*, indicating that the 3q26.1 region conserved a centromeric competence that was present before hominoids/ OWM splitting.

These findings further support the idea that some chromosomal domains have inherent centromeric capability. Two peculiar neocentromere cases, on chromosome 3 and on chromosome 4, have been recently reported in *literature*. Both arose in otherwise normal *chromosomes* of *normal individuals,* and were *transmitted* to the *subsequent generation*.

They can be considered as *living* examples of centromere repositioning events at their very early stages. The centromere is a specialized structure responsible for *chromosome segregation* at *mitosis* and *meiosis*, and is conserved throughout evolution. Unexpectedly, sequences underlying the centromere function are not evolutionary *conserved*, and show *variations* even among closely related species. Pericentromeric regions also show a high degree of evolutionary plasticity. They *accumulate* large amount of segmental duplications, which appear intrinsically more unstable than the bulk of euchromatic DNA. A detailed analysis of *pericentromeric* segmental *duplications* and their role in evolution can be found in Chapter enywhere chapter in this book.

Evolutionary studies aimed at *tracking* chromosomal marker order *conservation* in primates *unveiled* an additional *peculiar* feature of centromeres: the centromere repositioning, consisting in the evolutionary movement of the centromere along the chromosome in the absence of any *pericentric inversion* that would account for the discrepancy in centromere location between two species. In this chapter, we focus on this biological *phenomenon* that is relatively frequent during the evolution of *mammals*.

EVOLUTIONARY APPROACH

We discovered this phenomenon while studying the evolutionary history of human chromosome 9 in primates. Figure elsewhere in this chapter summarizes the results obtained using a *panel* of yeast artificial chromosome clones used to establish marker order along this chromosome in 9 different primate *species*. If the position of the centromere is not taken into account, a relatively small number of *rearrangements* can be invoked to *reconstruct* the marker order in the *studied* species.

Conversely, if the centromere is included in the analysis, a paradox emerges: the centromere appears to have *undergone* an independent evolutionary history. We *hypothesized* that the movement of the centromere toward the new *location* was not owing to classical chromosomal rearrangements (essentially inversions), but, rather, to the inactivation of the old centromere and the *simultaneous* appearance of a new centromere in a different *location*.

The complexity of the evolutionary history of chromosome 9 did not allow certain *alternative hypotheses* to be discarded, such as two successive inversions, the first involving the centromere and the second restoring the previous marker order with the exception of the centromere. For these reasons we undertook similar studies on other chromosomes. According to Ohno' s law, the gene content of the chromosome X has barely changed throughout mammalian development in the last 125 million years.

The X chromosome was, therefore, an ideal chromosome in this respect. The position of the centromere of this chromosome is substantially unchanged in primates, with two exceptions in *prosimians*: the black lemur (Eulemur macaco [EMA]), and the ring-tailed lemur *(Lemurcatta [LCA])*. The chromosome X appears acrocentric in black lemur and almost metacentric in *ring-tailed* lemur. We investigated the marker order organization using a variety of different molecular *cytogenetic tools*, but no variation in marker order was found.

We concluded that the move of the centromere in EMA and LCA was because of the centromere repositioning event. A centromere repositioning implies the inactivation of an ancestral centromere and the seeding of a new centromere. Neocentromere appearance is not a rare event in human. Human neocentromeres are defined as "rare human chromosomal aberrations where a new centromere has formed in a previously non-centromeric location".

The first well-documented case of a human neocentromere was reported by du Sart et al. Since then, more than 50 neocentromere cases have been described. Some neocentromeres are

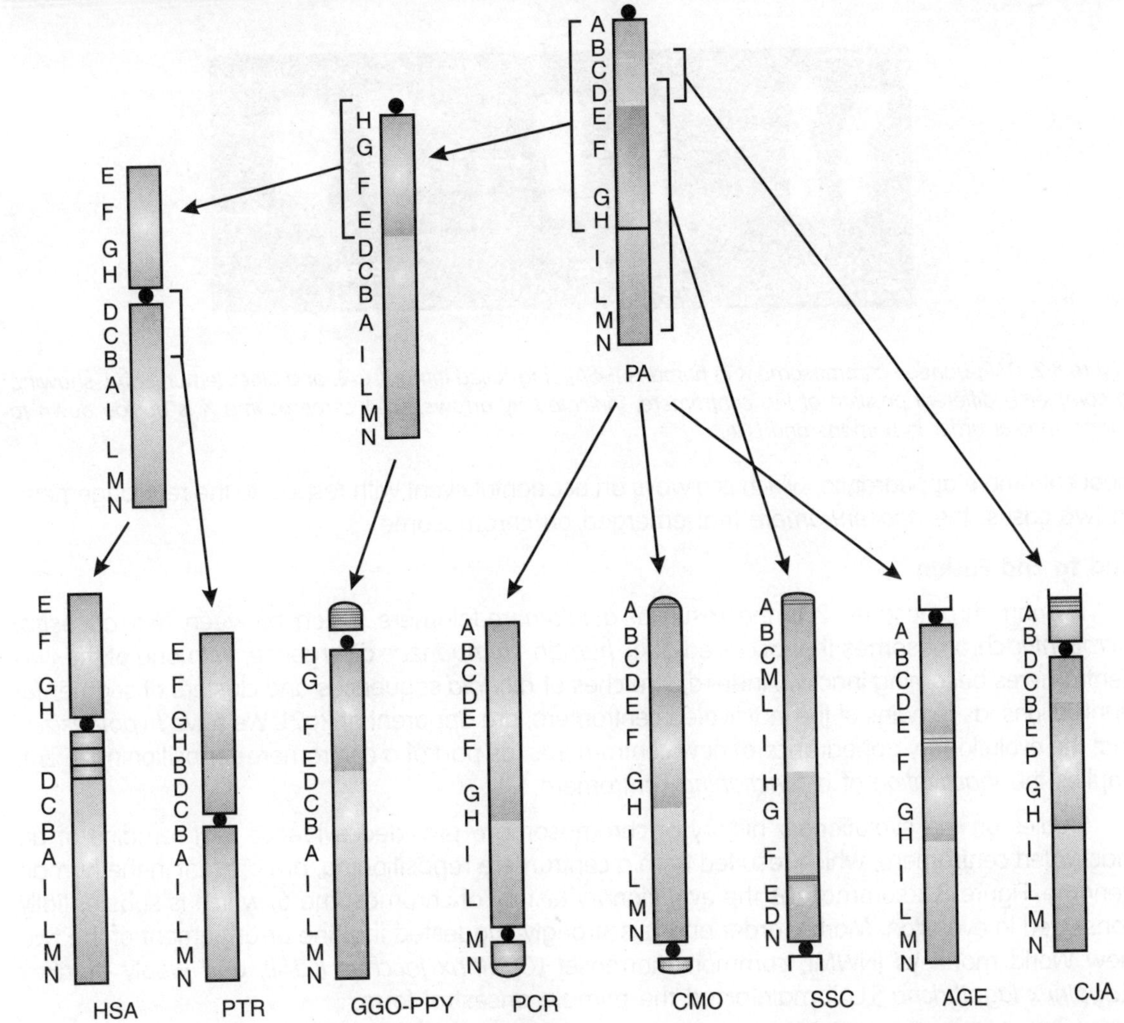

Figure 8.1: Summary of fluorescence in situ hybridization results delineating the evolutionary history of human chromosome 9 in primates. The hypothesized pericentric or paracentric inversions are indicated by brackets spanning the inverted segment. The brackets at the bottom of the figure groups species in which the difference in centromereposition cannot be explained on the basic of classical rearrangements. Species studied: common chimpanzee (Pan troglodytes, PTR), gorilla (Gorilla gorilla, GGO), and orangutan (Pongo pygmaeus, PPY). Old World monkeys, silvered leaf-monkey (Presbytis cristata, PCR). New World monkeys, dusky titi (Callicebus inolloch, CMO), spider monkey (Ateles geoffroyi): common marmoset (Callithrix jacchus, CJA), and squirrel monkey.

clustered in "hotspots" for neocentromere formation, which include 3q26-qter, 8p,13q21-q32, and 15q24-q26 regions. Most of the neocentromeres arise in acentric fragments because of a cytogenetic rearrangement. Acentric chromosomal fragments are usually lost.

Y was not accompanied by a rearrangement. In one of these cases, the chromosome was *segregating* in the family. However, in some cases, their *mitotic* survival is *rescued* by

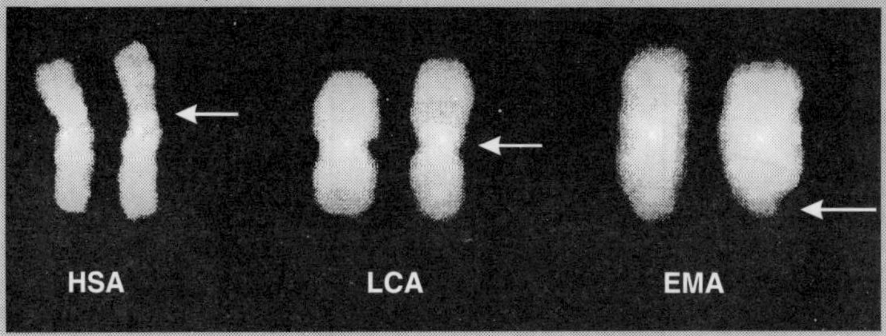

Figure 8.2: DAPI-banded chromosome X in humans (HSA), ring-tailed lemur (LCA), and black lemur (EMA), showing a completely different position of the centromere, indicated by arrows. EMA chromosome X is upside down to match marker order in humans and LCA.

neocentromere appearance, which is always an accidental event with respect to the rearrangement. In two cases, the *neocentromere* that emerged on chromosome

End to End Fusion

Human chromosome 2 is the result of a *telomere*-telomere fusion between two *ancestral acrocentric* chromosomes that occurred after *human-chimpanzee* divergence, with one of the two centromeres becoming inactive. Indeed, stretches of *alphoid* sequences and clusters of segmental duplications, as remains of the inactivated centromere, are apparent at 2q21. We have *hypothesized* that the evolutionary appearance of new centromeres, as part of a centromere-repositioning event, implies the *inactivation* of a *functioning* centromere.

Studies on the evolutionary history of chromosome 6 provided evidence that remains of an inactivated centromere, which resulted from a centromere repositioning, are present in the human genome. Figure 8.3 summarizes the evolutionary history of chromosome 6, which is substantially conserved in evolution. Marker order analysis strongly suggested that the arrangement of the two New World monkeys (NWM), common marmoset (*Callithrix jacchus [CJA]),* and wooly monkey *(Lagothrix lagothricha* [LLA]) maintained the primate ancestral form.

The marker arrangement of cat, chosen as the outgroup, supported this hypothesis. The centromeres of OWM and great apes, therefore, are *evolutionary* new centromeres. The major point of this analysis was the finding, in humans, of a *duplicon* cluster at 6p22.1 exactly *corresponding* to the region where the ancestral centromere was inactivated. No remains of alphoid *centromeric* sequences, however, were *detected.*

It can be hypothesized that, following the centromere *inactivation*, the *pericentromeric* duplicons *dispersed* in a relatively wide area (approx 10 Mb), whereas the large alphoid block, typical of active primate centromeres, completely disappeared. A second example of remains of an ancestral inactivated centromere is presented next.

Cytogenetics of Centromeres

Centromeres are visible cytogenetically as the primary constriction and in *primates* are always associated with the presence of an *array* of υ-satellite DNA. α-satellite, however, it is not an

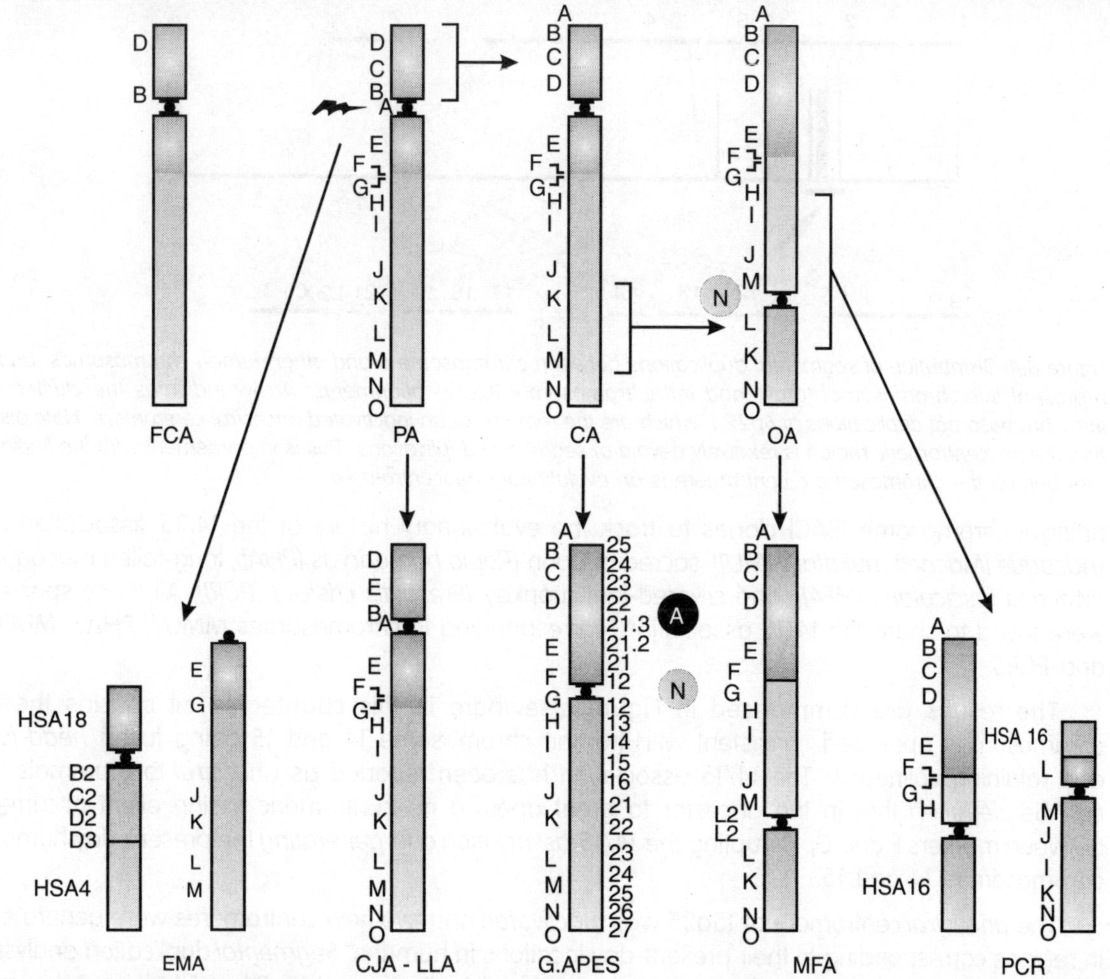

Figure 8.3: Evolutionary history of human chromosome 6 established on the bases of marker order comparison among the orthologous chromosomes in great apes, Old World monkeys, New World monkeys, and prosimians. MFA, long-tailed macaque, Macaca fascicularis; *PCR, silvered leaf-monkey.* Preshytis cristata; *CJA, common marmoset,* Callithrix jacchus; *LLA, wooly monkey,* Lagothrix lagothricha; *EMA, black lemur,* Eulemur macaco; *FCA, cat,* Felix catus. *N in a gray circle stands for a new centromere; A in the black circle stands for an ancestral centromere. PA, primate ancestor; CA, catarrhini ancestor; OA, OWM ancestor.*

absolute requirement for centromere function, as evident in *neocentromeres*, which are devoid of satellite DNA.

The mechanisms *underpinning* neocentromere *emergence* is obscure. du Sart et al. have hypothesized the existence of latent centromeres with a finite capacity for *neocentromerization*. Our investigations of chromosome 15 evolution provided evidence of an obscure but *intriguing* relationship between neocentromeres and ancestral inactivated centromeres.

Chromosomes 14 and 15 in humans and in great apes are separate chromosomes. In OWM, they form a unique chromosome (chromosome 7 in macaque). We used a panel of bacterial

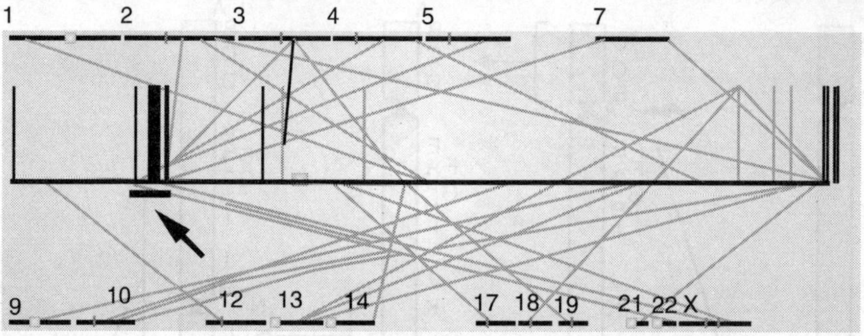

Figure 8.4: Distribution of segmental duplications between chromosome 6 and other human chromosomes. Lines represent interchromosomal (gray) *and intrachromosomal* (black) *alignments. Arrow indicates the cluster of intrachromosomal duplications at 6p22.1, which are the remains of an inactivated ancestral centromere. Note also that the pericentromeric region is relatively devoid of segmental duplications. This is in agreement with the finding that human the chromosome 6 centromere is an evolutionary neocentromere.*

artificial chromosome (BAC) clones to track the evolutionary history of the 14/15 association in *macaque (Macaca mulatta [MMU]),* sacred baboon *(Papio hamadryas [PHA]),* long-tailed macaque *(Macaca fascicularis [MFA]),* and silvered-leaf monkey *(Presbytis cristata [PCR]).* All these species were found to share the 14/15 association, corresponding to chromosomes MMU7, PHA7, MFA7, and PCR5.

The results are summarized in Figure elsewhere in this chapter. In all species these chromosomes appeared consistent with human chromosome 14 and 15 being fused *head-tail* and retaining colinearity. The 14/15 association has been reported as *ancestral* to mammals. It can be *deduced* that in the ancestor to great apes, a noncentromeric fission event occurred between markers F and G, disrupting the 14/15 *association* and *generating* the present-day human chromosomes 14 and 15.

The *ancestral* centromere at 15q25 was *inactivated* and two new centromeres were generated in regions corresponding to their present-day locations in humans. *Segmental* duplication analysis showed a large cluster of intrachromosomal *segmental* duplications at 15q24-q26 (data not shown). This cluster very likely derives from the inactivated ancestral centromere.

Fluorescence *in situ hybridization* (FISH) experiments using BAC probes containing duplicons of this region flanked the centromere in macaque, strongly indicating that they are dispersed remains of the ancestral inactivated centromere. The region encompassed by *duplicons* that flanked the ancestral centromere is approx 11 Mb in size. Similarly to the *inactivated* ancestral *centromere* at 6p22.1, no remains of alphoid sequences were found.

The structural organization of the centromeres of the *present-day* human chromosomes 14 and 15 are indistinguishable from any other human centromere. The conclusion is that the evolutionary new centromeres *progressively* acquired the *complexity* of a normal centromere, which very likely stabilize the centromere function.

Interestingly, one of the hotspot site of *neocentromeres* appearance in clinical cases is at 15q24-q26. We studied two of these neocentromeres that were found several *megabases* apart but inside the region encompassed by duplicons of the ancestral centromere. Alonso et al. have

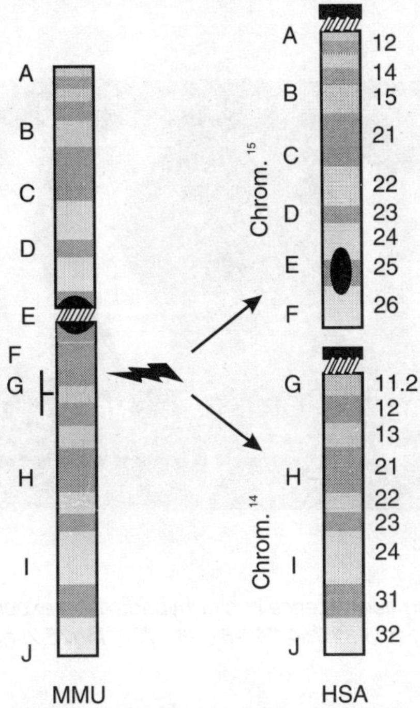

Figure 8.5: The diagram shows the organization of human chromosomes 14/15 association in macaque (MMU. Macaca mulatta*). A fission event gave rise to chromosomes 14 and 15. The fission was followed by the emergence of two new centromeres and by the inactivation of the old centromere at 15q25.*

reported that *neocentromeres* mapping at the 13q32 hotspot span a 6.5-Mb region. Both sets of data disproved the hypothesis that the neocentromere hotspots refer to a single specific locus. Interestingly, however, both 15q25 neocentromeres mapped to duplicons that are the remains of the *ancestral* centromere, therefore establishing an intriguing but unclear correlation between duplicon dispersal and neocentromere *emergence*. If this correlation is correct, one would expect the appearance of neocentromeres at 6p22.1 and at 2q21, where abundant *duplicon* remains of an *inactivated* centromere are present.

We think, however, that other important factors are involved in neocentromere appearance. It is worth noting that most of the *acentric* fragments are small in size and associated with phenotypes *compatible* with embryo development. It could be hypothesized that aneuploidies *generated* by other latent centromeres are *incompatible* with life and will never be detected among *liveborn*.

PHYLOGENY OF THIRD CHROMOSOME

Because the evolutionary history of the neocentromere hotspot at 15q25 revealed a *connection* to an ancestral *inactivated* centromere, we undertook a *comprehensive* study of the evolutionary history of chromosome 3, *harboring* an *additional neocentromere* hotspot at 3q25-q26.

A panel of human BAC clones was used to reconstruct the order of chromosome 3 markers in representative extant primates. The analysis clearly showed the *emergence* of new centromeres

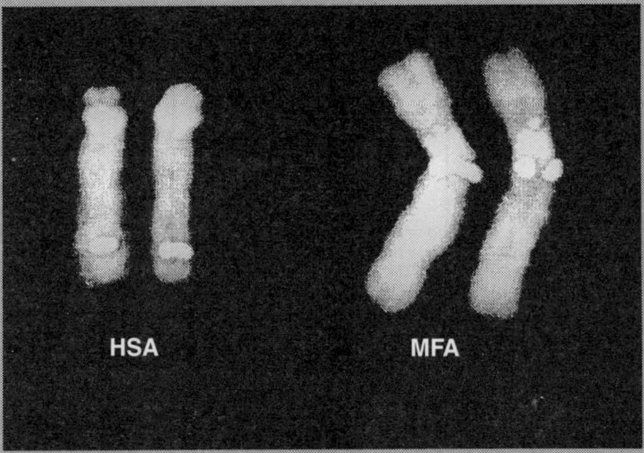

Figure 8.6: Partial metaphases showing fluorescence in situ *hybridization experiments using the human bacterial artificial chromosome clone RP11-182J1 (chr15:82,764,242-82,935,727; 15q25.2) on human chromosome 15 (left) and on macaque chromosome X.*

in OWM and great ape ancestors. Two human neocentromere cases at 3q25-26 were *investigated.* These two cases were found to have distinct but important implications toward a better understanding of new centromere emergence phenomenon both in *evolution* and in clinical cases. In the first case, the neocentromere emergence occurred at 3q26.1 as part of a *complex rearrangement.*

The functional centromere was *excised,* and a neocentromere seeded in the *acentric derivative* chromosome. Human BAC clone RPI 1-498P15 was identified to be the closest to the neocentromere. This same clone was found a few *kilobases* apart from the evolutionary new centromere that appeared in OWM ancestor. This coincidence suggested that the region at 3q26.1 maintained a latent centromere competence that was present before OWM/*Hominidae divergence* that occurred approx 25 million years ago.

Furthermore, the new centromere that appeared in the great ape ancestor was seeded very close to marker G. Interestingly, a new centromere was also seeded in this region in NWM ancestor (chromosome GF21) suggesting, again, that the same latent centromeric region was activated twice, in great ape and NWM ancestors. Pevzner and *Tesler, comparing mouse* and human evolutionary breakpoints, have suggested the "*reuse*" of breakpoint regions in evolution. Similarly, the examples of neocentromeres we have discovered, both in evolution and in *clinical cases,* seem to extend the "reuse" concept to centromeres.

Phylogeny of Centromeres and Telomeres

In four NWM species examined, human chromosome 3 is split into three chromosomes *annotated* as (I), (II), and (III) in the Figure elsewhere in this chapter Chromosomes (I) and (II) have an identical marker order in common *marmoset* (CJA), wooly monkey (LLA), and dusky titi *(Callicebus moloch [CMO]),* if centromeres are not considered.

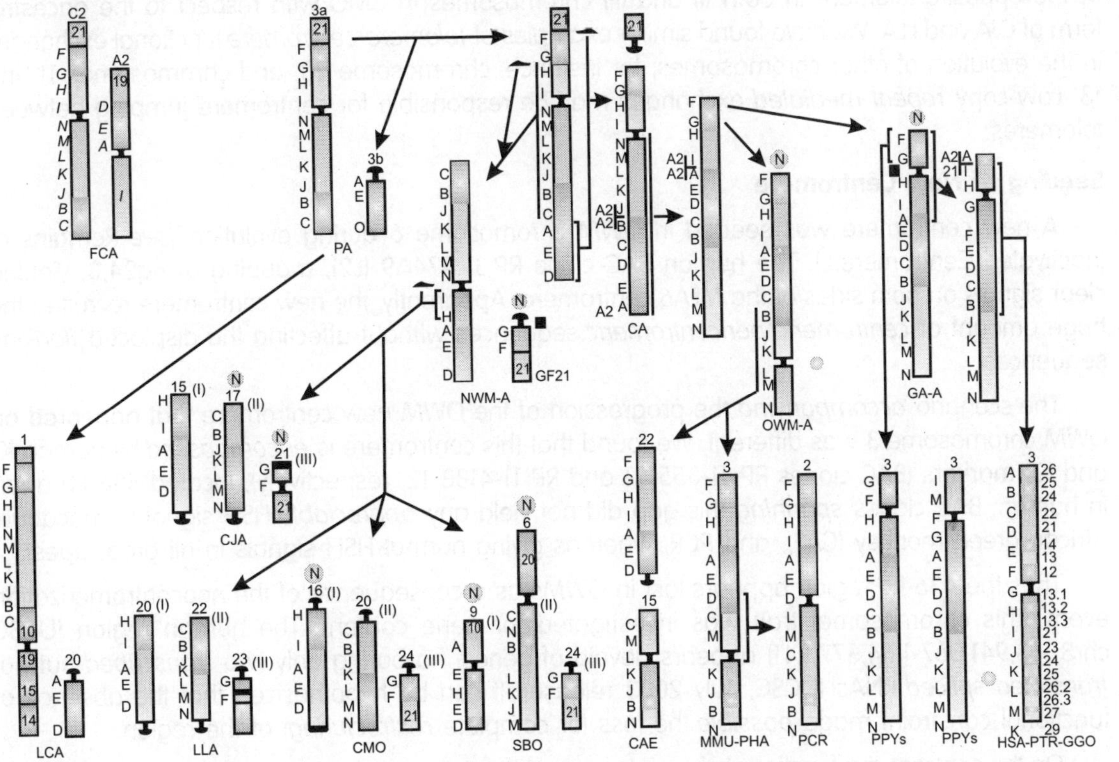

Figure 8.7: Summary of fluorescence in situ *hybridization results delineating the evolutionary history of human chromosome 3 in primates. The hypothesized pericentric or paracentric inversions are indicated by square parentheses spanning the inverted segment. Some chromosomes are upside down to facilitate comparison. PA, primate ancestor; CA, catarrhini ancestor; OWM-A, Old World monkeys ancestor; NWM-A, New World monkeys ancestor; GA-A, great ape ancestor; N (in a gray circle), new centromere. The number that identifies the chromosome in each species is reported on top of the chromosome. Arrows with hatched lines point to species for which an intermediate ancestor has not been drawn. A white N in a gray circle indicates the chromosomes where new centromeres appeared during evolution. The black circles (HSA and OWM-A) and the black squares (GA-A and NWM-A) are positioned at loci where seeding of an evolutionary new centromere was detected. Cat (FCA) probes were identified using the following strategy. The sequence encompassed by each human bacterial artificial chromosome (BAC) was searched for conservation against mouse and rat genome. "Overgo" probes were designed on the most conserved region of each BAC and were then used to screen the cat BAC library RPCI-86. Cat probes. therefore, map to loci orthologous the loci encompassed by the corresponding human BACs. Species analyzed: great ape: common chimpanzee* (Pan troglodytes, *PTR), gorilla* (Gorilla gorilla, *GGO), Borneo* (Pongo pygmaeus pygmaeus, *PPY-B) and Sumatra* (Pongo pygmaeus abelii, PPY-S) *orangutan; Old World monkeys (OWM): rhesus monkey* (Macaca mulatta, *MMU, Cercopithecinae), sacred baboon* (Papio hamadryas, *PHA, Cercopithecinae), African green monkey* (Cercopithecus aethiops, *CAE, Cercopithecinae), silvered leaf-monkey* (Presbytis cristata, *PCR, Colobinae); New World monkeys (NWM): wooly monkey* (Lagothrix lagothricha, *LLA, Atelinae), common marmoset* (Callithrix jacchus, *CJA, Callitrichinae), duskytiti* (Callicebus moloch, *CMO, Callicebinae), squirrel monkey* (Saimiri boliviensis, *SBO); prosimians: ring-tailed lemur* (Lemur catta, *LCA).*

The position of the centromeres, on the contrary, appears peculiar: the centromere is located at the opposite telomere in both (I) and (II) chromosomes in CMO with respect to the ancestral form of CJA and LLA. We have found similar examples of *telomere-centromere* functional exchanges in the evolution of other chromosomes, for instance, chromosome 20, and chromosomes 11 and 13. Low-copy *repeat-mediated* exchanges may be responsible for centromere jumping between telomeres.

Seeding of New Centromere

A new centromere was seeded in OWM chromosome 6 during evolution *(see* Remains of Inactivated Centromeres;). The human BAC clone RP 11-474A9 (L2), mapping at 6q24.3, yielded clear signals on both sides of the MFA6 centromere. Apparently, the new centromere recruited the huge amount of *centromeric/pericentromeric* sequences without affecting the displaced *flanking* sequences.

The scenario *accompanying* the progression of the OWM new centromere that appeared on OWM chromosome 3 was different. We found that this centromere is *encompassed* by human K1 and K2 markers (BAC clones RP 11-355121 and RP 11-418B 12, respectively), located 436 kb apart in human. BAC clones *spanning* this gap did not yield any *appreciable* FISH signal in macaque, African green monkey (CAE), and PCR, whereas giving normal FISH signals in all great apes.

Thus the 436-kb region appears lost in OWMs as a consequence of the neocentromerization event. This chromosomal trait was investigated for gene content. The human region (UCSC chr3:163,941,067-1.64,377,941) appears devoid of genes, harboring only two *transcribed* but not *translated spliced* RNAs (UCSC, July 2003 release). It can be hypothesized that the absence of functional constraint made *possible* the loss (or complete *restructuring*) of the region.

On the contrary, the functional elements around the new centromere insertion on chromosome 6 in OWM behaved as a strong selective barrier to the *degenerative* activity of the evolving centromeric region. Gene expression around the neocentromere that emerged at 10q25 was investigated by *Saffery* et al. The *authors* concluded that the *neocentromere* formation did not affect *underlying* gene expression.

The likely scenario of the progression of an evolutionary new centromere appears as a competition between the *degenerative* activity of the centromere in the seeding area, and the selection acting against disturbance of *functional* elements present in the region. Nagaki et al. have reported recently the organization of rice chromosome 8 centromere *featuring* an unusual organization, with genes embedded in the centromeric *heterochromatin*, as if the centromere were at an *intermediate* stage between new centromere seeding and a final stage in which all genes are excluded from the completely *heterochromatized* centromere, as occurs in the remaining rice centromeres.

Segmental Duplication

Segmental duplications accumulate at greater density in pericentromeric and telomeric regions. However, large blocks of duplicons exist also in euchromatic locations. We have previously shown that *duplicon clusters* located, in humans, at euchromatic regions 6p22.1 and 15q25 are the remains of *inactivated* centromeres. Other chromosomal rearrangements can *relocate pericentromeric* duplicons in noncentromeric areas.

One of the two breakpoints of both pericentric and paracentric evolutionary inversions falls, rather frequently, in pericentromeric regions. In these cases part of a pericentromeric duplicated region can be repositioned far apart from the centromere. Two nonpericentromeric olfactory receptor clusters lying on chromosome 3 are precisely located at domains identified by A2 and 11 markers.

The evolutionary history of chromosome 3 we have reconstructed indicated that these two clusters are examples of duplicons that were pericentromerically located during evolution *(see* marker organization in the primate ancestor [PA]). Most mutations of the *CYP21* locus (steroid 21-*hydroxylase gene*) are owing to a close pseudogene *(CYP21P)* that triggers gene conversions during unequal crossovers. Bush *CYP2land CYP21 P* are located inside the duplicon cluster at 6p22.1 that represents the remains of an evolutionary inactivated centromere. Gene *conversion*, therefore, is an additional, indirect mechanism by which duplicon cluster contribute to human genome *morbidity*.

GENERATION OF ACENTRIC FRAGMENTS

Most of human neocentromere cases arise as a consequence of a rearrangement that generated an acentric fragment. The presence of the acentric fragment as a supernumerary chromosome results in phenotypic abnormalities, which indirectly prevent the transmission of the neocentromere-bearing chromosome to successive generations. They, therefore, have no evolutionary relevance. Very recently, however, Amor et al. and *Ventura* et al. have documented two unprecedented neocentromere cases, at 4q21.3 and 3q24, respectively, that arose in an otherwise normal and mitotically stable chromosomes that were *transmitted* to subsequent *generations*.

These individuals had normal *phenotypes*. They can be regarded as present-day episodes equivalent to new centromere appearance that occurred during the course of evolution and that were fixed in the population. Individuals with chromosomal changes not *accompanied* by clinical signs or reproductive problems usually escape detection. It can be *reasonably* supposed that the "*normal*" cases reported so far are only a minimal fraction of the existing ones.

In individuals heterozygous for the neocentromere, an odd number of meiotic exchanges internal to the chromosomal region encompassed by the two active centromeres leads to the formation of dicentric and acentric chromosomes. These problems could affect the fitness of heterozygous individuals and act as a *negative* selective pressure. Meiotic drive in females in favor of the repositioned chromosome is among the possible *mechanisms* favoring the fixation of the neocentromere. Meiotic drive favoring *Robertsonian translocations* has been reported by Pardo-Manuel de Villenna and Sapienza.

SEQUENCING PROJECT

One of the first steps in a sequencing project is the construction of detailed genetic and physical maps obtained using linkage analysis and radiation *hybrids* studies, respectively. *Radiation hybrid* maps have been constructed, i.e., for cattle, cat, horse, and pig. These studies could represent, in theory, an efficient tool to pinpoint *centromerere positioning* events when maps of different species are compared. However, most of these studies were not focused on *evolutionary* aspect of genome *organization*.

The paper on the bovine radiation hybrids map published by Band et al., on the contrary, took into account the position of the cattle centromeres with respect to humans. The authors concluded that "the comparative map suggests that 41 translocation events, a minimum of 54 internal rearrangements, and repositioning of all but one centromere can account for the observed *organizations* of the cattle and human genomes."

Most likely, they have overestimated the centromererepositioning phenomenon in cattle. However, it indicates that the repositioning phenomenon may be much more widespread than expected. Horse and donkey (equidae) diverged about 2 million years ago. They produce hybrids (mule and hinny) that are mostly sterile owing to the several chromosomal rearrangements that differentiate horse and donkey chromosomes.

Yang et al., using a combination of classical G-banding and whole chromosome paints, have noticed the discordant position of the centromere in four *orthologous* chromosomes, i.e., four distinct centromere *repositioning* events occurred in a relatively short evolutionary time.

Chapter 9

Protein Sequence Alignment

Protein sequence alignment has become an essential part of biomedical research. It is one of the standard approaches to explore potential functional activity of a newly discovered protein by identifying sequence homologues that may be evolutionarily related. One can often infer the structural and functional information of a new protein from the knowledge of well-characterized homologous proteins. In general, closely related protein sequences are relatively easy to align using the existing sequence-based methods.

However, the success rate of these methods in finding correct alignment is significantly reduced when the sequence identity between two aligned sequences is lower than 30 %, a threshold often referred to as the twilight zone. The performance of a sequence alignment algorithm largely depends on its employed substitution matrix. PAM and BLOSUM are some of the most commonly used substitution matrices in sequence alignment algorithms. They are mainly derived from the frequencies of amino acid substitutions in a series of compiled families of protein sequences.

Algorithms employing BLOSUM62 and PAM250 are most effective in identifying and aligning homologous proteins. Efforts to develop improved sequence alignment of remote homologues have focused on incorporating additional information that improves the diversity of the substitution matrices. These methods include the use of position-specific substitution profiles derived from multiple sequence alignment of protein families. Structure-based substitution matrices have also been developed. Such matrices are derived from the frequency of amino acids occupying similar positions in a series of structurally aligned proteins.

Algorithms employing structure-based substitution matrices appear to have improved success in detecting and aligning remotely related protein sequences. Another approach that has achieved promising improvements in pairwise sequence alignment, particularly for sequences with low homology, is the sequence-template alignment method. This algorithm incorporates the structural knowledge of the template, as well as the amino acid propensities for secondary structures, into a substitution matrix for sequence alignment.

Although methods relying on structure-based sequence profiles are effective in identifying specific functional motifs in proteins, the overall improvement of these methods over the purely sequence-based methods in sequence alignments is limited. It appears that the inherent limitation of these methods is their dependence on the structure and the sequence diversity of the derived profile. If the profile construction emphasizes too much on sequence diversity, it would result in profiles with limited signature value.

On the other hand, profiles derived from conserved proteins are not sensitive in detecting distantly related homologs. It has always been a challenge to reach an optimal balance in selecting protein sequences for constructing profiles.

Some studies have suggested that the most effective profiles could be derived from sequences sharing 30-50 % homologue. Another key weakness of the structure-based sequence profiles is the lack of information in the loop regions, which are often ignored in structural alignments.

For sequences with long and functional loop regions, the structure-based profile alignment methods could be ineffective. A method incorporating both sequence- and structure-based profiles (the hybrid sequence profiles) has shown an improved performance in aligning proteins with distant homologues.

NdPASA is a new pairwise protein sequence alignment algorithm that incorporates the structure information of the aligning sequences. By incorporating the neighbor-dependent propensities (NDPs), the method has shown significant improvements over conventional methods in aligning proteins with remote sequence homologies. The NDP reflects the likelihood of an amino acid pair adopting a particular secondary structure conformation (α-helices, β-strands, and loops) in proteins.

The rationale for applying the NDPs in sequence alignment is easily recognized. Methods employing sequence-based substitution matrix often have limited success in aligning sequences sharing low percentage of sequence identity. The incorporation of NDPs has allowed us to estimate the probability of an amino acid pair to be aligned with a corresponding amino acid pair adopting a specific secondary structure in the template sequence.

For example, an amino acid pair in the query having a low neighbor-dependent propensity for α-helical conformation would be less likely aligned with an amino acid pair in an α-helix of the template. A structure-dependent gap opening and extension penalty scheme is also implemented. A higher gap penalty is applied for gaps within the regular secondary structures than for gaps in the loops.

The performance of the NdPASA is compared with the PSI-Blast, the standard global alignment (GA) using BLOUSUM62, and the individual propensity assisted sequence alignment (IPASA), a global sequence alignment algorithm that incorporates individual amino acid propensity for protein secondary structures.

NdPASA performs most effectively when the structural information of the template sequence is available. It shows significant improvements over conventional methods when aligning sequence pairs sharing less than 20 % sequence identity.

REQUIREMENTS

NdPASA is written in JAVA with CGI interface. It is compiled on a LINUX-based workstation. The web server can be freely accessed online at www.fenglab.org/bioserver.html.

All analyses are performed using a rational database derived from a nonredundant set of PDB (PISCES). The limit of redundancy chosen in this study is at 25 % sequence identity. Only high-resolution structures are selected with a resolution cut-off at 2.5 Å. NMR structures are not included in this study. A total of 1430 proteins are selected from the PDB databank.

The database containing these PDB entries (the sPDB) is used in the subsequent parsing protocols. Two independent secondary structure assignment criteria are used to extract the sequences and the secondary structure information from the sPDB entries: the assignments given by the experimentalists and the assignments generated by the DSSP algorithm, which assigns secondary structure based on the analysis of backbone dihedral angles and hydrogen bonds.

For the sake of comparing the two assignment methods, the more sophisticated DSSP structure assignments are converted as follows: helices, 3_{10} helices and the π helices are all considered as helices; the single-stranded α-strand and the multiple-stranded α-sheets are simply considered as strands; and a loop is defined as a region of a protein that is assigned as neither an α-helix nor a β-strand by either methods.

The database system used in this study is the PostgreSQL packaged in the Redhat Linux 7.2. The sequence and secondary structure information of the sPDB entries are parsed into rational tables. Two sets of tables are compared. One derived from parsing the author assigned secondary structure information and the other is parsed according to the DSSP calculations. Considering that the authors' assignments are experimentally observed, we choose to use them as the standard for defining secondary structures in the protein structures.

A double-check mechanism is installed to avoid erred assignments: the author assigned structural elements are compared with the DSSP assignment. The selected secondary structural elements are consistent with both assignment methods. The parsing exercise of the sPDB database results in libraries containing specific secondary structural elements (SSLs).

Studying the Secondary Structures of Amino Acid

The NDP reflects the neighboring probability of a pair of amino acids, in any combination, in the three classes of protein secondary structures S*j* (*j* = α-helix, β-strand, and loop). Neighbors can be defined as the first neighbor, where the amino acids in the pair are immediately next to each other in sequence; the second neighbor, where the pair of amino acid is separated by one amino acid residue in sequence; the third neighbor, where the pair of amino acid is separated by two amino acids; or the fourth neighbor, where the pair of amino acids is separated by three amino acids.

First introduced by Chou and Fasman, the amino acid propensity (ε_a) is defined as the relative distribution of amino acids in between the SSLs and the sPDB. It reflects the likelihood of amino acids adopting secondary structures. For neighbor-dependent propensity analysis, the frequency of occurrence of the residue type x at neighboring positions of the residue a in a specific secondary structure S_j is calculated as follows:

$$\varepsilon x(a \pm i), j = \frac{\left[\Sigma x(a \pm i)_{s_j}\right] / n_{(pair)S_j}}{\left[\Sigma x(a \pm i)_p\right] / n_{(pair)P}} \tag{9.1}$$

where $\Sigma x(a \pm i)S_j$ and $\Sigma x(a \pm i)P$ are the occurrences of the residue type *x* at the *±ith* positions of the residue *a* in the SSLs (*S*) and in the sPDB (*P*), respectively; $n(_{pair})S_j$ and $n_{(pair)P}$ are the total number of residue pairs in S_1 and *P*, respectively. The numerator in Eq. 9.1 calculates the frequency of occurrence of residue *x* neighbors the residue type *a* in the secondary structure S_j, whereas the denominator in the equation calculates the frequency of occurrence of the residue *x* neighboring the residue type a in the sPDB (P).

The ratio of these values would be the propensity of residue x in S_i when it is neighbored with residue type *a*. We applied the neighbor-dependent sequence analysis on the residues of immediate neighbors in secondary structures of proteins. The NDPs are represented as $\varepsilon_{x(a\pm1)}$; An $\varepsilon_{x(a\pm1),j}$ value of 1.0 means that the occurrence of the residue pair, *ax* (or *xa*), in the secondary structure *j* is the same as its frequency of occurrence in proteins. A value greater than 1.0 means the pair has occurrence in the secondary structure *j* higher than that in proteins, i.e., the pair has preference for adopting the *j* secondary structure conformation.

A $\varepsilon_x(_{a\pm1})_{,j}$ value lower than unity would suggest that the pair is less likely to adopt the *j* secondary structure than random distributions. For example, a $\varepsilon_p(_{A-1})_{a^*}$ = 1.52 in short helices means that Pro has 62 % more chance to be found in short helices than it in the proteins when it precede Ala, i.e., Pro at -1 position of Ala. On the other hand, a $\varepsilon_{p(A+1),a^*}$ = 0.47 suggests that Pro is less likely to be found in short helix when it follows Ala, i.e., Pro at +1 position of Ala.

The neighbor-dependent sequence analysis of proteins revealed that the amino acid pair preference for secondary structures has its unique pattern and that such pattern are not always predictable by assuming proportional contributions from the propensity values of the individual amino acids. Some of those sequence patterns are most pronounced in some subgroups of secondary structures, such as the short loop and helix subgroups.

Our analysis also yielded a series of amino acid dyads that showed preference for a particular secondary structure. The dyads are defined as the amino acid pairs *"ax"* having strong preferences for the particular secondary structure, whereas the amino acid pairs of *"xa"* having weak preferences for the secondary structure. It is evident that the neighbor-dependent protein sequence analysis method revealed some "hidden" sequence codes in proteins.

Pairwise Sequence Alignment

The first step to developing a pairwise sequence alignment algorithm is to setup a data set of sequence pairs for training and testing. The dataset of sequence pairs is constructed as follows. Protein sequences sharing less than 90 % sequence identity are first extracted from the nonredundant PISCES database. The resolution cut-off is at 2.5 Å. The homologous sequence pairs selected for this study satisfies two criteria: (1) the pair shares certain level of sequence homology and (2) the protein pair adoptes same structural fold.

Proteins homologous to the selected sequences in the PISCES database are identified by PSI-BLAST searches running against a nonredundant nucleotide database, the NCBI-nr. Sequences returned with *e*-values in the range of 10^{-6} to 1000 are retained. Sequence pairs with multiple high segment pairs are removed. The sequence pairs are then checked against the SCOP classification database to ensure that they belong to the same structural fold. Sequence pairs of the same protein family, protein super family, and fold in the SCOP database are selected.

All selected sequence pairs also have comparable sizes where the shorter sequence of the pair is at least 80 % of the length of the longer sequence. The data set of sequence pairs contains a total of 2521 sequence pairs that shares sequence identities ranging between 13 and 25 %. Five hundred of these protein sequences pairs are selected as a training set, and 2021 pairs are selected as a testing set. The training set is used to optimize parameters in the alignment algorithm.

Alignment Algorithm

NdPASA incorporates the information of secondary structure propensity into the Needleman-Wunsch global alignment algorithm with affined gap penalties. The Needleman-Wunsch dynamic programming algorithm is effective in finding the optimal scoring alignment or a set of alignments. To incorporate NDPs, we formulate *M*(*i, j*) as follows:

$$M(i,j) = \max\begin{Bmatrix} M(i-1,j-1) + s(x_i,y_j) + scale^* p(x_{i-1},x_j,x_{i+1}|ss_j) \\ I_x(i-1,j-1) + s(x_i,y_j) + scale^* p(x_{i-1},x_i,x_{i+1}|ss_j) \\ I_y(i-1,j-1) + s(x_i,y_j) + scale^* p(x_{i-1},x_i,x_{i+1}|ss_j) \end{Bmatrix} \quad (9.2)$$

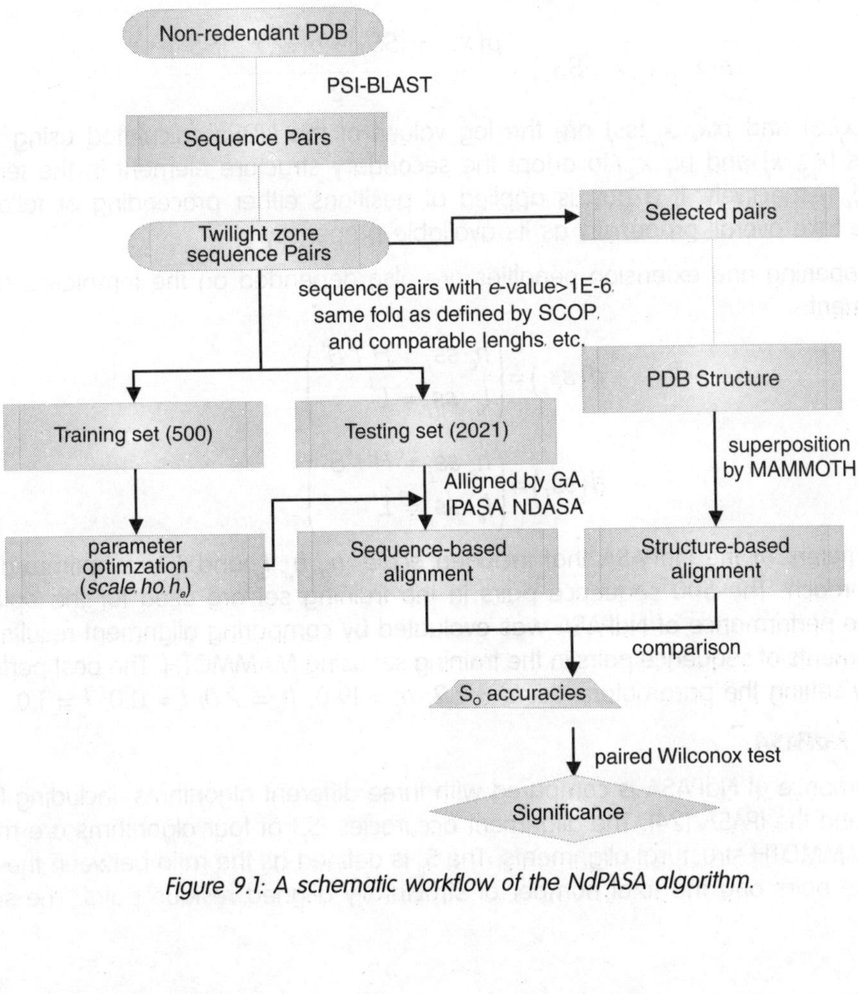

Figure 9.1: A schematic workflow of the NdPASA algorithm.

$$I_x(i,j) = \max\begin{Bmatrix} M(i-1,j) - d(ss_j) \\ I_x(i-1,j) - e(ss_j) \end{Bmatrix} \quad (9.3)$$

$$I_y(i,j) = \max\begin{Bmatrix} M(i-1,j) - d(ss_j) \\ I_y(i-1,j) - e(ss_j) \end{Bmatrix} \quad (9.4)$$

where $M(i, j)$ is the best score up to (i, j) given that x_i is aligned to $_{y}j$; and $I_x(i, j)$ *is* the best score given that x_i is aligned to a gap; $I_y(i, j)$ *is* the best score given that y_j is in a insertion with respect to x_i and $s(x_i, y_j)$ is the amino acid substitution score of x_i and y_i given a substitution matrix (BLOSUM62 for this study). Scale is the scaling factor, denoting a relative weight between propensity score and substitution score. $p(x_{i-1}, x_i x_{i+1} | ss_j)$ is the neighbor-dependent secondary structure propensity score for adopting the secondary structure element of the template sequence at position *j*. It depends on the amino acid at *ith* position and the amino acid types of its neighboring positions both proceeding $(i - 1)$ and following $(i + 1)$ the position *i*.

$$p\left(x_{i-1}, x_j, x_{i+1} \middle| SS_j\right) = \frac{p\left(x_{i-1}, x_i \middle| SS_j\right) + p\left(x_i, x_{i+1} \middle| SS_j\right)}{2} \quad (9.5)$$

where $p(x_{i-1}, x_i | ss_j)$ and $p(x_i, x_{i+1} | ss_j)$ are the log values of the NDPs calculated using Eq.9.1 for residues pairs (x_{i-1}, x_i) and (x_i, x_{i+1}) to adopt the secondary structure element in the template at position j (ss_j), respectively. If a gap is applied at positions either proceeding or following the residue x_i, we take overall propensity as its available propensity.

The gap opening and extension penalties are also depended on the template's secondary structure element:

$$d(ss_j) = \begin{pmatrix} h_0 & ss_j = 'H'/'S' \\ l_0 & ss_j = 'L' \end{pmatrix} \quad (9.6)$$

$$d(ss_j) = \begin{pmatrix} h_e & ss_j = 'H'/'S' \\ l_e & ss_j = 'L' \end{pmatrix} \quad (9.7)$$

The parameters of the NdPASA that included *scale*, h_o, h_e, l_o and l_e are optimized using a bootstrap approach. The 500 sequence pairs in the training set are used for the optimization procedure. The performance of NdPASA was evaluated by comparing alignment results with the structure alignments of sequence pairs in the training set using MAMMOTH. The best performance is obtained by setting the parameters: *scale* = 2.2; $h_o = 19.0$; $h_e = 2.0$; $l_o = 11.0$; $l_e = 1.0$.

Performance NdPASA

The performance of NdPASA is compared with three different algorithms, including PSI-Blast, pairwise GA, and the IPASA (24). The alignment accuracies (S_0) of four algorithms are measured against the MAMMOTH structural alignments. The S_0 is defined as the ratio between the correctly aligned residue pairs and the total number of structurally aligned residue pairs. The sequence pairs selected for the evaluation share 13-25 % sequence identity. NdPASA demonstrated an average of 5-32% improvement over other algorithms. Its improvement over the GA algorithm was 5-23%, 2-10% over the IPASA algorithm, and 17-48% over the PSI-BLAST. The significantly larger improvement over PSI-BLAST could be attributed in part to the fact that PSI-BLAST aligns sequence pairs locally. NdPASA is most effective in aligning sequence pairs with low sequence homology. For sequence pairs with approx 23-25% identity, NdPASA has an average accuracy of 3-5% better than that of the ISAPA and GA algorithms, respectively.

For sequence pairs sharing approx 15-17% identity, NdPASA performed by an average of 6-14% better than that of the IPASA and GA algorithms, respectively. These results suggest that the sequence patterns derived from the neighbor-dependent sequence analysis of protein structures have more significant contribution to sequence alignments for sequence pairs that are remotely related.

Algorithm Implementations

The NdPASA server is designed mainly to aid homologous protein structure modeling of query sequences (see notes). Figure 9.2 shows a schematic diagram of the algorithm implemented in the server. The first step to homologous modeling is to identify a template that shares the same fold as that of the query sequence. An input page is designed with three options. Option 1: user-identified template with an input of either a sequence or a PDB ID. With an input of PDB ID, NdPASA will automatically retrieve the corresponding sequence from the PDB and assign secondary structure elements using DSSP before performing sequence alignment with the query.

When a template sequence is entered, NdPASA server will perform a PSI-BLAST search against the PDB database and return results containing PDB entries that share at least 80 % sequence identity with the template. The user is asked to select the desired template from the returned results. When a template is identified, the NdPASA server will perform secondary structure

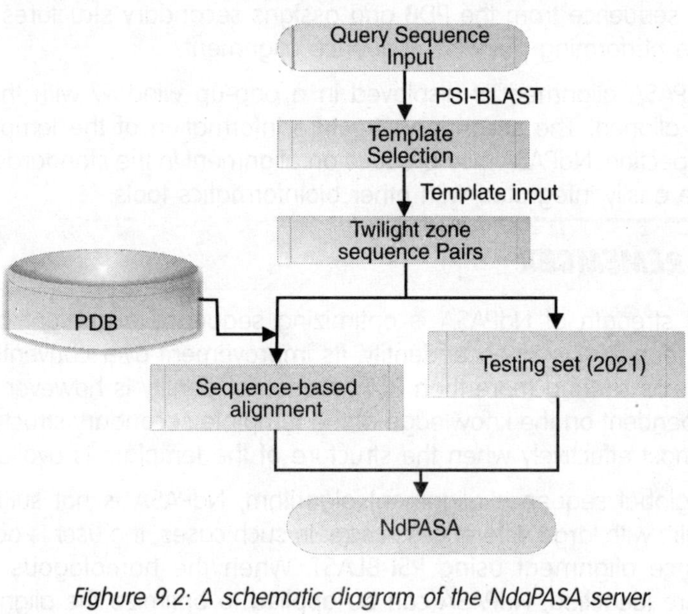

Fighure 9.2: A schematic diagram of the NdaPASA server.

assignments and sequence alignment with the query sequence. Option 2: input template sequence with user-assigned secondary structures.

When the structure of the template is not known, user can submit template sequence, to a secondary structure prediction server, such as the PSIPRED server (http://bioinf.cs.ucl.ac.uk/psipred/psiform.html). NdPASA accepts the returned secondary structure assignments for subsequent alignment. Option 3: template unknown. In this case, NdPASA server performs a PSI-BLAST search against the nonredundant protein structure database (PDB) for sequences homologous to the query using the BLOSUM62 matrix.

The user also has the option to choose different scoring matrices, including PAM250, PAM300, PAM120, BLOSUM35, BLOSUM45, BLOSUM50, BLOSUM60, BLOSUM62, and BLOSUM80. The gap opening and extending penalty parameters can also be changed. The default options of BLOSUM62, the gap opening (-11) and extending (-1) penalties are selected based on experimental test results. All returned results from the PSI-BLAST searches are displayed with their sequence names, PDB entry-ID, PSI-BLAST scores and the percentage sequence identities as determined by the PSI-BLAST when compared with the query protein.

To limit the scope of template selection, we specified the PSI-BLAST output to contain only the sequences with either the top 5-ranked scores or the top 15-ranked scores for inspection. An optional filter is also implemented where the user may limit the output of the PSI-BLAST search to those sequences that share sequence homology above a defined threshold. When a template candidate is identified, the user may select the radial button next to the sequence and click "submit" for optimized pairwise sequence alignment by NdPASA.

Alternatively, the user may choose one of the more sophisticated fold recognition servers, such as the GenTHREADER, 3D-PSSM, or Pcons5, to identify template. Upon receiving command to align the query sequence against one of the templates identified by PSI-BLAST, the program fetches the template sequence from the PDB and assigns secondary structures for the template by using DSSP before performing pairwise sequence alignment.

The result of NdPASA alignment is displayed in a pop-up window with the query and the template sequences aligned. The secondary structure information of the template sequence is also displayed for inspection. NdPASA also produce an alignment in the standardized FASTA format so the results can be easily integrated with other bioinformatics tools.

POINTS TO REMEMBER

1. The greatest strength of NdPASA is optimizing sequence alignment between proteins sharing less than 20% sequence identity. Its improvement over conventional methods in aligning proteins sharing more than 30% sequence identity is however not significant. It is. highly dependent on the knowledge of the template secondary structure assignments. It performs most effectively when the structure of the template is available.

   Adopting a global sequence alignment algorithm, NdPASA is not suitable for aligning sequence pairs with large differences in size. In such cases, the user is advised to perform local sequence alignment using PSI-BLAST. When the homologous regions in both sequences are identified, NdPASA can be applied to optimize the alignment.

Chapter 10

Analyzing the Copy Number

Cancer genomes undergo a variety of changes and modifications during neoplastic development. Because tumorigenesis should be ultimately ascribed to the alterations of the genetic programs caused by these changes and modifications, it is of particular importance to establish a technology to analyze such genetic and epigenetic changes in a genome-wide manner, to get better understanding of human cancers.

In this respect, recent development of a number of genome-wide copy number detection systems using array technologies represents one of the major advances in the field of cancer genetics. Complementing conventional cytogenetics and FISH analysis, or replacing chromosome-based CGH techniques, microarray-based copy number detection methods are now widely used to analyze complex cancer genomes as well as to study large-scale copy number variations found in normal human populations.

Among different copy number detection array systems now commercially available, the Affymetrix GeneChip® system has a unique feature that it can detect copy number alterations in an allele-specific manner using a set of single nucleotide polymorphisms (SNP)-specific probes. In fact, currently, this is one of the most sensitive systems to detect allelic imbalances or loss of heterozygosity* (LOH) in cancer genomes using allele-specific copy number determination.

In addition, the total number of probes on a single array now exceeds 250K, which allows very high resolution copy number detection (http://www.afymetrix.com/products/arrays/specific/500k.affx). A main drawback about these SNP arrays arises from PCR amplification before microarray analysis, which biases copy numbers as measured from raw signal ratios between test and reference experiments.

However, well-designed experiments and a set of algorithms to compensate the biases which we developed in the software tool Copy Number Analyzer for Affymetrix GeneChip (CNAG) could overcome the drawback and produce satisfactory results in most situations. In this chapter, we describe the basics of genome-wide copy number/LOH analysis using CNAG2.0 on the GeneChip platform, focusing on the practical use of the software with the underlying principles of the algorithms CNAG2.0 uses.

Before going on to the next section, some key issues in the "wet" part of the analysis are discussed anywhere else in this chapter for those who are unfamiliar with copy number analysis using Affymetrix platform, but these issues are also essential to get high-quality results and the readers are recommended to see the section in advance.

REQUIREMENTS

Cancer Cell Lines

NCI-H2171, H1437, and H2009, and NCI-BL2171, BL 1437, BL2009 are lung cancer cell lines and immortalized lymphoblastoid cell lines established from the corresponding patients, which were obtained from American Type Culture Collection (Rockville, MD; http://www.atcc.org/ common/ cultures/NavByApp.cfm). A glioma cell line U251 was distributed from Riken (Tsukuba, Japan).

Peripheral blood samples and primary tumor specimens were collected from 96 normal volunteers and 33 patients with a variety of malignancies, and subjected to DNA extraction. Microarray experiments were performed completely according to the manufacturer's protocols, using Affymetrix GeneChip 50K *XbaI,* 50K *Hind*III, 250K *StyI,* and 250K *NspI* arrays.

.cel and .chp Files

CNAG2.0 uses both cel and chp files for the copy number calculations and LOH inference. These files are created by GCOS and GTYPE programs, respectively, when array signals are scanned (.cel files) and analyzed genotypes (.chp files). The wet part details are not provided in this article, because they are essentially the same as what is described in the manuals of GeneChip.

See the Affymetrix manuals about how to do array experiments and create these files (http:/ /www.affymetrix.com/products/application/whole_genome.affx). Note that they need to be placed within the same directory that is accessible from the PC in which CNAG2.0 is installed.

Installation

CNAG2.0 is distributed from our website (http://www.genome.umin.ac.jp). After downloading the installer, set up the program according to the manual. By default, it is installed under the directory, \C\Program Files\Genome Lab\CNAG. Requirements for machines are more than 500 Mb RAM and more than 1 GHz Pentium processors.

Recommended speck is, however, more than 1 Mb RAM and more than 3 GHz processors. Portable PC performs well but desktop PC equipped with enough RAM and faster processor is recommended when a large number of 500 K samples are analyzed.

Chip Description Files (CDF)

CNAG2.0 uses Chip Description (CDF) files to extract signal data from .cel files. CDF files should be downloaded from the Affimetrix website (http://www.affymetrix.com/support/technical/ libraryfilesmain.affx).

PROCEDURES

1. Launch CNAG2.0 from Desktop CNAG icon or from within the Windows Startup Menu- > Programs- > CNAG- > CNAG.

2. Parameter->Default Setting to open the "setting" dialog, where all the settings of CNAG2.0 are specified.
3. Choose "Folders" tab and set indicated directories as follows.
4. Set "CHP CEL files" folder to that directory where all the cel and chp files are stored to be analyzed.
5. Set "Library files" folder to that directory where the CDF files are installed..cdf files describe probe information in cel files and prepared for individual array types. They usually come with Affymetrix programs or are available from the Affymetrix website. CDF files are essential for CNAG to extract data from cel files.
6. Set "Array data files" folder to that directory where you want to save cfh files in which CNAG2.0 stores the signal and genotype data extracted from cel and chp files as well as other information used for copy number calculations. Any directory can be used for this purpose but we recommend creating a new folder named, for example, "array-data," under CNAG directory.
7. Similarly, create another folder to save the analyzed results, for example, "Results folder" and set "Results files" folder to that directory. CNAG holds all the results within this folder.
8. Set "Output files" folder, where CNAG output copy number as well as other data in text files. After all the folders are properly set, close the dialog by clicking "OK."

Extracting Data From.cel and .chp Files

1. Data-> Extract Data to open a dialog for extracting data from cel and chp files.
2. Place each sample that appears in the middle window in one of the four categories, by choosing each sample name and click "Add" on the appropriate window on the right. Paired tumor and reference files are put in the "Paired TEST" and "Paired REF" windows. Otherwise, non-paired samples are put into either the "non-Paired TEST" or "non-Paired REF" window. Each paired samples should appear at the same level in the corresponding windows.
3. After all samples are properly placed, click "OK" to start extraction. Through this step, CNAG2.0 looks into each cel file to get array information and separately calculate signal sum of all the perfect match probes for each SNP type for all SNP loci, whereas genotyping and the gender information are extracted from the corresponding cfh file. Extracted data is then stored in the cfh file, which is used for all the CNAG analyses hereafter.

Modes of Analyzing Copy Number

Two modes of copy number analysis are available in CNAG2.0. In nonallele-specific analysis, array signals for both SNP types are combined and compared between tumor and reference files.

1. Data->Sample Manager to launch Sample Manager.
2. Apply appropriate filters based on the date of array experiments, array types, sample types, and other user-defined attributes.
3. Copy number analysis in CNAG is iteratively performed. First, choose the sample to be

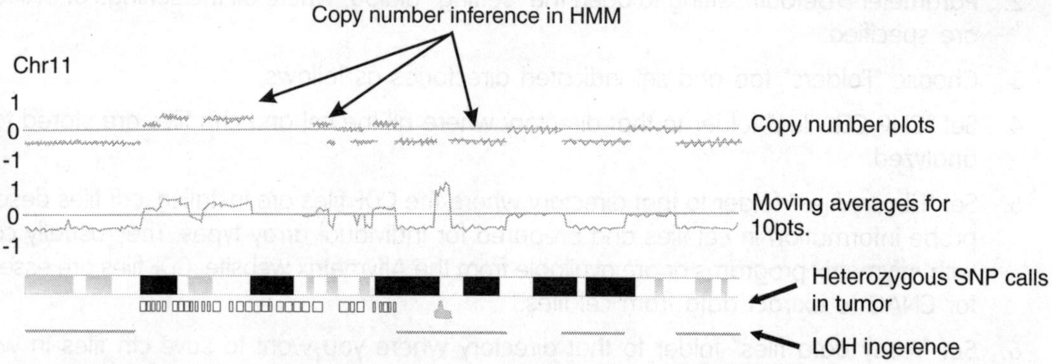

Figure 10.1: Nonallele-specific copy number analysis in chromosome view. Copy Number Analyzer for Affymetrix GeneChip v2.0 calculates log2 ratios of signals for all single nucleotide polymorphism (SNP) loci and plots them on chromatograms. It also displays inferred copy numbers (brown lines), moving averages for 10 adjacent SNPs (blue lines), heterozygous SNP calls (green bars), and loss of heterozygosity prediction (light blue lines). The horizontal axis is interchangeably depicted in SNP order with uniform intervals or in proportion to real genetic distance (not shown).

analyzed and the mode of analysis, and do "non-self analysis".

For every available reference, CNAG takes log2 ratios of signals between tumor and reference (measured copy numbers) for all .SNP loci, applies regressions to these log2 ratios, and calculates their standard deviation assuming all autosomes are diploid. In "automation" mode, CNAG calculates the best combination of the references that provides the lowest SD. In "manual" mode, it sorts and displays all the references in SD's ascending order and prompts one to select the references to be used for analysis.

CNAG averages the log2 ratios calculated for each selected references and plots them on chromatogram. By default, all copy number plots (red points), their moving average for 10 loci (blue lines), and heterozygous SNP calls (green bars) in tumor samples are displayed. One may change items to be displayed and their appearances.

4. In the next step, do Parameter-> Set range of SD calculation to select a region to be assigned as diploid by moving the mouse. CNAG recalculate copy numbers and applies regressions assuming the region as diploid.

Inferred Values

By default, CNAG infers real copy numbers, which take integer values, from the observed log2 ratios based on the HMM and overlays the inferred values on the log2 plots. When processing multiple samples, some kind of automation is required to extract copy number abnormalities. Because the default parameters of the HMM is optimized for pure tumor cells, they should be adjusted for fresh tumor samples having significant degrees of normal tissue components. To do this:

1. Parameter->ContaminationRate to select a region used for setting the parameters. It should have copy number 1 (loss) or 3 (gain).
2. Assign a copy number for that region (1 or 3).

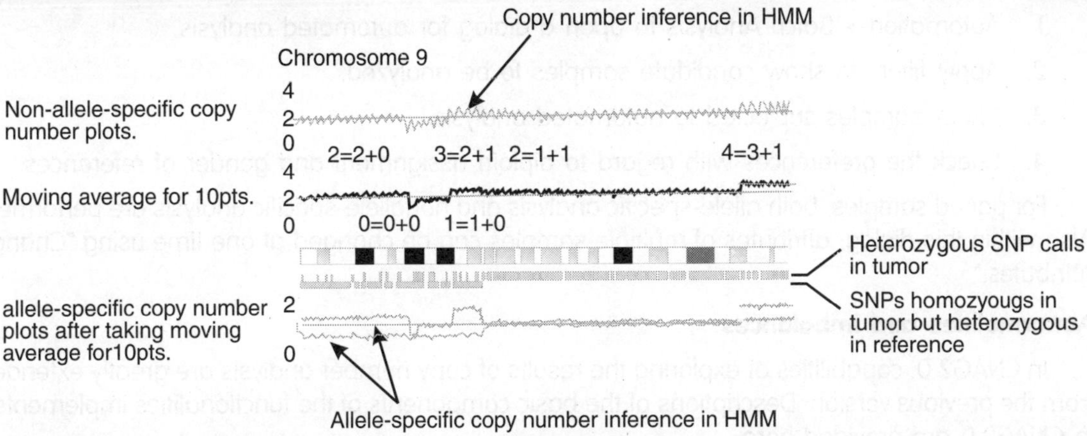

Figure 10.2: Allele-specific analysis on Styl 250K array. Allele-specific copy number analysis on chromosome 9 reveals allelic imbalances in a lung cancer cell line (NCI-H1437) chromosome 9, where signal ratios are separately calculated for different SNPs at heterozygous SNP loci in paired normal DNA (bottom panel). Allele-specific copy number inferences are provided by dark red and light green lines. Combined signal ratios are plotted for all SNP loci together with inferred copy numbers in HMM (upper panel). Moving averages for 10 adjacent SNPs and heterozygous SNP calls in the tumor sample are also depicted.

3. CNAG recalculates the real copy numbers based on the adjusted parameters.

In addition to the inference of real copy numbers, CNAG predicts the presence of LOH based on the regional reduction of heterozygous SNP calls.

Putative LOH regions are indicated by blue bars, which are drawn in varying height that is proportional to the logarithm of LOH likelihood.

Calculation of log2 Ratio

For those samples where paired normal DNA is available, copy numbers in log2 ratios can be calculated in an allele-specific manner.

1. Choose a tumor sample in Sample Manager as described anywhere else in this chapter. When the sample has a paired reference, "Allele-specific analysis" becomes enabled.
2. Do "Allele-specific analysis."
3. Assign a diploid region just in the same manner as anywhere else in this chapter.

In this mode, CNAG displays allele-specific copy numbers by separately plotting larger and smaller log2 ratios for each heterozygous SNP locus. The parameters of allele-specific copy number inference by HMM should be adjusted.

The latest CNAG version (CNAG 3.0) enables allele-specific copy number analysis without depending on the availability of paired normal DNA, but using a few anonymous references (AsCNAR algorithm).

Analysis of Multiple Samples

Multiple samples can be analyzed in background, although diploid assignment is the responsibility of researchers and cannot be automated.

1. Automation-> Batch Analysis to open a dialog for automated analysis.
2. Apply filters to show candidate samples to be analyzed.
3. Check samples subjected to automated analysis.
4. Check the preferences with regard to diploid assignment and gender of references.

For paired samples, both allele-specific analysis and nonallele-specific analysis are performed. Also within this dialog, attributes of multiple samples can be changed at one time using "Change attributes."

Abnormalities and Imbalances

In CNAG2.0, capabilities of exploring the results of copy number analysis are greatly extended from the previous version. Descriptions of the basic components of the functionalities implemented in CNAG2.0 are provided here.

Open "Display Samples"

1. Data->Display Samples to open a dialog for displaying the results.
2. Apply filters to show up the samples.
3. Check samples to be explored.

Results

1. Single display to open a pop up window that shows selected samples. Choose one and CNAG display the sample.
2. Multi display to explore multiple samples.
3. Ihtegral view to summarize the results of multiple samples. "Set the Integration Parameters" dialog is open to prompt the input of the required parameters. Then do "OK."

Microarray Data

Copy number data from different array types can be analyzed in combination. This is possible between 50K *Xba* and *50K Hind* arrays, and 250K *Sty*I and 250K *Nsp*I arrays.

1. Data Display Samples to open a dialog for displaying the results.
2. Apply filters to display specific samples.
3. Check the paired samples to be combined.
4. Combine. The combined result is displayed and stored in a file, named A_and_B.CFS(CFN) for a combination of A.CSF(CFN) and B.CFS(CFN) files.

Copy Number Data in Text format

1. Data Display Samples to open a dialog for displaying the results.
2. Apply filters to show up the samples.
3. Check the paired samples whose data is going to be output.
4. Output raw data and "Output raw data" dialog will open.

5. Check the data to be output.
6. Also check "combined to one file for each array type" if all the data should be output into a single text file. Otherwise, data of each sample goes into a separate file.

Chromosomes to be Explored

1. Display the sample and move the chromosome to be explored.
2. Parameter measure.
3. Assign the region under interest by drugging the mouse.
4. "Statistics" dialog will open to show the information and statistics about the assigned region, including the start and end position and corresponding SNP numbers and cytobands, length of the region, and the mean and SD of log2 ratios therein. The data can be directly copied to the clipboard.
5. Click "go to web" to jump to the UCSC browser according to the selected positional parameters.

POINTS TO REMEMBER

1. The wet part of the copy number analysis on Affymetrix platform is out of the scope of this chapter but still needs to be briefed on not only for completeness of description but also for understanding how profoundly the wet part influences the quality of the copy number analysis. SNP typing on the GeneChip platform is performed based on specific hybridization of PCR-amplified genomic DNA to large numbers (~10, 000 to ~250, 000 depending on array types) of SNP-specific probe sets chemically synthesized on microarrays.

   Each probe set consists of varying numbers of 25-mer oligonucleotide probes perfectly matched to each SNP type together with the same number of mismatch base-introduced probes. The key step on this platform is the whole genome selection assay. Briefly, 250 ng genomic DNA is digested with an indicated restriction enzyme, attached with a common primer, and PCR-amplified using a single primer.

   The PCR products are subjected to ultrafiltration for desalting followed by DNasel fragmentation, and biotinylated DNA fragments are hybridized to microarrays. Through this assay, less than 1.5- to 2-kb restriction fragments are effectively amplified and thus, selective amplification of target DNA fragments is accomplished while reducing overall complexity of genomic DNA.

   On the other hand, however, the whole genome selection assay produces unfavorable side effects on copy number calculations, because it is the signals from amplified DNA that are compared between test (tumor) and reference. For example, degraded DNA produces less PCR templates after restriction digestion and adaptor ligation for longer than shorter restriction fragments.

   As a result, when the quality of DNA is different between test and reference samples, copy numbers are biased according to the length of the restriction fragments. More complicated is the different PCR efficiency between' test and reference experiments. Of course, efficiency of PCR varies for different restriction fragments even within the same sample.

However, the difference in PCR efficiency between different fragments needs to be the same degree between test and reference experiments, if ever accurate comparison of copy numbers as to the nonamplified DNA is made possible. Unfortunately, this is not always the case and copy numbers as determined from amplified DNAs are frequently biased according to the parameters that affect PCR efficiency. Thus it is critically important to perform test and reference experiments as uniformly as possible.

The quality of test and reference DNA should be of the same degree, if not excellent. All the experimental procedures need to be kept uniform, using the same thermal cycler that can achieve uniform thermal cycles across different wells. All the reagents must be prepared as premix. Performance of polymerase is another parameter. According to our experience for the 100K protocol, TAKARA LA Taq outper-forms the Platinum Taq recommended by Affymetrix.

It is more difficult to keep experimental conditions uniform when the experiments are performed on different occasion. As discussed in the following sections, it is possible to compensate the effects from different PCR conditions using computations to a certain extent, but keep in mind that these compensations never replace well-performed experiments.

2. As described next, CNAG uses multiple references to improve copy number calculations. However the random selection of references from both sex prevents the accurate copy number calculations in X chromosome. When the "same sex" box is checked, CNAG only uses the references having the same sex as the tumor sample, and therefore correctly detect copy number gains and losses in X chromosome. Note that in this mode, however, the quality of over all analysis may be affected, because availability of references is restricted.
3. On any array based platform, copy numbers are calculated as relative values to a reference sample, usually from normal individual. Given that the PCR amplification preserves relative copy numbers across different restriction fragments and that the array signals are proportional to the amount of PCR products for each fragment, the copy number of the ith SNP fragment in log2 ratio is expressed as:

$$C^i = K \times \frac{\Sigma\left(PA^i_{tum} + PB^i_{tum}\right)}{\Sigma\left(PA^i_{ref} + PB^i_{ref}\right)} \tag{10.1}$$

or

$$\log C^i = \log_2 K + \log_2 \frac{\Sigma\left(PA^i_{tum} + PB^i_{tum}\right)}{\Sigma\left(PA^i_{ref} + PB^i_{ref}\right)}, \tag{10.1'}$$

where PA^i and PB^i are signal intensities for type A and type B SNPs at the ith SNP probe, respectively, and lower suffixes are to distinguish tumor *(turn)* and reference (ref) values. Signal sums are taken for all the perfect match probes for the ith SNP. K is a constant determined from the mean array signals of tumor and reference samples and also from the mean ploidy of tumor sample.

CNAG separately calculates ΣPA^i and ΣPB^i and stores them into the cfh file at the data extraction time and at the initial step of non-self copy number analysis, it preliminary set the *K*-value so that the mean intensities of the two arrays are the same and mean ploidy for all autosomes is Eq. 10.2, or,

$$K_0 = \frac{\sum_i \sum \left(PA^i_{ref} + PB^i_{ref}\right)}{\sum_i \sum \left(PA^i_{tum} + PB^i_{tum}\right)} \tag{10.2}$$

where sum is taken for all autosomal SNPs.

4. As discussed elsewhere in this chapter, the K-value is not a constant but varies for different SNP fragments, and copy numbers calculated as Eq. 1 are deviated from the real values according to the variable K_i-value, which is determined by the base composition of the ith fragment and the difference in experimental parameters between test and reference experiments.

   In fact it can be demonstrated, empirically though, that the K_i explicitly depends on the

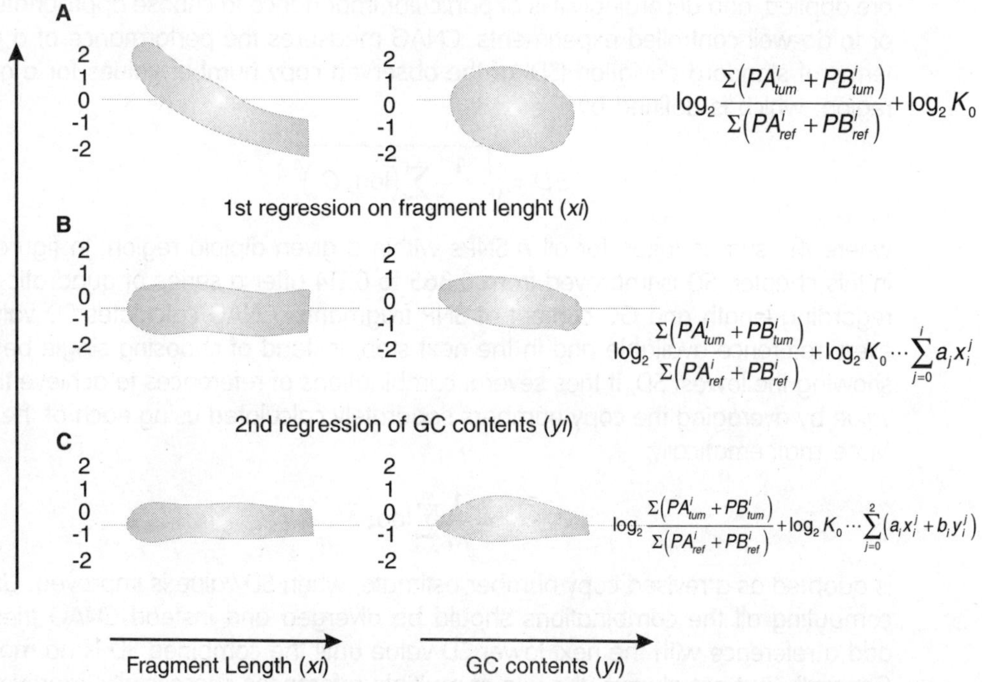

Figure 10.3: Dependence of Log2 ratios on fragment length and GC content. Log2 ratios of normalized array signals were calculated between two normal samples and plotted against the two parameters, fragment length (left) and GC content (right). Biased log2 ratios (A) are serially compensated by a couple of quadratic regressions with regard to both parameters (B,C), which contributes to improved SD value.

length and GC content of the ith SNP fragment, when $\log_2 C$'s in expression Eq. 1′ under a constant K given by Eq. 10.2 are plotted against these parameters. Based on this finding, we obtain a new expression using the second order approximation for $\log_2 K_i$:

$$\bar{\Lambda}^i = \log_2 \frac{\Sigma\left(PA^i_{tum} + PB^i_{tum}\right)}{\Sigma\left(PA^i_{ref} + PB^i_{ref}\right)} + \log_2 k_0 - \sum_{j=0}^{2}\left(a_j x_i^j + b_j y_i^j\right) \tag{10.3}$$

where x_i and y_i are the length and GC content of the ith SNP fragment and a_j and b_j are determined by the quadratic regression from the measured PAs and PBs for the SNPs in a given diploid (C_i = 1) region. Different DNA quality between both samples also affects k-value but the effect is already incorporated in expression (Eq. 3) because it is related to the length of the fragment. After a series of quadratic regressions, the biases from different PCR conditions and DNA quality are partially compensated and better inference of copy numbers is obtained. CNAG initially calculates the coefficients of the regression (a_js and b_js) using all the autosomal SNPs and in the next step recalculates these coefficients as well as K_0 using the SNPs within the region assigned as diploid and revises all the copy numbers according to the recalculated parameters.

5. Quality of the copy number analysis strongly depends on the selection of the reference, or "control," to be used for copy number calculations, even after a series of regressions are applied, and accordingly it is of particular importance to choose appropriate references or to do well-controlled experiments. CNAG measures the performance of a reference in terms of standard deviation (SD) of the observed copy number values for a given diploid region, which is defined by:

$$SD = \sqrt{\frac{1}{n-1}\sum_i^n \left(\log_2 C^i\right)^2}$$

where the sum is taken for all n SNPs within a given diploid region. In figure elsewhere in this chapter, SD is improved from 0.365 to 0.114 after a series of quadratic regressions regarding length and GC content of SNP fragments. CNAG calculates SD values against every reference available and in the next step, instead of choosing single best reference showing the lowest SD, it tries several combinations of references to achieve the lower SD value by averaging the copy numbers separately calculated using each of the references. More mathematically,

$$\bar{\Lambda}^i = \frac{1}{m}\sum_{k=1}^{m} \log_2 \bar{\Lambda}^k$$

is adopted as a revised copy number estimate, when SD value is improved. Unfortunately computing all the combinations should be diverged and instead CNAG tries to serially add a reference with the next lower SD value until the combined SD is no more reduced. Generally, but not always, the use of multiple references successfully improves SD values on average. Two factors could contribute the improvement in SD.

First, random error components in reference experiments are expected be reduced by taking averages. Note, however, that including dissimilar references,,in terms of SD values is of little use to reduce the overall SD value or even deteriorates the result. We recommend that experiments should be planned to include more than five normal controls to get enough references with similar SD values.

The other factor that contributes to improved SD is the fact that the difference in genotype between test and reference tends to be neutralized by taking averages. Because SNP specific probes may show significantly different hybridization kinetics, $\Sigma(PA^i_{tum} + PB^i_{tum})/ >(PA_{rer} + PB_{ref})$ could be considerably affected depending on the combination of genotypes in tumor and reference samples Figure 10.3 demonstrates the effect of comparison between different genotypes.

6. To be rigorous, it is impossible to unambiguously determine the *K*-value based only on array experiments. It is essentially related to the ploidy of the tumor genome, which can be determined only by cell-based analysis, for examples, cytogenetics, FISH, or measurement of DNA content using a flowcytometer. We can only guess the value but unfortunately frequently fail to do that correctly, especially for solid tumor samples, as revealed by FISH confirmation.

   Also note that assignment of copy number gains (more than two copies) and losses (less than two copies) for a region easily changes depending on the mean ploidy. In allele-specific analysis, purely triploid tumor genome can be distinguished from diploid genome in that the former shows allelic imbalance over the whole genome without showing disappearance of heterozygous SNP calls.

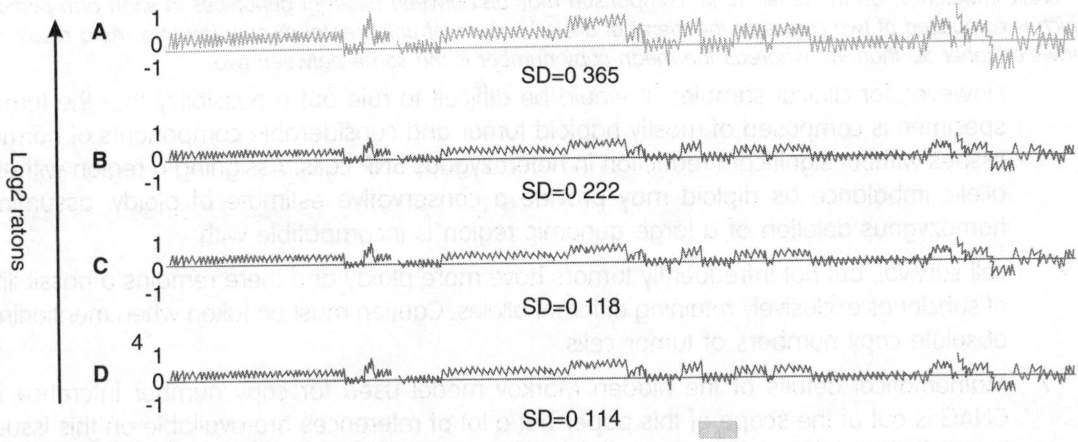

Figure 10.4: Improvement of SD after a series of compensations. Genomic DNA from a glioma cell line U251 was analyzed for copy number alterations with or without compensations for different experimental conditions using different compensation algorithms. The raw log 2 ratios in the chromosomal order show widely fluctuated baselines (A; SD = 0.365), which are greatly reduced after compensation of experimental conditions and using the best-fit single (B; SD = 0.222) and multiple references (C; SD = 0.118). Regions of homozygous and hemizygous deletion are indicated by small and large arrows. The SD value is further decreased (SD = 0.114) when the second round compensation procedures are applied by specifying a region of diploid alleles (D, red bar). With this approach, diploid regions as well as other regions showing varying ploidy states are correctly assigned to their real ploidy. For proper comparison, SD values are calculated for the SNPs within the diploid range (red bar).

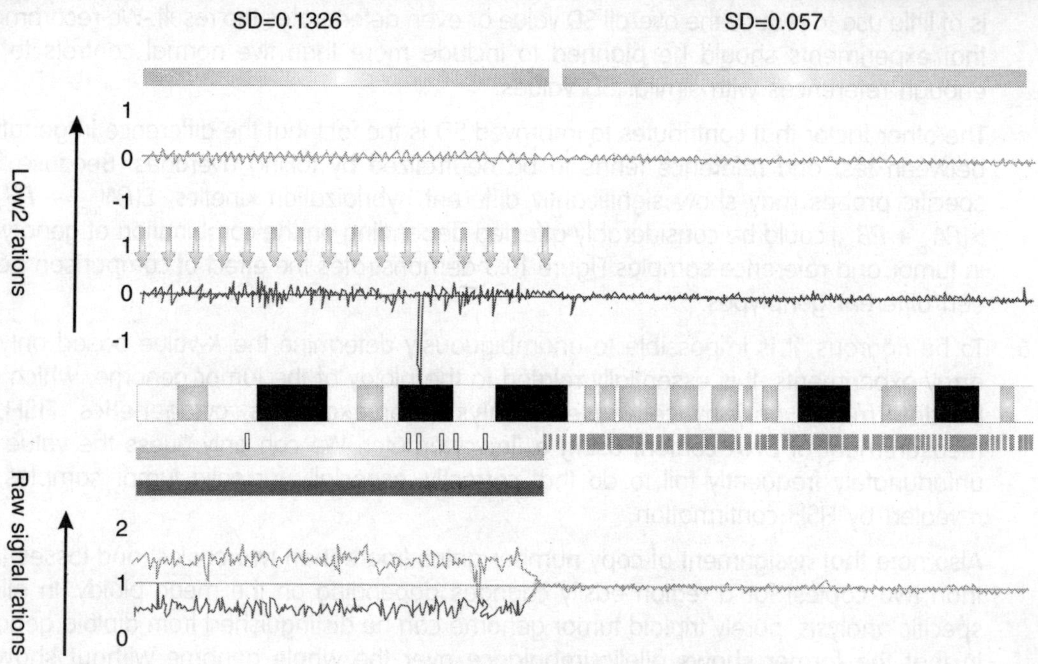

Figure 10.5: Effects of difference in genotype on SD. In this example of allele-specific analysis on chromosome 9, signals from test and reference are compared between the identical genotypes on the long arm where there is no allelic imbalance. On the other hand, comparison may be between different genotypes in short arm portion, which is composed of two copies of the identical allele because of uniparental disomy (arrows). As a result, 9p shows a higher SD than 9p, whereas the mean copy number is the same between two.

However, for clinical samples, it would be difficult to rule out a possibility that the tumor specimen is composed of mostly haploid tumor and considerable components of normal tissues without significant reduction in heterozygous SNP calls. Assigning a region without allelic imbalance as diploid may provide a conservative estimate of ploidy, assuming homozygous deletion of a large genomic region is incompatible with

cell survival, but not infrequently tumors have more ploidy and there remains a possibility of subclones exclusively retaining different alleles. Caution must be taken when mentioning absolute copy numbers of tumor cells.

7. Mathematical details of the hidden Markov model used for copy number inference in CNAG is out of the scope of this paper but a lot of references are available on this issue. It was first introduced for copy number inference on Affymetrix GeneChip platform by Li et al.. In this model, each measured copy number is assumed as an emission from one of the definite copy number states, which is now unknown or hidden.

   Each SNP locus has its own copy number state and default parameter values for HMM analysis in CNAG were empirically determined from the measurements in tumor cell lines where no normal cells are contaminated.
8. Prediction of LOH in nonallele-specific analysis is based on a kind of maximum likelihood method, in which LOH regions are locally determined so that the probability of observing

the current distribution of heterozygous SNP calls takes a maximum value.

For paired samples, signals from the two discrete alleles can be discriminated at those SNP loci where the paired reference has heterozygous SNP calls. Thus, copy numbers in each allele can be individually evaluated by separating signals from the different alleles.

$$\Lambda_A^i = \log_2 \frac{\Sigma PA_{tum}^i}{\Sigma PA_{ref}^i} + \log_2 k_0 - \sum_{j=0}^{2}\left(a_j x_i^j + b_j y_i^j\right)$$

$$\Lambda_B^i = \log_2 \frac{\Sigma PB_{tum}^i}{\Sigma PB_{ref}^i} + \log_2 k_0 - \sum_{j=0}^{2}\left(a_j x_i^j + b_j y_i^j\right)$$

A ploidy constant K_0 and coefficients of the regressions are determined so as to minimize

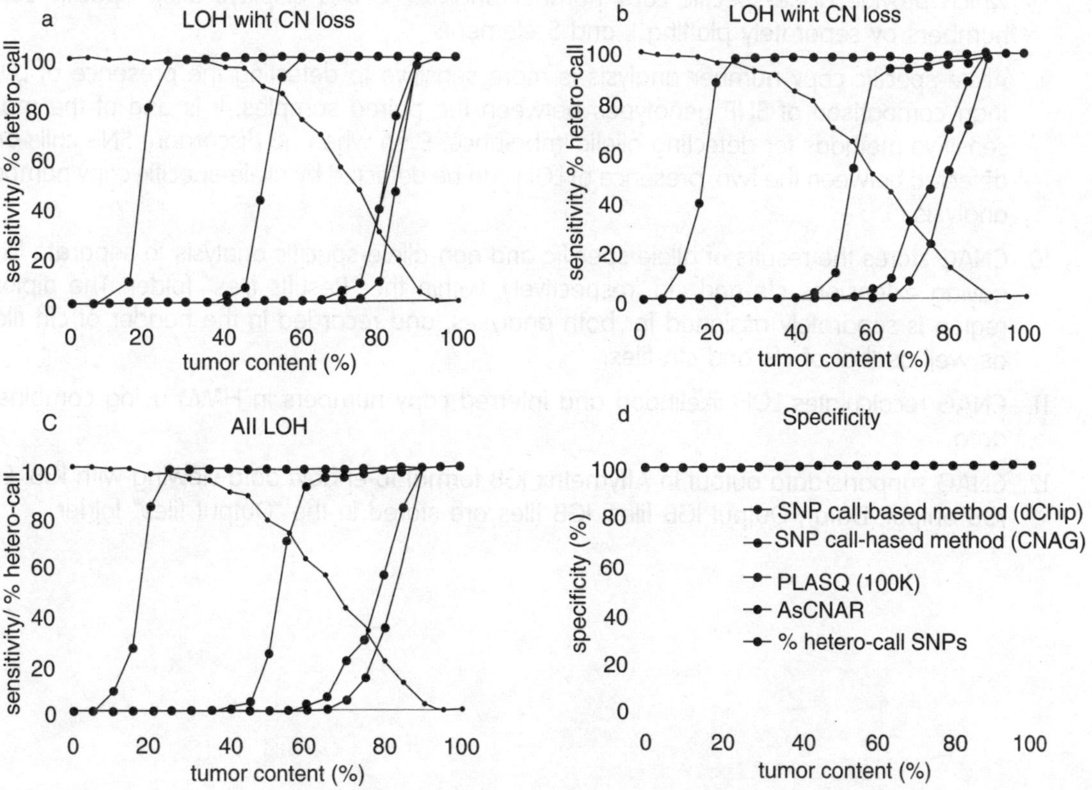

Figure 10.6: Sensitivity of detecting LOH either with or within CN loss (a, b) in different algorithms are compared using mixture of tumor (NCI-H2171) and the paired LCL sample (NCI-BL2171), where SNPs which are heterozygous in the paired LCL and homozogous in the tumor sample are considered to be truly positive for LOH, and SNPs which are heterozygous both in the paired LCL and the tumor are to be truly negative. SNPs typing were carried out three times and discordant SNPs were exclude from calculation. The result for all LOH regions are shown in c and specificities of LOH detection are depicted in d. Proportions of heterozygous call SNPs remained in LOH regions of each sample are also shown in a-c.

$$\sum_{i}^{n}\left\{\left(\log_2\frac{\Sigma PA^{i}_{tum}}{\Sigma PA^{i}_{ref}}+\log_2 K_0-\sum_{j=0}^{2}\left(a_j x_i^j+b_j y_i^j\right)\right)^2\right.$$
$$\left.+\left(\log_2\frac{\Sigma PB^{i}_{tum}}{\Sigma PB^{i}_{ref}}+\log_2 K_0-\sum_{j=0}^{2}\left(a_j x_i^j+b_j y_i^j\right)\right)^2\right\}$$

where sum is taken for all heterozygous SNP fragments within a given diploid region. Two log2 ratios are separated into two groups,

$$L=\left\{\max\left(\Lambda_A^i,\Lambda_B^i\right)\middle| i\in hetero\_SNPs\right\}$$

$$S=\left\{\min\left(\Lambda_A^i,\Lambda_B^i\right)\middle| i\in hetero\_SNPs\right\}$$

which provide allele-specific copy number analysis. CNAG displays allele-specific copy numbers by separately plotting L and S elements.

9. Allele-specific copy number analysis is more sensitive to detecting the presence of LOH than comparison of SNP genotypes between the paired samples. It is one of the most sensitive methods for detecting allelic imbalance. Even when no discordant SNP calls are detected between the two, presence of LOH can be detected by allele-specific copy number analysis.
10. CNAG stores the results of allele specific and non-allele-specific analysis to separate files having extensions cfs and cfn, respectively, within the "Results files" folder. The diploid region is separately assigned for both analyses, and recorded in the header of cfh files as well as that of cfs and cfn files.
11. CNAG recalculates LOH likelihood and inferred copy numbers in HMM using combined data.
12. CNAG supports data output in Affymetrix IGB format to enable data viewing with IGB. For IGB output, Data->Output IGB files. IGB files are stored in the "Output files" folder.

Chapter 11

Culture of Embryonic Cells

Functional analysis of the mammalian genome will be a major task of biologic science in the future. Gene trap is a method designed to provide functional information for novel genes. It offers tagging of a gene (facilitating subsequent cloning), analysis of the expression pattern by a simple staining technique, and generation of a mutant mouse strain in uric experimental approach. According to our experience, about two thirds of the genes isolated by his method are either completely unknown or published only as expressed sequence tags.

The reporter gene reflects the expression pattern of the endogenous gene faithfully in about three quarters of all cases. Overt phenotypic abnormalities can be expected in about half of the mutant mouse lines generated by gene trap. However, note that frequently the trapped allele is not a null allele. but rather a hypomorphic allele.

This may be beneficial in cases in which preimplantation lethality would ordinarily preclude the analysis of functions of the gene later in development or in adult life. The most serious criticism to date is that there is little possibility to direct mutagenesis by gene trap to a specific organ system, although attempts in that direction have been made.

In this chapter, we describe methods used in producing mouse lines containing genes tagged by gene trap construct insertions. The approach used involves electroporation into murine embryonic stem ES cells.

REQUIREMENTS

Culture Techniques

1. Dissection instruments: scissors and forceps.
2. Petri dishes: 15, 10, and 3.5 cm (Falcon).
3. 24 and 96-Well plates (Falcon).
4. Erlenmeyer flask with 100-ml- glass beads.
5. Wire meshes.

6. Stirring rod.
7. General cell culture glassware.
8. 0.1% Gelatin (G-1890: Sigma). Make up the stock in H_20 and autoclave.
9. Feeder medium: Dulbecco's modified Eagle's medium (DMEM) (Gibco). 10% fetal bovine serum (FBS), virus and mycoplasma screened.
10. Primary embryonic fibroblasts isolated from embryonic day 13.5 (e13.5) murine embryos or SNL cells.
11. Mitomycin C (M-0503; Sigma): Dissolve 1 vial (2 mg) in 4 mL of phosphate-buffered saline (PBS). filter sterilize, and store protected from light at 4°C. Use within 1 wk; check for precipitates before use.
12. PBS (10X stock): 80 g/L of NaCl, 2 g/L of KCl, 11.5 g/L of Na_2HPO_4 2 g/L of KH_2PO_4. Sterilize by autoclaving. Dilute 10X PBS 1:10 with water, check pH (7.2-7.4), and autoclave again for storage.
13. Trypsin/EDTA (5X stock): 8 g/1, of NaCl, 0.4 g/L of KCl, 0. 1 u/1, of Na_2HPO_4, 1 g/L of glucose, 3 g/L of Tris-base, 0.4 g/L of Na_2-EDTA, 5 g/L of trypsin (Seromed 1:250). Adjust the p11 to 7.6 and sterilize by filtration (0.2 lam). Store at 4°C and use within 1 wk (or freeze aliquots at-20°C). Dilute the stock 1:5 with sterile PBS, store at 4°C, and use within 1 wk.
14. ESC medium: DMEM (high glucose. 4.50 mg/L; Gibco), 0. 1 *mM b**-mercaptoethanol (Sigma 47522), 1X nonessential amino acids (Gibco), 1X L-glutamine (Gibco), 1X sodium pyruvate (Gibco), 2W'(FBS (Gibco offers FBS tested for ES cell culture). 1000 U/mL of ml-IF (Gibco).
15. Incubator, 37°C and 5% CO_2.
16. Sterile laminar flow hood that fits a dissection microscope.
17. Inverted microscope, magnification ×50-×320, phase-contrast optics.
18. Cell culture centrifuge.
19. Liquid nitrogen container to store cells.
20. 2X Freezing medium: mix 30 parts ESC medium. 9 parts dimethylsulfoxidc (DMSO) (D-8779: Sigma), and 6 parts FBS. The final concentration after treatment of cells for freezing is 10% DMSO, 20% fetal bovine serum and 16.7% trypsin/EDTA.
21. 2-20 µL micropipet plus tips: 20-200 µL micropipet plus tips.

ES Cell Clones

1. Gene Pulser electroporator (Bio-Rad).
2. Electroporation cuvet: 4-min electrode distance, 0.8-ml- vol (Bio-Rad).
3. G418 (GeneticinR, Gibco).
4. Dissection microscope, magnification ×6-×40 (Wild-Leitz).

Staining Method for β-Galactosidase

1. 37-39% Formalin (Fluka).

2. 5-Bromo-4 chloro-3-indolyi-β-D-galactopyranoside (X-gal).
3. 25% Glutaraldehyde.
4. $K_3Fe(CN)_6$ (0.5 *M* stock in water).
5. $K_4Fe(CN)_6$, (0.5 *M* stock in water).
6. $MgCl_2$ (0. 1 *mild* stock in water).
7. NP40 (10% stock solution in water).
8. Sodium deoxycholate (1% solution in water).
9. 30°C Incubator.
10. 20X PBS buffer.
11. Glycerol stocks (20. 40. 60. and 80% glycerol in PBS).

Production of Chimeras

1. 70%: Ethanol.
2. Avertin: Prepare the stock by dissolving 1 g of 2,2.2-tribromo ethyl alcohol in 1 ml- of tertiary amyl alcohol. Keep the stock protected from light at 4°C. Dilute to 2.5% in 42°C, 0.9% sterile saline for use (injection solution). Observe complete mixing. Cool the injection solution down to room temperature slowly. Keep protected from light at 4°C, and use within 4 wk.
3. Acid Tyrode's solution: 0.8 g of NaCl, 0.02 g of KC1, 0.024 g of $CaCl_2{\bullet}2H_2O$, 0.01 g of $MgCl_2.6H_2O$, 0.1 g of glucose, and 0.4 g of polyvinylpyrrolidone. Make up the volume to 100 mL, filter, and aliquot. Store at-20°C.
4. M2 and M16: Prepare from stock solutions stored at -20°. Prepare medium for one week and store at 4°C. Replace stocks after approx 3 mo. For a more extensive description see elewhere in this chapter. For 10 mL:

   M2: 1 mL Stock A, 160 μL Stock B, 100 μL Stock C, 100 μL Stock D, 840 hl. Stock E, 7.8 mL H_2O, 40 mg BSA. Sterile filtrate.

   M 16: 1 mL Stock A, 1 ml. Stock B, 100 μL, Stock C, 100 μL Stock 1), 7.8 ml. H_2O. 40 mg serum bovine albumin (BSA). Sterile filtrate.
5. 10X Solution A: 5.534 g/100 mL, NaCl, 0.356 g/100 mL KCl, 0.162 g/100 ml, KH_2PO_4, 0.293 g/l00 mL $MgSO_4{\bullet}7\ H_2O$ 2.610 g/100 mL sodium lactate. 1 g/100 mL- glucose, 0.060 g/100 mL, penicillin, and 0.05 g/100 mL streptomycin.
6. 10X Solution B: 2.101 g/100 mL $NaHCO_3$, and 0.01 g/100 mL phenol red.
7. 100X Solution C: 0.036 g/10 mL, sodium pyruvatc.
8. 100X Solution D: 0.252 g/10 ml- $CaCl_2.2\ H_2O$.
9. 100X Solution E: 5.958 g/100 mL, HEPES and 0.01 μ/100 ml. phenol red. Adjust the pH to 7.4 with 0.2 *N* NaOH.
10. BSA, Fraction V, minimum 96% (A-9647: Sigma).

11. Light mineral oil (M-8410; Sigma).
12. Surgical swabs.
13. Aluminum foil.
14. 1-mL Syringes.
15. 2-ml. Syringes.
16. Short 27-gage injection needles.
17. Flushing needle, 33-gage needle with the tip cut off and slightly bent.
18. 3.5-cm Petri dishes.
19. 10-cm Petri dishes.
20. Mouth pipet assembled from a mouthpiece. Soft silicon tubing, a sterile filter inserted between two parts of the tubing, and a pipet tip for 200-1000 µL (cut off the tip and invert; the wide opening will fit the wide opening of a Pasteur pipet).
21. Short Pasteur pipets,
22. Medium tapestry needles to make depressions in the bottom of a 3.5-cm Petri dish: For aggregation wells, try several to find a suitable needle, and then equip with a handle of either metal or plastic (heated and formed in place).
23. Bunsen burner.
24. Tissue culture incubator. 37°C and 5%f CO_2 in air.
25. Low-magnification binocular microscope.
26. Cold fiberoptic light source.
27. Fine clippers.
28. Dissection instruments: large scissors (for cutting skin), small scissors, blunt small forceps. and no. 5 watchmaker's forceps.
29. Surgery instruments: small surgical forceps, small bunt forceps, small scissors, small Serrefine clamp, small wound clips (Michel's clips), and wound clip applicator.

PROCEDURES

We used two splice acceptor trap (pGT1.8geo and pKCl99b*geo) and two exon insertion trap constructs (pEIT2 and pEIT3). All four constructs were promoterless. pGTl.8geo consists of intron and exon sequences spanning a splice acceptor of the murine engrailed-2 gene fused in frame to codon 8 of the b*-galactosidase coding sequence and in frame to codon 2 of the neomycin phosphotrartsferase coding sequence.

The neomycin resistance coding region is concluded by an SV40 polyadenylation signal. pKCl99b*geo contains an Hoxc9 splice acceptor fused in frame to the b*geo fusion of the b*-galactosidase coding region and a mutated neomycin phosphotransferase coding region as in pSAb*geo. pEIT2 and pEIT3 are derivatives of pGTl.8geo.

Instead of the intron sequences of the engrailed-2 splice acceptor. they contain base 865-905 (pEIT2) and base 490-905 (pEIT3) of the microtubule-associated protein 4 coding region fused in frame to the engrailed-2 exon of pGT1.8geo resulting in 25w- and 634-bp buffer sequence 5′ of the β-galactosidasc coding region for pEIT2 and pEIT3, respectively. A more specialized gene trap construct has been designed to select for insertions into genes coding for secreted molecules.

Promoterless constructs rely on activation by an endogenous promotor and reliably integrate into bona fide transcriptional units. To generate G418-resistant ES cell colonies, the trapped endogenous gene must be active in ES cells. Our data suggest that genes with very low transcriptional activity in ES cells can be trapped. However, the population of genes that can be targeted by this method is almost certainly limited by the need to be expressed in ES cells.

Initially, gene trap constructs were designed containing a splice acceptor fused to the *lacZ* gene with a polyadenylation signal followed by a promotor driving a neomycin phosphotransferase gene concluded by a polyadenalytion signal. Although eliminating the need to insert into a transcriptionally active locus in ES cells. these constructs produced G418-resistance inserting outside of transcriptional units, and, therefore, actual gene trap events were rare (2-5%).

To overcome this problem, polyadenylation trap constructs were designed. Polyadenylation trap constructs do not contain polyadenylation signals at the end of the neomycin phosphotransferase gene and rely on linking up to an endogenous polyadenylation signal to produce a stable mRNA.

This alone is thought to select for insertions into transcriptional units allowing the use of a promotor. A polyadenylation trap construct may consist of a splice acceptor, an internal ribosomal entry site (IRES), a β-galactosidasc coding region, a polyadenylation signal, and a promotor driving a neomycin phosphotransferasee gene concluded by a splice donor.

Such a construct will allow integrations into genes silent in ES cells. In this case, the ES cells will be G418 resistant but will not exhibit β-galactosidasc activity. Only on activation of the endogenous locus will the *lacZ* gene be activated. This could be achieved by an in vitro differentiation procedure. One drawback of this type of gene trap construct is that in the case of multiple copy tandem integrations, the splice donor 3′ of the neomycin phosphotransferase gene will be spliced to the splice acceptor 5′ of the *lacZ* gene of the following copy of the gene trap construct.

Thus, the *lacZ* gene can he driven by the promotor meant to drive the neomycin phosphotransferase alone. The IRES will render such an artificial transcriptional unit polycistronic so that functional b*-galactosidase is produced in every case. Splice acceptor trap constructs such as pGT1.8geo or pKC 199βgeo function when inserted into intron or exon sequences. In the case of an intron insertion. the splice donor of the preceding endogenous exon is spliced to the splice acceptor of the gene trap construct.

In the case of an exon insertion the splice donor of the endogenous exon preceding the exon carrying the insertion is spliced to the splice acceptor of the gene trap construct. In our hands. insertion into exons seem to be more likely to mutate the endogenous locus functionally. A gene trap construct inserted into an intron can be spliced out apparently without affecting the function of the endogenous locus.

Feeder Cell Monolayers

Murine ES cells can be cultured on gelatin-coated tissue culture plasticware or on monolayers of feeder cells. We prefer to culture ES cells on feeder cells, because, in ow hands, the rate of germline chimera production is higher after culture on feeder cells. For the last three passages of ES cells before chimera production, we prefer primary embryonic fibroblast feeder cell monolayers over SNL feeder cells.

Isolation of Primary Embryonic Fibroblasts

Primary embryonic fibroblasts (PEFs) are isolated from e 12.5 to 04.5 mouse embryos. Transgenic mice carrying a neomycin resistance gene have to be used if PEFs are intended for 6418 selection.

1. Kill one to two pregnant mice by cervical dislocation, and open the abdominal cavity by changing scissors between the skin layer and muscle layer.
2. With a sterile pair of forceps and scissors, dissect out the uterus, and wash with three changes of sterile PBS.
3. Transfer into a sterile laminar flow hood and wash again by passing through three Petri dishes containing sterile PBS.
4. Dissect embryos from uterus and wash 3X with PBS.
5. Remove the head, heart, liver, and other internal organs and wash the remainder three times in PBS.
6. Cut into small pieces with fine scissors in a small volume of PBS (5 mL).
7. Transfer to a prewarmed (37°C) sterile Erlenmeyer flask containing 50 mL of 1X trypsin/ EDTA solution, a stirring rod, and 100 mL of glass beads.
8. Stir the solution gently at 37°C for 30 min.
9. Recover the first 50 nil, of trypsin solution by passing through a wire mesh, and add an equal volume of feeder medium. Keep this batch at room temperature until the second digest is recovered.
10. For the second digest, add 50 mL of prewanned trypsin solution to the tissue pieces in the Erlenmeyer flask and incubate at 37°C with gentle stirring for 30 min.
11. Pass the second digest through a wire mesh and add feeder medium.
12. Spin both digests at 200g at for 10 min.
13. Discard the supernatant and resuspend the pellet in 20 mL of feeder medium per six embryos.
14. Plate the cells at three to six embryos per 15-cm dish. Change the medium the next day.
15. Carefully examine the cells for signs of contamination and cells without fibroblast morphology.

Freezing of Cells

Freeze the clean preparations of PEF when confluent at three vials per 15-cm plate.

1. For freezing, wash PEFs twice with PBS.

2. Add 5 mL of 1X trypsin/EDTA and incubate at 37°C for 3 in in.
3. Add 5 ml- of feeder medium, centrifuge at 2008 for 5 min. and remove supernatant.
4. Resuspend the pellet in 1.5 mL of feeder medium and add an equal volume of 2X freezing medium.
5. Mix the cells carefully and transfer to cryovials (1 mL each).
6. Slowly freeze to -70°C, and transfer quickly to liquid nitrogen. In liquid nitrogen, they can be kept for years. Defrost frozen vials in a 37°C water bath, no longer than it takes for the last ice crystals to disappear. Pipet out the contents slowly into 10 mL of feeder medium, mix by inverting, centrifuge at 200g for 5 min, and remove supernatant.
7. Disrupt the pellet carefully by tapping and resuspend in 20 ml., of feeder medium.
8. Plate onto a gelatin-coated 15-cm Petri dish. PEFs should not be passaged more than once before cleavage inactivation and use.

Routine Culture of SNL Cells

SNL cells can be obtained from A. Bradley. They are a permanent fibroblast cell line (STO) transfected with neomycin phosphotransferase and leukemia inhibitory factor (LIF) expression vectors.

1. Defrost frozen vials in a 37°C water bath, no longer than it takes for the last ice crystals to disappear.
2. Pipet the content of a cryovial slowly into 10 mL of feeder medium, mix by inverting slowly, centrifuge at 200g for 5 min, and remove supernatant.
3. Disrupt the pellet carefully by tapping. and resuspend in the appropriate volume of feeder medium.
4. Plate onto 15-cm tissue culture dish.

Inactivation of Feeder Cells

PEF or SNL cells are grown under sterile conditions in feeder medium. SNL cells have to be passaged before they reach confluence. PEFs are passaged only once before use.

1. To cleavage inactivate feeder cells, treat approx 80% confluent cells with medium containing 10 µg/mL of mitomycin C for 2 h.
2. Meanwhile, coat tissue culture dishes with 0.1 % gelatin solution for 30 min.
3. Wash mitomycin C-treated feeders twice with PBS.
4. Trypsinize for 3 min and transfer to feeder medium.
5. Centrifuge, remove supernatant, and resuspend 20 mL of feeder medium per 15-cm Petri dish.
6. Count and plate at a density of 5×10^4 cells/cm^2. Medium is changed the next day and from then on twice a week. Feeders should be used within 1 wk.

Defrosting

1. To defrost ES cells, place a cryovial into a 37°C water bath for as long as it takes to

dissolve the last ice crystals.

2. Carefully transfer the cells to a centrifuge tube containing 10 mL of ESC medium and centrifuge at 200g for 2 min.
3. Discard the supernatant and resuspend carefully in 2 mL of ESC medium.
4. Plate onto freshly inactivated feeders. Vigorous pipetting at this stage will kill cells and result in low plating efficiency. Change the medium daily, and keep it from turning acidic.

Passaging of ES Cells

1. To passage the ES cells, wash the dishes twice with PBS.
2. Incubate in 1X trypsin/EDTA at 37°C for approx 5 min (each hatch of trypsin/EDTA solution has to he checked for the appropriate tine of FS cell dissociation).
3. Pipet the cells in trypsin/EDTA about three times.
4. Add 3 vol of ESC medium.
5. Centrifuge at 200g for 2 min.
6. Discard the supernatant and disrupt the cell pellet carefully.
7. Resuspend in an appropriate volume of ESC medium.
8. Plate the cells onto a monolayer of inactivated feeder cells.

Freezing of ES Cells

1. For freezing. trypsinize ES cells (0.2 mL trypsin/EDTA per well of a 24-well plate, 0.5 mL per 3.5-cm dish, or 2 mL per 10-cm dish).
2. After dissociation of the cells in trypsin/EDTA solution, add twice the volume of ESC medium and mix by pipetting (0.4 mL per 24-weli. 1 mL per 3.5-cm dish, or 4 mL per 10-cm dish).
3. Finally, add an equal volume of 2X freezing medium (0.6 mL per 24-well, 1.5 mL per 3.5-cm dish or 6-mL per 10-cm dish).
4. Mix the cell suspension by gentle pipetting, and transfer 1 mL of each to prelabeled cryovials.
5. Cool the cryovials slowly in a styrofoam box to -70°C, and transfer quickly to liquid nitrogen.

Electroporation

1. Wash exponentially growing ES cells (usually two 10-cm plates will yield 4 to 6 × 10^7 ES cells) twice in *PBS* solution.
2. Trypsinize and add to ESC medium.
3. Centrifuge at 200g for 5 ruin, remove supernatant and resuspend in *PBS*.
4. Count and centrifuge again at 200g and remove supernatant.
5. Resuspend in PBS to a density of 10^7 Cells/0.8 ml for each electroporation cuvet.
6. Add 20 µg of DNA to the cells and mix by gentle pipetting avoiding the formation of air bubbles.

7. Incubate the cell suspension plus DNA for 5 min at room temperature.
8. Resuspend and transfer to an electroporation cuvet (4-mm electrode distance, 0.8-ml, vol) avoiding the formation of air bubbles, and expose to a pulse of 250 V and 500 µF.
9. Add 1 mL, of ESC medium immediately after electroporation.
10. Incubate for 5 min at room temperature and resuspend in 4 mL of medium.
11. Plate onto a 1.5-cm plate of feeder cells.

Selection

Usually gene trap constructs contain a neomycin phosphotransferase gene that confers neomycin resistance. As the selection, Geneticin® (G418) is used. G418 can be kept in aliquots as a sterile filtered 1000X stock solution at –20°C. Repeated defrosting and freezing of G418 reduces activity.

1. Select stable integration events with 250 µg/mL of G418 (by weight, corresponding to approx 125 µg/mL of active substance) in ESC medium starting 24 h after electroporation for 7-12 d.
2. Change the medium daily.
3. To test the activity of a new batch of G418, subject sham electroporated ES cells (without DNA) to 250 µg/mL of G418.

Picking of ES Cell Clones

1. Pick colonies of a 500-µm diameter or larger starting after 7 days of selection using a 2-20 µL pipet set to 2 and a dissection microscope placed in a laminar flow hood.
2. Transfer the colonies to 100 µL of 1 X trypsin solution in a 96-well plate.
3. Incubate at 37°C for 5 min.
4. Dissociate the cells by pipetting up and down using a 20-200µL pipette.
5. Transfer to 24-well dishes of feeder cells with 1 mL of ESC medium.
6. Incubate for 24 h and change medium to 0.5 mL of fresh ESC medium.
7. Propagate the individual colonies as individual cell lines.

Freezing

After 2-3 d either freeze one cryovial per well of a 24-well plate or passage each well onto a 3.5-cm dish. Freeze three cryovials from each confluent 3.5-cm dish. After freezing add more ESC medium and grow remaining cells for b*-galactosidase staining.

Basic Differentiation Protocol

1. After aliquots of each G418-resistant ES cell clone are frozen, grow remaining cells in each well at very low cell density for 14 d maintaining G418 levels at 250 µg/ml and 200 U/mL of LIF.
2. Stain the cultures after 14 d for β-galactosidase activity.
3. Evaluate the cultures microscopically at ×100 and ×320 magnification. Cell clones with a

variety of morphologic features will be observed (undifferentiated, neuron-like, epithelioid, mescnchymal, fibroblast-like, endoderm-like, vesicle forming and tubule forming). Clones exhibiting the cell type of interest can be selected for further studies.

Fixing Cells or Embryos

1. Wash the cells once in PBS.
2. Fix cells for 2 min in 1% formaldehyde and 0.2% glutaraldehyde in PBS, and wash twice in PBS. Fix embryos in 1% formaldehyde, 0.2% glutaraldehyde, 0.02% NP40, and PBS for 10 (e8.5) to 60 min (e15.5).
3. Wash in two changes of PBS for 10-60 min.

Process of Staining

1. Transfer the cells to staining solution (PBS containing 1 mg/mL of X-gal, 5 mM $K_3Fe(CN)_6$, 5 *mM* $K_4Fe(CN)_6$, 2 mM $MgCl_2$).
2. Stain overnight at 37°C.
3. Wash stained cells in PBS. They can be kept in PBS at 4°C for a few days. If later reference is needed, sodium azide can be added to the PBS to prevent fungal growth.

Staining of Embryos

1. Transfer the embryos to staining solution as for cells (PBS containing 1 mg/mL of X-gal. 5 mM $K_3Fe(CN)_6$, S mM $K_4Fe(CN)_6$.2 mM $MgCl_2$. (Add 0.2% NP40 and 0.1 % sodium deoxycholate for embryos older than e 11.5.)
2. Incubate overnight at 30°C.
3. After staining, wash the embryos in PBS briefly.

Clearing of Embryos

Embryos are either processed for histologic sections or cleared in ascending concentrations of glycerol in PBS (20. 40, 60, and 80%). When first placed into the next higher glycerol concentration, the embryos float in the solution.

Once an equilibrium is reached, the embryos sink to the bottom of the dish and can be transferred to the next higher concentration of glycerol. If embryos are transferred to higher concentrations of glycerol too quickly, some parts of the embryos shrink, e.g., the roof of the hindhrain.

Production of Chimeras

Fertile females of the mouse strains NMRI, CD1, or C57B16 (coat color of embryo donor mouse strain must he distinguishable from that of the mouse strain from which the ES cells were. derived) are superovulated by ip injection of pregnant mares' serum gonadotropin (PMSG) (7.5 IU for NMRI and CD 1: 5 IU for C57B16) followed by an injection of human chorionic gonadotropin hCG (5 IU) 46 h later.

Hormone-treated females are mated 1:1 with stud males of the same mouse strain overnight. Successful matings are recognized by the presence of a vaginal plug the next morning. Noon on the day of the vaginal plug is called e0.5. Eight-cell stage embryos are recovered between 9:00AM and 11:00AM on e2.5 from the oviduct. We usually use between 10 and 20 females per operator.

1. Before starting to flush the oviducts, fill a 3.5-cm Petri dish with 2 mL of M16.
2. Place 6 μL drops of M16 in another 3.5-cm Petri dish (10-20 drops) and cover with mineral oil.
3. Push six to nine depressions through the mineral oil into the plastic of the bottom of the Petri dish within each drop of M 16 using a tapestry needle (sterilize in 70% ethanol and allow to dry before use).
4. Leave three drops without depressions for ES cell clump selection. This is the aggregation dish.
5. Place both dishes containing M 16 into a 37°C, 5% CO, in air tissue culture incubator for equilibration.
6. Warms M2 to room temperature before use in subsequent steps.
7. For collection of embryos and further manipulations, use finely drawn out Pasteur pipets.
8. Kill superovulated and plugged female mice by cervical dislocation.
9. Open the abdominal cavity and hold the uterus with a pair of blunt forceps.
10. Remove the mesometrium from the uterus by perforating it with a pair of small scissors and opening the scissors parallel to the uterine horn and oviduct.
11. Excise the oviducts leaving 0.5 cm of uterus, and cut carefully between the ovary and oviduct in order to prevent damage to the oviduct and the infundibulum.
12. Transfer the oviducts to drops of M2 in a 10-cm dish, one drop per animal.
13. When all oviducts are collected, transfer to fresh drops of M2 in a 10-cm dish, this time one drop per oviduct.
14. Flush the oviducts under a low magnification microscope. Load a 2-mL syringe with M2 and attach a 33-gage flushing needle. Fix the infundibulur using no. 5 watchmaker's forceps. Insert he flushing needle into the infundibulum and hold with the forceps in place.
15. Flush about 200 μL of M2 through the oviduct with little pressure. Leave the uterine end of the oviduct in the drop of M2. During this flushing procedure embryos can be observed exiting the oviduct through the remaining uterine stump.
16. Attach a pipet to the mouth pipet and preload with M2. Collect embryos (10–20) by gently sucking into the pipet, and transfer them to a fresh 3.5-cm Petri dish filled with 2 mL of M2.

    During collection, good 8-cell stage or morulac embryos should be counted vs degenerated embryos. embryos delayed in development, and unfertilized eggs to assess success of superovulation and mating.

    Usually similar numbers of embryos can be expected from each of the two oviducts recovered from one animal. Eight-cell stage embryos and more or less compacted morulae can be used for the following steps of the procedure.

Separation of Zona

1. To remove zona pellucida, collect the embryos with a pipet preloaded with acid Tyrode's solution and transfer to a 3.5-cm Petri dish filled with 2 mL of acid Tyrode's solution.
2. Monitor the process of zona disappearance closely under the low-magnification microscope. The zona disappears within 1 to 2 min. To avoid damage to the zonaless embryos, collect them swiftly with a pipet preloaded with M2.
3. Wash zonaless embryos twice through 3.5-cm Petri dishes containing 2 mL of M2.
4. Wash the embryos (zonaless) again through the equilibrated dish containing 2 mL of M 16; and then place singly into the depressions in the previously prepared aggregation dish.
5. Return the aggregation dish to the incubator.

Isolation of ES Cells

1. Passage the ES cells for aggregation at least two times after defrosting, and keep in logarithmic growth phase throughout culture to enrich for healthy undifferentiated cells. The optimal time for aggregation is about 20-24 h after the last passage.
2. Change medium on the cultures on the morning of the aggregation. i.e.. several hours before aggregation, to ensure that the cells are growing under optimal conditions.
3. Wash a 3.5-cm Petri dish of ES cells twice with PBS and partially trypsinize in 0.5 mL of 1X trypsin/EDTA solution. The trypsinizing time will be similar to that for passaging the ES cells, but mechanical dissociation is reduced so that small numbers of ES cells stay loosely attached to each other in small clumps or branching chains, whereas all the feeder cells fall off the ES cells.
4. Add 1.5 mL of ESC medium to the cells and check the number of ES cells per clump or chain. If the clumps are still too large, pipet once or twice again with a 2-mL pipet and check the cell number.
5. Dilute partially trypsinized ES cells with 4-6 mL of ESC medium.

Aggregation

1. Select ES cell clumps or chains under x40 magnification and transfer to the depression-free drops of aggregation dish.
2. Select the ES cells again and transfer single clumps or chains of the cells to each zonaless embryo. and position such that a maximal contact between the embryo and the ES cells is achieved.
3. After supplying all embryos with ES cells, check again that they are in contact with ES cells and carefully transfer to the incubator.

Transfer of Embryos

After about 26 h of culture most of the aggregates will have developed into blastocysts and are ready for embryo transfer into the uterus of pseudopregnant recipient mice. Aggregates that did not develop to the blastocyst stage call still be transferred after a further 24 h of culture.

Pseudopregnant embryo transfer recipient trice are generated by mating fertile females to vasectomized males. After successful mating, a vaginal plug can be observed the next morning. The pseudopregnant females are used on e2.5. but minus 24 h asynchronous to the aggregated embryos that were recovered on e2.5 and then cultured for 1 d.

1. Transfer blastocysts from the aggregation dish to a 3.5-cm dish containing 2 ml, of M2.
2. Transfer again to a 3.5-em dish containing five 20-μL drops of M2. Transfer all embryos (four to eight) for one side of a recipient uterus to one drop. From these they can easily be recovered for embryo transfer during surgery.
3. Keep the embryos (in drops) outside the incubator at room temperature. In initial experiments, embryos for only one transfer should be transferred to drops because the embryos should not be kept outside of the incubator for too long.
4. Put the recipients into general anaesthesia by injecting approx 20 μL, of the injection solution of avertin 1 g of body weight. Avertin is more potent the first 2 d after the preparation of the final solution. Each new preparation of avertin may vary and has to be tested for best results.
5. Shave the lateral body walls between the last ribs and pelvis with fine clippers. Disinfect the surgical instruments with 70% ethanol and allow to dry. Make "surgery nay" by folding aluminum foil (four layers. 10 × 6 cm) and place it under the low-magnification microscope.
6. Place the anesthetized mouse onto the surgery tray.
7. Disinfect surgery area of mouse with 70% ethanol.
8. Cut a small incision, roughly halfway between the last rib and the tuber coxae of the pelvis and about 1.5 cm below the spine, into the skin in an angle of 45° at the last rib with small scissors while holding the skin with surgical forceps.

   In this area, almost always the fat pad of the ovary forms a small mound that is visible through the skin. This can be used for orientation. The incision can be made on top of this mound. The opening is bluntly widened by inserting the scissors and opening them.
9. Cut the muscle layers of the abdominal wall. Make the incision in an area lacking blood vessels and nerves. This cut will often include the peritoneum. For initial experiments, it is advisable to cut a larger opening to facilitate a better view of the organs. Experienced operators will cut an opening of < 1 cm. The ovarian fat pad is usually substantial and found easily.

   It can be used as a handle to expose the fat pad, ovary, the oviduct, and cranial part of the attached uterine horn through the body wall opening. A Serrefine clamp can now be attached to the fat pad and placed onto the back of the animal to weigh down the fat pad and hold the uterine horn in place. The animal can be moved out of the field of view for the next step by moving the aluminum surgery tray.
10. Collect four to eight embryos into a drawn-out Pasteur transfer pipet attached to mouth pipetting tubing preloaded with two drops of M2 with an intermediate air bubble and an air bubble at the end. In the end, the pipet should contain from the tip: four to eight embryos in a minimal amount of M2, air, M2, air. M2, air. The loaded pipet is carefully set

aside so that it can be picked up easily.

11. Move the animal back into the view field, and adjust the focus to the surface of the uterine horn about 1 cm from the oviductal-uterine connection.
12. Hold the uterine horn with small blunt forceps at the oviductal end.
13. Perforate the uterine wall with a 27-gage injection needle near the forceps.
14. Make a note of this place, pick up the transfer pipet carefully, and insert it into the uterus. Gently blow the embryos into the uterine cavity with minimal medium while avoiding blowing air into the uterus.
15. Under microscopic control, suck M2 into the pipet and blow out again to check, that the embryos are successfully transferred.
16. Release the Serrefine clamp from the fat pact, and replace the organs into the abdominal cavity. Now the peritoneum and the musculature of both sides of the abdominal wall opening are juxtaposed such that peritoneum touches peritoneum and is held together using two pairs of forceps.
17. Use one pair of forceps to hold the tissues in this position while using the other to pull the skin of both sides of the wound over the musculature. The resulting wound ridge consists of six layers (skin, muscle, peritoneum, peritoneum, muscle, and skin) and is fixed using small surgical wound clips.
18. Repeat this procedure on the contralateral side.
19. Wrap the mouse in paper tissue and place onto a warm plate at 37°C to prevent hypothermia. As the animal starts to recover from anesthesia, place it into a clean cage.
20. Examine the animal and the surgery wounds the next day.

Mouse Breeding

An estimate of the overall contribution of the ES cells to the chimera can be made from the ES cell contribution to the pigment cell population (i.e., percentage of coat color chimerism). In our experience, all mice with about 90-100% ES cell contribution, as assessed by coat color, were germline chimeras, if they were fertile. About half of the lower percentage chimeras were germline chimers with no correlation to the contribution to the pigment cell population.

Very low percentage chimeras were rarely germline chimeras. The resulting chimeras are bred to females of a mouse strain chosen such that the phenotype of the coat color genes of the ES cells can be recognized. This first mating is to test for germline transmission of the ES cell genome. Once germline transmission is established, we routinely cross chimeras back to the mouse strain that was used for ES cell production (in Our case I29/SvPas).

This mating will generate mutant mice on a defined genetic background. Coat colors are no indication in this mating. The inbred mice are usually less fertile but always much more consistent in all parameters of interest. Therefore, we usually perform an initial phenotypic analysis on a mixed genetic background and then repeat these experiments on an inbred background for better consistency. Unlike in homologous recombination experiments, gene trap only generates one ES cell clone per mutation of a specific gene.

Therefore, care must be taken to eliminate the possibility of an unrelated mutation in the genome of the ES cells causing the observed phenotype. The final phenotypic analysis should be performed after backcrossing the gene trap allele to wild-type mice for several generations.

This will help remove unlinked mutations. We recommend that 100 meioses should be checked for cosegregation of the gene trap allele and the phenotype. Tightly linked mutations are not likely to segregate.

However, the most likely linked mutation would be a rearrangement at the insertion site. This possibility can be eliminated by testing for the integrity of the endogenous locus outside of the gene trap insertion by Southern analysis of regions outside of the insertion site.

Genotyping by PCR

DNA is isolated from tail biopsies. To genotype offspring of heterozygous by wild-type coatings, polymerase chain reaction (PCR) amplification of 508 by of the *lacZ* gene can be performed. The PCR buffer consists of: 50 mM KC1, 10 mM Tris-HCl, 2 mM $MgCl_2$, 0.1% Triton X-100. and 0.2 mM dNTPs.

1. Perform PCR using *5′ TTGGCGTAAGTG,AAGCGAC3′* as a forward and 5′ *AGCGGCTGAT GTTGAACTG3′* as a reverse primer to delete *LacZ* gene.
2. PCR cycle: Denaturing at 94°C for 5 min, 30 cycles of 94°C for 1 min, 60°C foi 1 min 30 s, and 72°C for 1 min; final extension 72°C, 10 min. Separate the PCR products on a 2% agarose gel and visualize with ethidiuni bromide.

Method for Southern Hybridization

During the initial screen for mutant phenotypes, an endogenous probe for a gene-specific Southern analysis may not be available. Because only about 50% of the gene trap mouse lines exhibit a mutant phenotype, it may be advisable to postpone laborious cloning procedures until after an interesting phenotype has been observed.

To genotype offspring of heterozygous by heterozygous matings in the absence of a specific probe, we use a quantitative Southern analysis distinguishing one and two copies (or multiples of these) of the gene trap construct per genome.

1. Cut the DNA with appropriate restriction enzyme.
2. Separate by agarose gel electrophoreses and transfer onto nylon filters.
3. Hybridize with [a*-^{32}P] dCTP-labeled eDNA probes for the b*-galactosiitase or the neomycin phosphotransferase gene and an endogenous marine gene as an internal control for two copies per genome.
4. Compare the intensity of the *LacZ* or the *neomycin phosphoirarrsferase* gene signals and the endogenous gene signals on the same filter to each other and among animals to determine animals with one or two copies of the gene trap vector per genome.
5. Test suspect homozygous males by mating with the wild-type females for their ability to transmit the trapped allele to 100% of their offspring.

Identification of Tagged Gene

The quickest way to obtain sequence information from the tagged locus is to use a 5′-RACE technique. We have used two different methods for this and both have worked well. Both these methods are also available as kits from Gihco-BRL and Clontech. When only a small number of 5¢-RACE products are required, using one of these kits is probably the most cost-effective option.

In general, we have found 5¢ RACE on RNA purified from ES cells more successful than from adult tissues. This is probably because the sequence complexity is lower and we routinely select the cells in G418-containing medium first, which selects for cells expressing the fusion mRNA in case of promoterless gene trap constructs. We use the repeated differential precipitation RNA purification method. It is important to use an RNA purification method that gives low DNA contamination.

1. Synthesize first-strand eDNA using *lacZ* gene-specific primer (5′-ATTCAGGCTGCGC AACTGTTGG-3′).
2. Amplify by two rounds of PCR using nested oligonucleotide primers (5′-CTG-CAAGGCGATTAAGTTGG-3′ and 5′-TAACGCCAGGG'I'TTTCCCAG-3′) from the *lacZ* sequence. Size select the products of the first PCR by agarose gel electrophoresis before second round of PCR.
3. After ligation into a plasmid vector, perform bacterial colony hybridization on filter lifts probing with a nested primer complementary to sequences in the gene trap construct 5¢ of the *lacZ* gene, e.g., the engrailed-2 exon (5¢-AAACTCAGCCTTGAGCCTCTGGAGCTGCTCA GCAGTGAAGGC 3¢) in the case of *pGT1.8,geo*. This last step greatly increases the number of specific clones.
4. Confirm 5′-RACE products by reverse transcriptase PCR.

POINTS TO REMEMBER

1. Low functional mutation efficiency depends on the gene trap construct used. Each new gene trap should construct. therefore. be tested sufficiently before use in a large-scale screen.
2. We have no indication that multiple tandem copy integrations increase the chance of producing a functional mutation. They are problematic when using certain types of polyadenylation trap gene trap constructs.
3. STOs should riot be allowed to reach confluence: instead, they should be passaged before they reach confluence by washing them twice with PBS, adding 1X trypsin/EDTA, incubating them at 37°C for 3 min, adding feeder medium (4-9 vol of trypsin/EDTA), and plating them onto gelatin-coated tissue culture plastic. SNL cells can he frozen as described above for PEFs.
4. Mitomycin C is very unstable and should be stored in the dark. Different batches from the same supplier can vary in activity, resulting in feeders that continue to proliferate or feeders that are so damaged that they start dying after plating.

   Ideally, feeders should not proliferate, but otherwise look healthy the day after inactivation. The presence of large numbers of dead cells may indicate that mitomycin C is more active than usual and poor-quality feeders will result.

5. The split ratio depends on the density of cells in the original dish and can vary between 1:3 and 1: 12 (if there are only very few colonies present in a dish, they may have to he passed 1:1 into a fresh dish of the same side as the previous well) The cells should be split in such a ratio that they can be passaged every other day with daily medium change in between.

   The cell density should he maintained such that the medium does not turn yellow before the next 48-h cell passage is due. It is important to keep track of the passage number, since it becomes increasingly difficult to produce germline chimeras efficiently with ES cell lines of higher passage numbers (more than 16 passages).
6. ES cells can be kept for years in liquid nitrogen without affecting their potential to form germline chimeras.
7. Low plating efficiency may result if ES cells have not been in maximal growth phase before electroporation or the gene trap construct DNA contains bacterial endotoxins. In our hands, using $CsCl_2$ gradient purified DNA for electroporation yielded the most consistent results.
8. After 7-9 d all colonies should have died. After 3 to 4 d of selection, the death of nonresistant cells is clearly visible. The dead cells lose the usually tight contact to neighboring cells in their colony, detach from the feeder cells, and float in the medium.
9. G418 concentration will have a significant effect on the number of resistant clones. G418 is known to vary from hatch to hatch in active component. It may be advisable to test the minimum concentration that will kill ES cells that have not been electroporated in 7 d. Avoid repeated thawing and freezing of G41 8.
10. We obtain an average of about 25 G418-resistant colonies electroporating 10' ES cells with 20 µg of pGT1.8geo and 10 colonies using pKC199βgee
11. In cases where insufficient colonies have grown after picking, it may be necessary to passage the cells 1:1 onto fresh feeders.
12. These culture conditions favor the cell type that expressed the selection gene very strongly while allowing slow differentiation into this cell type. Differentiation by complete LIF withdrawal is not as successful. because the cells tended to differentiate rapidly along the pathways of least resistance and a complex culture does not develop.
13. For embryos older than e 11.5 0.2% NP40 and 0.1% sodium deoxycholate should be added to the fixative. Embryos older than e 12.5 must be cut into halves longitudinally.
14. Whole-mount β-galaetosidase staining gives satisfactory results only up to e 11.5. In older embryos, the fixative and stain do not penetrate well, giving irregular staining. This is normally sufficient for a rough overview, but for detailed histologic analysis. β-galactosidase staining of cryostat sections is necessary.
15. Background staining can occur because of endogenous β-galactosidase activity. Usually, ensuring that the glutaraldehyd component has been store correctly and/or lengthening the fixation time solves the problem.

    However, low intensity real staining can be abolished by overfixation. It is important that the volume of fixative is at least 10X the tissue volume and that the embryos arc gently rocked so all surfaces are exposed to the fixative.

16. The described superovulation protocol may need to be adjusted when using other mouse strains. For this protocol, we prefer postpubertal females to prepubertal females. Prepubertal females produce high numbers of fertilized oocytes but the embryos frequently do not develop well.
17. If the stud males are too old very few healthy embryos will be obtained. When using our males two to three times a week, we exchange them three times a year. Despite the use of a controlled environment (light cycle, temperature, and humidity control), we observed seasonal changes in reproductive performance of our mouse colonies, with a low in the coldest month and the hottest month.
18. Pasteur pipets are finely drawn out by hand after heating their thin section above the small flame of a Bunsen burner such that the diameter of the pipet is slightly larger than the diameter of an embryo including the zona pellucida. The fine part of the pipet is broken such that it is about 4 cm long and the broken edge is straight. We usually prepare a large number to choose from, which are kept upside down in a glass beaker.
19. When very different numbers of embryos are often obtained from both oviducts within animals, this is an indication of lack of skill on the part of the operator.
20. When too much M2 is transferred with the embryos into the acid Tyrode's solution. zona removal will take much longer.
21. We use associations of 12-20 cells for aggregation with embryos of the strains NMRI and CDI and 2-6 cells for C57B16. These numbers may vary with ES cell lines and with mouse strains used and have to be established empirically.
22. For best results, our aggregations have to be finished by 11:00 PM, but this time point may vary with the superovulation schedule and mouse strain used. However. embryos that are developed too far will not incorporate ES cells. Four hours after aggregation, the success can be assessed by examining the aggregates under the microscope. The ES cells should be incorporated into the embryos, which should be compacted morulae by this time.
23. If you notice problems with in vitro development of aggregates. check the water used to make up M2 and M16, the M16, the mineral oil, and the incubator. We regularly test new batches of autoclaved Milli-Q water, first M16 made from new stock solutions, and mineral oil on unmanipulated embryos.

    The aggregates also will have problems developing when too much ESC medium was introduced to their M16 while supplying them with ES cells. Embryos may also fail to develop if they are left too long in acid Tyrode's solution or not handled carefully.
24. Swellings at the site of abdominal surgery occur if the peritoneum and musculature are riot closed properly.
25. If low implantation rates are observed, check surgery techniques by transferring unmanipulated embryos.
26. Low ES cells contribution to chimeras may result if either ES cell culture before aggregation or the parental ES cell line is suboptimal. Check that the parental ES cell lines produce

germline chimeras with high efficiency before proceeding further.

27. We observed multiple integrations in 2 of 26 mouse lines crossed to homozygosity. They were unlinked and segregated within one generation of mouse breeding.

Chapter 12

Examine Mutations

The advent of restriction landmark genome scanning (RLGS) techniques permitting the two-dimensional separation of DNA fragments from a genomic digest is an important advance that has allowed researchers to study the occurrence of spontaneous and induced mutation in both somatic and germinal tissues.

A unique gel holder for the first-dimension agarose disc gel was made from a Teflon tube and a glass tube. The development of the disc gel system resulted in high-resolution separation of DNA fragments with enhanced reproducibility. A new device for the in-gel restriction enzyme digestion of the fractionated DNA fragments was also made from a Teflon tube. The development of this vertical disc gel system has made the RLGS a practical method.

The overview of the revised RLGS method is schematically represented in Figure elsewhere in this chapter In a revised protocol, the DNA samples were double digested with two restriction enzymes: *EcoRV* (GAT$^{\downarrow}$ATC) and *NotI* (GC$^{\downarrow}$GGCCGC). The *NotI* sites of the digests were specifically end-labeled with ^{32}P by fill-in reaction with a DNA polymerase and separated by high-resolution 2-DE.

The DNA samples were first separated on agarose disc gels by size. After the first electrophoresis, the DNA fragments were again digested *in situ* with the third restriction enzyme Hinf I (G$^{\downarrow}$ANTC) and cut into smaller fragments that were separated perpendicularly on polyacrylamide slab gels.

Following the second electrophoresis the sheet gels were dried and the *NotI* fragments were visualized by autoradiography. The autoradiographic image of labeled DNA fragments represents the RLGS profile and is reproducible enough that profile comparisons can be made between two different DNA samples.

This technique can be used to determine mutation events within a genome after exposure to an environmental contaminant. In addition, subsequent isolation and sequencing of the differential RLGS fragments found between profiles can be used to identify the specific locus that these mutations have occurred. Since earlier description of the methods we employ, several modifications have introduced. Current protocol and the equipment will be described in this chapter.

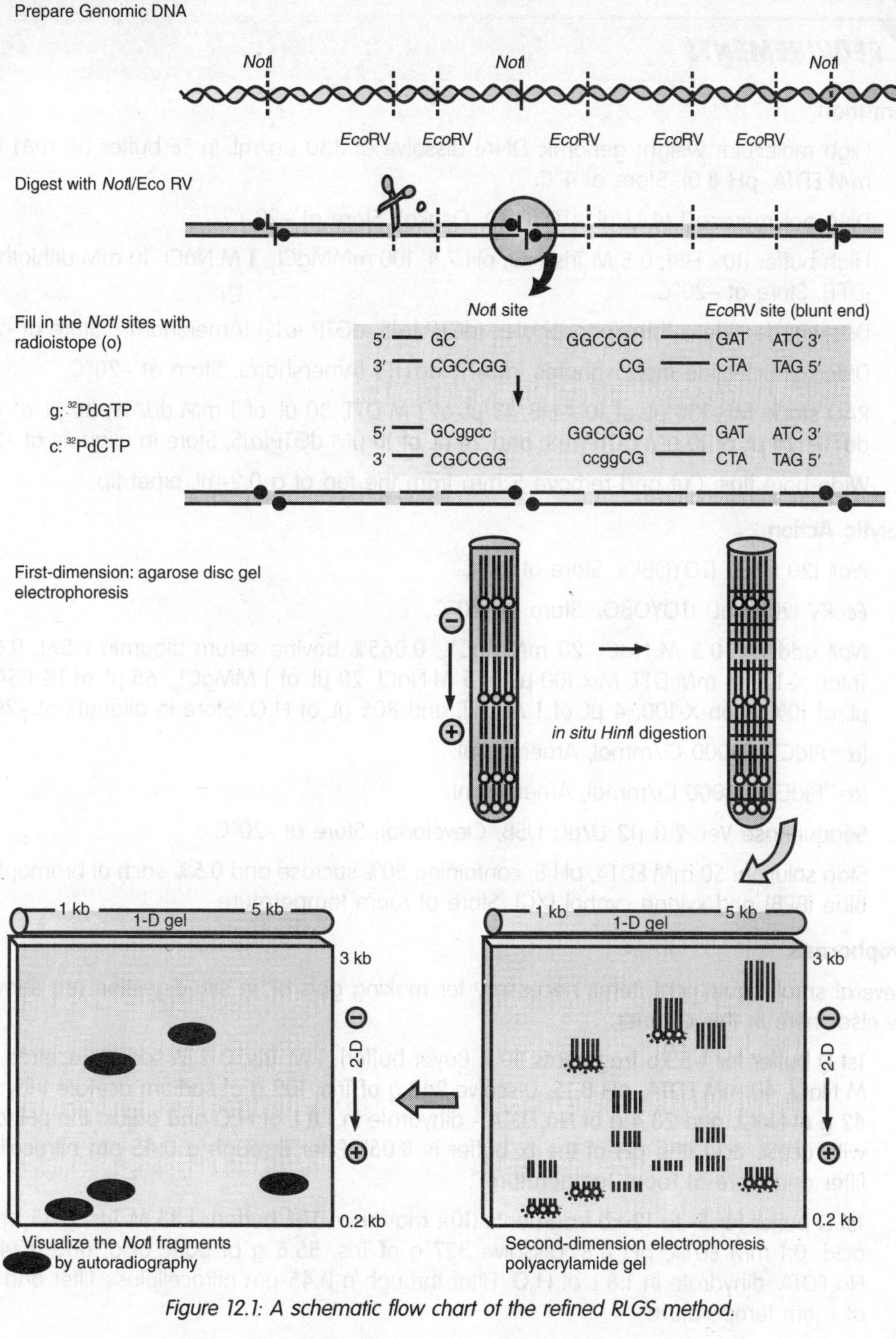

Figure 12.1: A schematic flow chart of the refined RLGS method.

REQUIREMENTS

Presentation

1. High molecular weight genomic DNA: dissolve at 330 μg/mL in TE buffer (10 m*M* Tris, 1 m*M* EDTA, pH 8.0). Store at 4°C.
2. DNA polymerase 1 (4 U/μL) (TOYOBO, Osaka). Store at –20°C.
3. High buffer (10× HB): 0.5 *M* Tris-HCl, pH 7.4, 100 mMMgCl$_2$, 1 M NaCl, 10 mM dithiothreitol (DTT). Store at –20°C.
4. Deoxynucleotide α-thiotriphosphates (dCTP [α]S, dGTP [α]S) (Amersham). Store at -20°C.
5. Dideoxynucleotide triphosphates (ddATP, ddTTP) (Amersham). Store at –20°C.
6. RAD stock: Mix 130 μL of 10 x HB, 13 μL of 1 M DTT, 30 μL of 1 mM ddATP, 30 μL of 1 mM ddTTP, 24 μL of 10 μM dCTP[α]S, and 24 μL of 10 μM dGTP[α]S. Store in aliquots at –20°C.
7. Wide-bore tips: Cut and remove 5 mm from the top of a 0.2-mL pipet tip.

Hydrolytic Action

1. *Not*I (20 U/μL) (TOYOBO). Store at 20°C.
2. *Eco*RV (20 U/μL) (TOYOBO). Store at -20°C.
3. *Not*I additive: 0.5 *M* NaCl, 20 mMMgC1$_2$, 0.065% bovine serum albumin (BSA), 0.065% Triton X-100, 4 mM DTT. Mix 100 μL of 5 *M* NaCl, 20 μL of 1 MMgC1$_7$, 65 μL of 1% BSA, 6.5 μL of 10% Triton X-100, 4 μL of 1 *M* DTT, and 805 μL of H_2O. Store in aliquots at –20°C.
4. [α-^{32}P]dCTP (3000 Ci/mmol; Amersham).
5. [α-^{32}P]dGTP (3000 Ci/mmol; Amersham).
6. Seaquenase Ver. 2.0 (12 U/μL; USB, Cleveland). Store at -20°C.
7. Stop solution: 50 mM EDTA, pH 8, containing 50% sucrose and 0.5% each of bromophenol blue (BPB) and xylene cyanol (XC). Store at room temperature.

Electrophoresis

Several small equipment items necessary for making gels or *in situ* digestion are shown in Figure elsewhere in this chapter.

1. 1st D buffer for 1-5 kb fragments (l0 × Boyer buffer): 1 M Tris, 0.4 M sodium acetate, 0.36 M NaCl, 40 mM EDTA, pH 8.15. Dissolve 242 g of Tris, 109 g of sodium acetate trihydrate, 42 g of NaCl, and 23.4 g of Na$_2$EDTA - dihydrate in 1.8 L of H_2O and adjust the pH to 8.15 with acetic acid (the pH of the lx buffer is 8.05). Filter through a 0.45-pm nitrocellulose filter and store at room temperature.
2. 1st D buffer for 5- to 12-kb fragments (10× marathon TBE buffer): 1.35 M Tris, 0.45 M boric acid, 0.1 mM EDTA, pH 8.8. Dissolve 327 g of Tris, 55.6 g of boric acid, and 0.74 g of Na$_2$EDTA. dihydrate in 1.8 L of H_2O. Filter through a 0.45-μm nitrocellulose filter and store at room temperature.

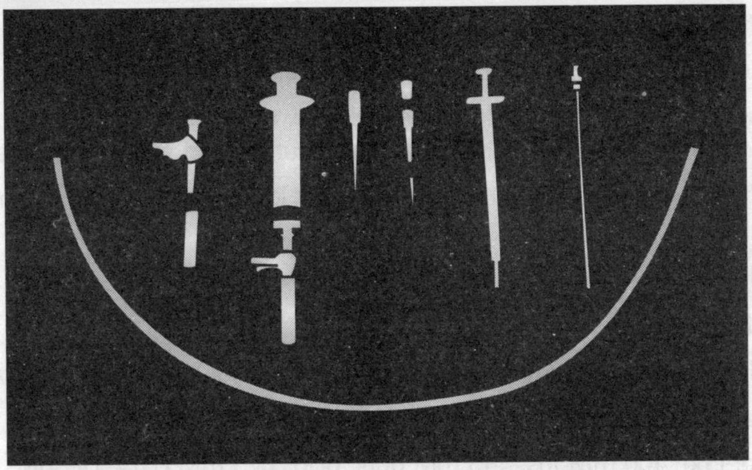

Figure 12.2: Small equipment items used for the agarose disc gel preparation and in situ *digestion.*

3. Stacking agarose gel (0.6%): Add 2.5 g of sucrose, 0.3 g of SeaKem Gold agarose (FMC), and 5 mL of corresponding 10× 1st D buffer (Boyer or marathon TBE) to 40 mL of H_2O. Boil the mixture in a microwave oven. After the agarose was completely melted, add warm water to obtain the initial weight. When the temperature of the solution drops to 65°C, remove undissolved particles via vacuum filtration through a 0.45-μm nitrocellulose filter. Divide 5 ML aliquot each in a Falcon 15-ml centrifuge tube and store at 4°C.
4. Running agarose gel for 1- to 5-kb fragments (0.9%): Mix 10 g of sucrose, 0.9 g of SeaKem Gold agarose (FMC), 0.9 g of SeaKem GTG agarose (FMC), and 20 mL of 10x Boyer buffer and adjust the volume to 200 mL with H_2O. Boil the mixture and melt the agarose in a microwave oven. Adjust the volume and filtrate the solution as described above. Divide 30 mL aliquot each in a Falcon 50-mL centrifuge tube and store at 4°C.
5. Running agarose gel for 5- to 12-kb fragments (0.7%): Mix 10 g of sucrose, 0.4 g of SeaKem Gold agarose (FMC), 1.0 g of SeaKem GTG agarose (FMC), and 20 mL of 10x marathon TBE buffer and adjust the volume to 200 mL with H_2O. Boil the mixture and melt the agarose in a microwave oven. Adjust the volume and filter the solution as described above. Divide 30 mL aliquot each in a Falcon 50-mL centrifuge tube and store at 4°C.
6. Support stands with double-buret clamps.
7. Disposable plastic syringes (1-mL and 6-mL).
8. Needles (10-cm long 19-gauge, 90° cut).
9. Three-way stopcocks: Fit the plastic syringes.
10. Silicon-tubes [inner diameter (i.d.) of 3 mm and 5 mm].
11. Cutoff tips: Cut and remove 5 mm from the top and 10 mm from the bottom of 0.2-mL pipet tips to fit the plastic syringes.
12. Size marker: mix 3 μL of 1-kb ladder (500 ng/μL; NEB), 30 μL of stop solution, and 67 μL of H_2O. Store at 4°C.

Role of Restriction Enzyme

1. BPB solution (0.01%). Store at room temperature.
2. Reaction tubes: 36-cm long Teflon tubes with i.d. of 2.7 mm connected to 3 cm of Silicon-tubes (i.d. 3 mm). Attach ID labels with consecutive numbers to the tubes.
3. *Hinfl* buffer (10x): 200 mM Tris-HCl, pH 8.3, 1 M NaCl, 100 mM $MgCl_2$, 10 m*M* DTT. Store at –20°C.
4. *Hinfl* (20 U/µL) (TOYOB). Store at –20°C.

Second Stage Technique

1. Repel-Silane ES (Amersham: dimethyldichlorosilane solution for glass plates coating).
2. Adhesive tape: Scotch 3M or compatible.
3. TBE buffer (5×): 400 m*M* Tris, 445 mM boric acid, 12.5 mM EDTA. Dissolve 108 g of Tris, 55 g of boric acid, 9.3 g of $EDTANa_2$ H_2O in H_2O. Adjust the volume to 2 L. Filter through a 0.45'µm nitrocellulose filter and store at room temperature.
4. Glycerol solution (50%): Mix equal volumes of glycerol (BRL, ultrapure) and H_2O. Store at room temperature.
5. Acrylamide stock solution (30%): Dissolve 290 g of acrylamide (99.9% pure) and 10 g of N, N-methylene bis-acrylamide in H_2O. Adjust the volume to 1000 mL and filter thorough a 0.45-µm nitrocellulose filter. Protect from light and store at 4°C.
6. Ammonium persulfate: Prepare fresh 10% solution.
7. N,N,N′,N′- Tetramethylethylenediamine (TEMED).
8. Water saturated 2-butanol: Mix 300 mL each of 2-butanol and H_2O, take out upper part and store in a spray bottle at room temperature.
9. Blue connecting agarose solution: Melt 0.5 g of SeaKem Gold agarose in 100 mL of l× TBE buffer containing each of 0.05% BPB and XC dye. Divide into three Falcon 50-mL centrifuge tubes. Store at 4°C.
10. Sampling platform: a plastic plate 3 mm thick, 33 cm × 10 cm.

Autoradiographyic Analysis

1. Gel drier with vacuum pump: Bio-Rad model 583 or equivalent.
2. Filter paper: 35 cm × 43 cm, 0.7 mm of thick. (Chromatography paper No. 526, Advantech, Tokyo).
3. Saran Wrap (50 cm of width).
4. X-ray films (35 × 43 cm): Fuji RX-U, Kodak BioMax MS or equivalent.
5. Intensifying screens (35 × 43 cm): Fujifilm GRENEX HR-12, Kodak BioMax MS or equivalent.
6. Exposure cassette for 35 × 43 cm film.
7. X-ray film developer.

PROCEDURES

Presentation

1. Melt the RAD stock and prepare a master solution for blocking; mix 0.3 μL of DNA polymerase I and 1.8 μL of RAD stock per DNA sample. Mix via a gentle pipetting and keep on ice. (Do not vortex!)
2. Add 2.1 μL of the RAD master solution to 7 μL of DNA (~2 μg). Mix via repetitive gentle pipetting with wide-bore tips. Incubate at 16°C for 30 min followed by 65°C for 30 min.

Restriction Enzyme Digestion and Isotope Labeling

1. Add 2.1 μL of *NotI* additive, 1 μL each of *NotI* and *EcoRV.* Mix thoroughly to achieve complete digestion by gently pipetting up and down (10 times) via a wide-bore tip. Incubate at 37°C for 3 h.
2. Radiolabel the NotI-derived 5 protruding ends of the sample DNA by a fill in reaction with a DNA polymerase (Sequenase Ver. 2.0). Prepare label master stock: mix 1.6 μL each of [α-^{32}P]dCTP and [α-^{32}P]dGTP, 0.2 μL of 1 M DTT, and 0.3 μL of Sequenase per reaction. Add 3.7 μL of the label master stock to the DNA digest and mix thoroughly by pipetting. Incubate at 37°C for 30 min.
3. Add 5 μL of stop solution and allow the mixture to stand on ice.

Electrophoresis Method

The development of the current first-dimensional electrophoresis procedures has involved considerable experimentation. The Biochemical Genetics Program at the Radiation Effects Research Foundation has in the past made extensive use of a disc gel electrophoresis system in the study of protein variants, and this system has been adapted for the DNA gels.

The gel consisted of a 1.5-cm 0.6% stacking agarose gel, and a 59-cm 0.9% running agarose gel. The vertical disc gel electrophoresis system is shown in Figure elsewhere in this chapter By using a single system, up to 10 DNA samples can be analyzed simultaneously.

The development of the vertical disc gel system resulted in high-resolution separation of large DNA fragments. This system dramatically reduced the space for electrophoresis as well as the amount of expensive enzymes (e.g., *HinfI;* the original system required 10,000 U of enzyme but now we need 500 U per reaction) used for *in situ* digestion and thus made the RLGS method more practical and less costly.

The first dimension agarose disc gel electrophoresis system and the latest version of the second dimension multiple giant gel vertical electrophoresis apparatus (for 10 or 8 gels) are shown in Figure's elsewhere in this chapter, respectively. Both systems are available from Ohtsuka-Rikagaku (Hiroshima, Japan) on request.

Electrophoresis Gels

The gel preparation of the first electrophoresis gel is schematically represented in Figure elsewhere in this chapter. Start the gel preparation before isotope labeling. Prepare two types of

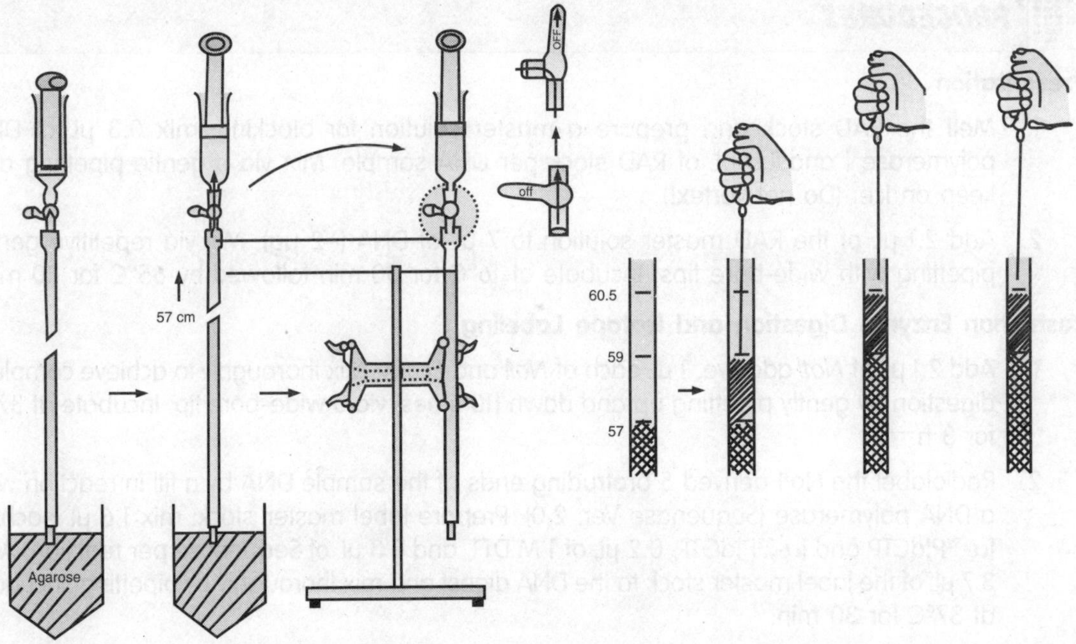

Figure 12.3: A schematic representation of the preparation of the agarose disc gel in a Teflon tube.

gels simultaneously if two patterns (1-5 kb and 5-12 kb) are to be studied. Preparation of softer gels for 5- to 12-kb long fragments is optional.

1. Melt the running agarose and stacking agarose in a microwave oven and place the container in a heater block at 70°C.
2. Connect a 6-mL plastic syringe fitted with a three-way stopcock and the gel holder with 3 cm of silicon tubing (5 mm i.d.).
3. Dip the bottom of the gel holder to a 1-cm depth and slowly suck up the running agarose solution to reach the first line, 57 cm from the bottom.
4. Hold it approx 30 s and stand it on the double buret clump on a support stand. Solidify for about 10 min.
5. Turn the stopcock 90° to the right to open the left inlet, and carefully remove the mounted syringe with the stopcock from the gel holder.
6. Overlay the running gel solution to the second line (59 cm) with a warmed l-mL syringe fitted with a 90°-cut long needle. Tap the upper part of the gel holder several times with a fingernail to make a flat surface. Solidify for about 3 min.
7. Overlay the stacking gel solution to the third line (60.5 cm). Tap the gel gently with a fingernail to make the top of the gel flat. Solidify for about 10 min.
8. Fill the tube with 1 × 1st D buffer.

9. Fit each gel holder into the anodal tank through a silicon plug with a 5-mm diameter hall
10. Add 350 mL of 1 × 1st D buffer to the bottom tank.
11. Place the top tank on the bottom tank and fill the tank with 300 mL of 1 × 1st D buffer.

Electrophoretic Analysis

1. Load 10 µL of the labeled sample, containing about 1 µg of DNA, onto each gel.
2. Load 10 µL of size marker (150 ng) onto a gel.
3. When all of the samples are loaded, put the cover on the top tank and connect the electrodes to the power supply.
4. Run the gel at constant voltage. For short DNA fragments (1-5 kb), run at 130 V for 19 h followed by at 140 V for about 24 h (~5800 V . h) until the center of the marker dye, BPB, migrates 50 cm. For long DNA fragments (5-11 kb), run at 70 V for a week (about 11,200 V . h) until the center of the marker dye, XC, migrates 50 cm.

Role of Restriction Enzyme

The *NotI/EcoRV* fragments separated on the agarose disc gels need to be further digested

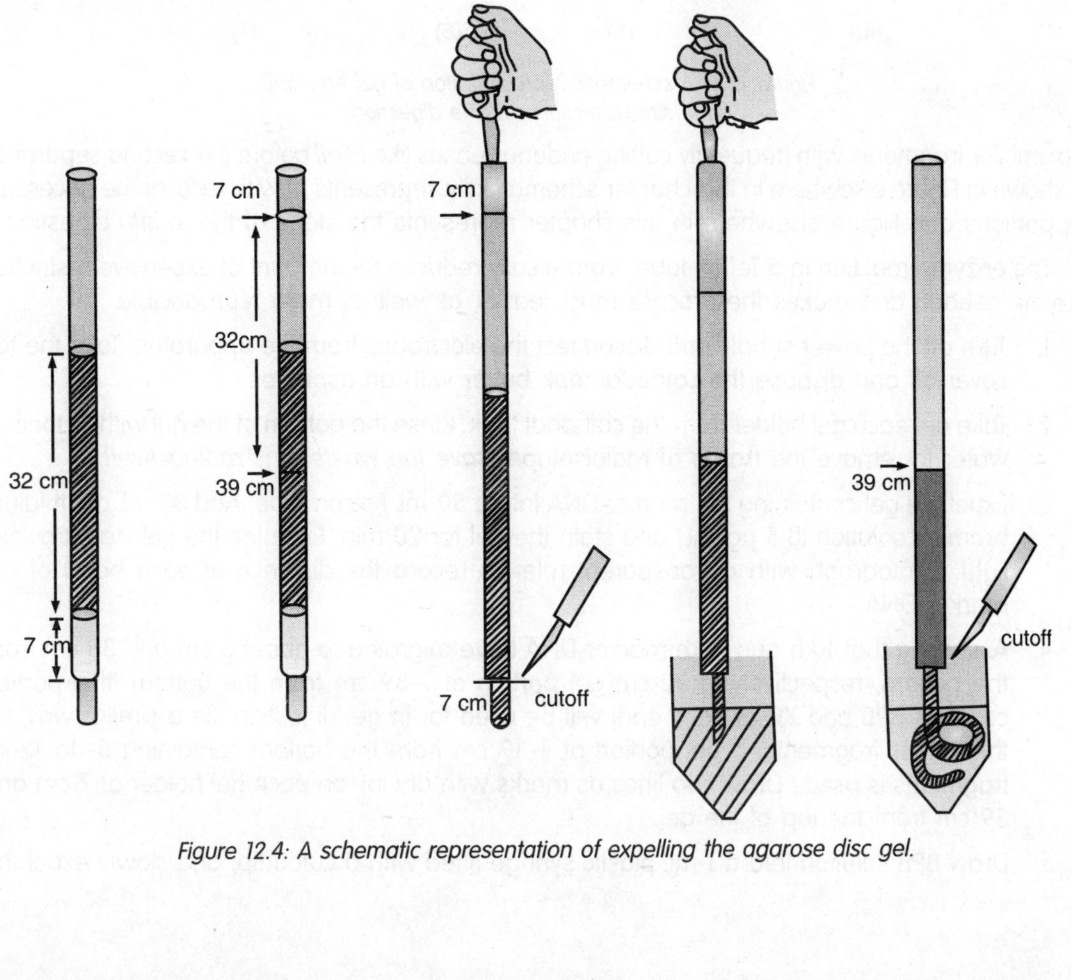

Figure 12.4: A schematic representation of expelling the agarose disc gel.

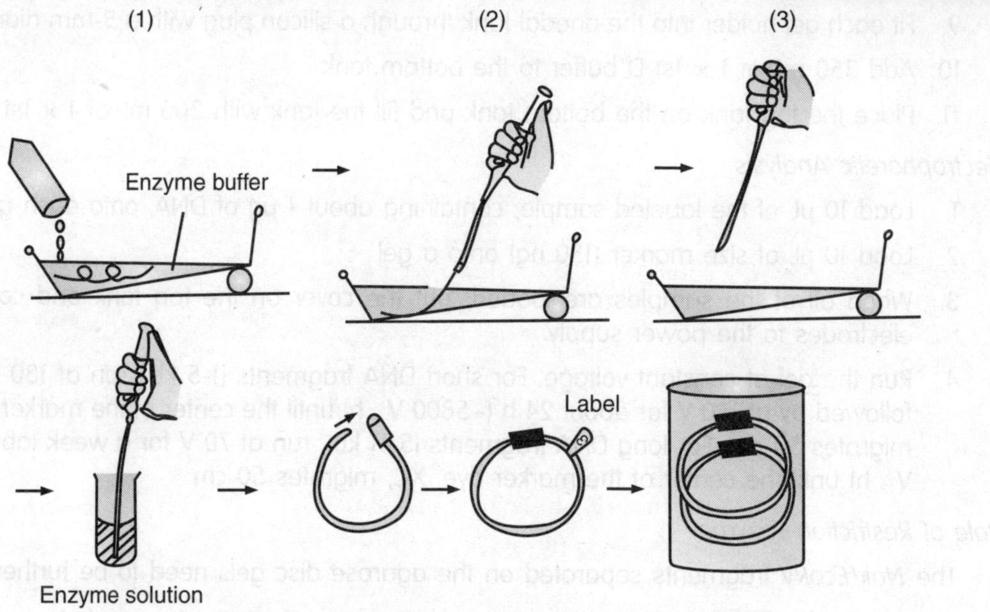

Figure 12.5: A schematic representation of gel handling for in situ restriction enzyme digestion.

into smaller fragments with frequently cutting endonucleases like *HinfI* before the second separation as shown in Figure elsewhere in this chapter schematically. represents how to recover the necessary gel portions and Figure elsewhere in this chapter represents the steps of the *in situ* digestion.

The enzyme reaction in a Teflon tube dramatically reduces the amount of expensive restriction enzyme needed and makes the process much easier as well as more reproducible.

1. Turn off the power supply and disconnect the electrodes from the apparatus. Take the top cover off and dispose the cathodal tank buffer with an aspirator.
2. Take out each gel holder from the cathodal tank. Rinse the bottom of the gel with deionized water to remove the traces of radioisotope *(save the waste, it is radioactive!).*
3. Expel the gel containing the marker DNA into a 50-mL Falcon tube. Add 30 mL of ethidium bromide solution (0.5 μg/mL) and stain the gel for 20 min. Examine the gel via ultraviolet light. Photograph with a transparent ruler to record the distance of each band of the marker DNA.
4. Assuming that 1-kb and 5-kb marker DNA have migrated to about 8 cm and 38 cm from the bottom, respectively, a 32-cm gel portion of 7-39 cm from the bottom (this portion contains BPB and XC at each end) will be used for in gel digestion. (In a similar way, for the longer fragments, a gel portion of 7-39 cm from the bottom containing 5- to 12-kb fragments is used.) Draw two lines as marks with dry ink on each gel holder at 7 cm and 39 cm from the top of the gel.
5. Draw BPB solution into a 1-mL plastic syringe fitted with a cutoff tip, and slowly expel the

gel by forcing through the blue BPB solution. Carefully let the blue solution to the first 7-cm line. Cut off the expelled 7-cm gel at a 45° angle of with a scalpel. Expel the gel portion of 39 cm into a 50-mL Falcon tube containing 20 mL of 1 × *Hinf*I buffer. When the BPB solution reaches the second 39-cm line, cut the gel at a 90° angle of.

6. Equilibrate the gel to 1 × *Hinf*I buffer at room temperature. Rock the tube occasionally and change the buffer two times at each 10 min. Incubate for 30 min in total.
7. Pour the gel noodle with buffer into a small tray placed at an angle about 10°.
8. Fit a cutoff tip to a 6-mL syringe and connect the tip to a 33-cm long 2.7-mm i.d. Teflon tube with a 3-cm long 3-mm i.d. silicon tube. Attach the Teflon tube to the edge of the noodle gel and slowly draw up the gel into the tube. Draw the buffer into the syringe and remove the buffer from the gel as much as possible.
9. Add 600 µL of lx *Hinf*I buffer containing 400 U of *HinfI* and 0.01% BSA to a 2-mL round-bottom microtube. Draw up the solution and fill the tube to cover the gel surface with the *Hinf*I mixture.
10. Remove the syringe and loop the Teflon tube by connecting one end to the other with the 3-cm silicon tube. Put the looped tube into a nylon bag and incubate at 37°C for 4 h.

Second Stage Technique

Prepare gels at least 1 day before starting the second electrophoresis. By the system shown in Figure elsewhere in this chapter 10 gels (33 cm wide, 46 cm long and 0.8 mm thick) are made simultaneously. The acrylamide solution is toxic; wear plastic gloves during the gel preparation.

1. Clean each glass plate meticulously and treat the surface of the beveled side with Repel Silane.
2. Set up the apparatus and seal the bottom holes with adhesive tape (Scotch 3M).
3. Mix 348 g of 30% acrylamide stock, 504 g of 5x TBE, 23.3 g of 50% glycerol, 15 g of 10% APS, and 1110 g of H_2O in a flask to make 5.15% acrylamide solution in 1.25× TBE. Divide into two 1-L vacuum flasks. We find that measuring these reagents by weight is more accurate than measuring by volume using graduated cylinders. This accuracy is important in ensuring the reproducibility of the electrophoresis steps.
4. Deaerate the solution by applying vacuum, gently at the beginning. Swirl the flask and continue deaeration until no more air bubbles are generated.
5. Add 210 µL of TEMED to each 1-L acrylamide solution and mix thoroughly.
6. Immediately fill the apparatus from the bottom entrance through a silicon tube. Clamp the entrance when the solution has reached to 1 cm from the top of glass plates.
7. Cover the surface by a gentle spraying of water-saturated 2-butanol.
8. After about 1 h of polymerization, replace the surface solution to 1 ×TBE.
9. Cover the TBE solution of the gel with Saran Wrap and store at room temperature until use. The gels are stable for about 5 days at room temperature. Avoid allowing the top of the gel to dry.

Electrophoretic Analysis

1. Melt the blue connection agarose and place the container in a heater block at 70°C.
2. After 4 h of digestion with *Hinf*I, expel each gel into a 50-mL Falcon tube containing 20 mL of 1 × TBE buffer.
3. After 10 min of equilibration, discard the buffer and pour the noodle gel onto a sampling platform laid on the top of the gels. Straighten the noodle with a ruler and slide to the edge of the platform.
4. Overlay the top of the gel with the connection agarose to reach to the beveled line of the glass plate.
5. Slip the noodle down along with the glass plate using a spatula and flow of melted blue connection agarose.
6. Cover the noodle enough connection agarose and solidify for 10 min. Remove the platform.
7. Add 2.5 L each of 1× TBE buffer to both reservoirs. Remove the 3M tape from the holes and contact the gel portion with the buffer.
8. Connect the electrodes to a power supply and run the gel at 150 V for about 40 h until the marker dye XC migrates to 35 cm.

Methods of Autoradiography

1. At the end of electrophoresis, turn off the power and disconnect the electrodes.
2. Dispose of electrophoresis buffer.
3. Lay the apparatus and remove the plastic cover.
4. Slowly pry apart the top glass plate from the gel using a metal spatula.
5. Place a piece of filter paper (35 cm × 43 cm, 2 cm wider than gel) on the gel. Apply gentle pressure so that the gel firmly attaches to the filter paper. Cut off the 3-cm portion of the gel with a scalpel.
6. Hold the edge of paper, and quickly flip the gel over and lay it on another piece of filter paper on a gel drier (place the gel on the top). Cover the gel surface with Saran Wrap. Try to avoid creases and bubbles.
7. Dry the gel under vacuum on a gel drier set at 80°C.
8. Put an ID number of the gel on the edge of the filter paper.
9. Establish an autoradiography by exposing the gel to X-ray film and intensifying screen in a metal cassette at −80°C for about a week.
10. Develop the autoradiogram.

Analyzing Results

A digitized image of a BALB/c mouse RLGS pattern is shown in Figure elsewhere in this chapter this method, end labeled DNA fragments are visualized by autoradiography. The spot intensity represents the copy number of the spot DNA. Except for sexchromosome-derived fragments and

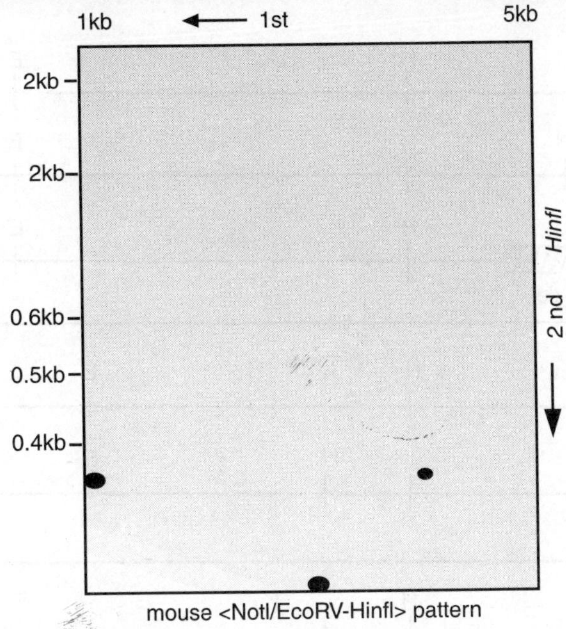

Figure 12.6: A digital image of RLGS pattern of BALB/c mouse. Fragment sizes shown for each dimension are based on sequence data of cloned spots.

multiple-copy fragments such as ribosomal DNA, the intensities of most spots in an RLGS pattern represent two copies of autosomal DNA fragments.

Genomic mutation detectable by this method consists of alterations in the length of the DNA fragments visualized, either by insertions or deletions within these fragments, or base changes at the recognition sites of the restriction enzymes used. Mutations can be detected by searching for new spots in patterns of progeny not present in either parent, or by detecting half-intensity (one-copy) spots in the progeny that are two-copy in the parents.

A mutated fragment will shift position on the gel if the distance from the *Notl* site to the closest *Hinfl* or *EcoRV* site is altered. The expected positions of mutant fragments are schematically represented in Figure elsewher in this chapter. In some cases, the altered fragment may not be found on this gel (although it might appear on a different gel) or may produce no spot if the *Notl* site is involved in a deletion or base substitution. In any event, the intensity of the normal spot should decrease by 50% in theory.

To analyze the complex 2-DE patterns, computer programs have been utilized. The computer-based image analysisdetect the presence or absence or any changes in intensity of each spot; thus, we can detect deletion, amplification, insertion, and rearrangement throughout the genome efficiently. Examples of mutations detected in our mouse study are shown in Figure elsewhere in this chapter

Additional Hydrolytic

The combination of restriction enzymes-Notl, *Eco*RV, and *Hinfl*-gave a mostly evenly distributed spot pattern for human and mouse (rat also) DNA. The most reliable landmark enzyme currently

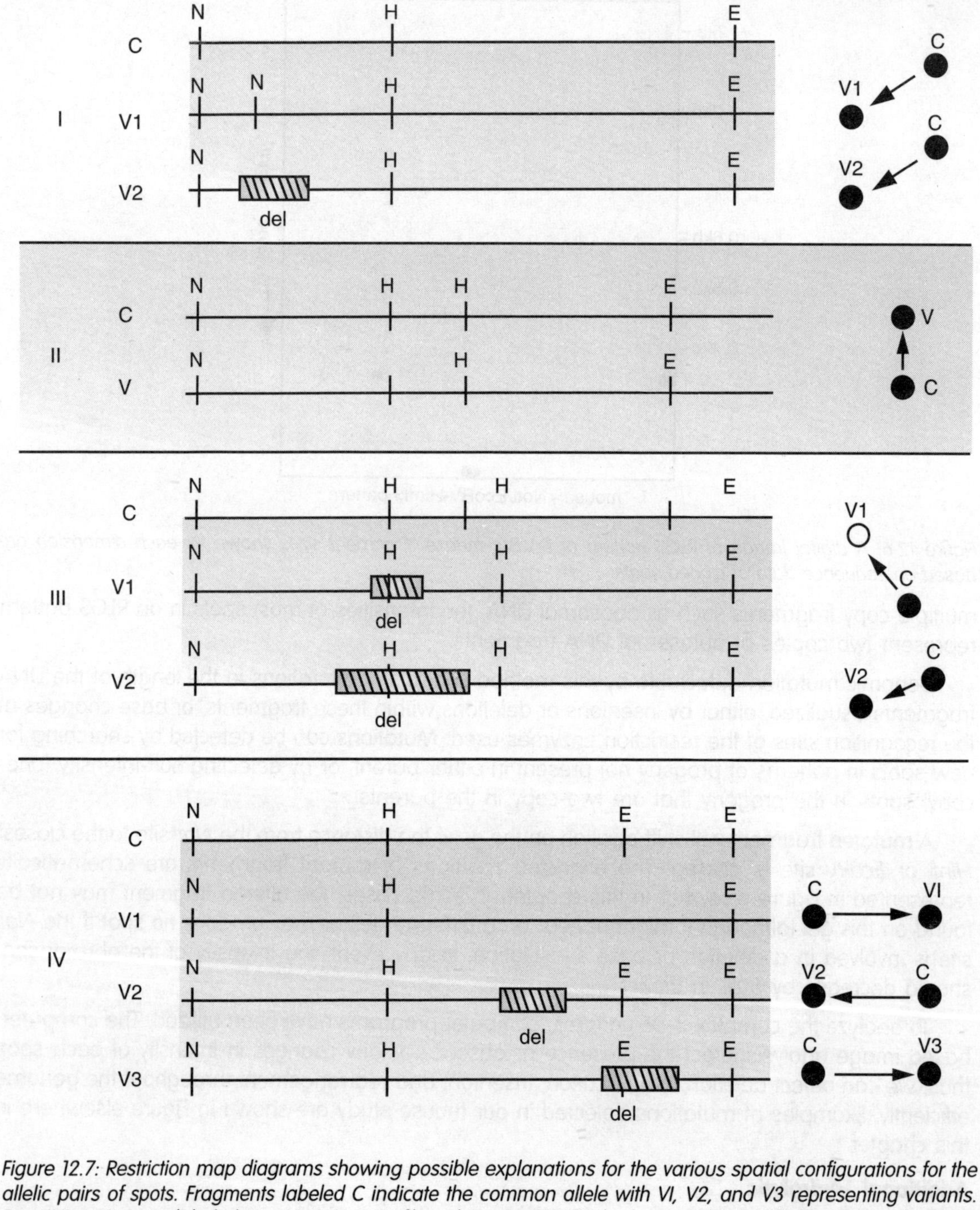

Figure 12.7: Restriction map diagrams showing possible explanations for the various spatial configurations for the allelic pairs of spots. Fragments labeled C indicate the common allele with VI, V2, and V3 representing variants. Restriction sites are labeled: N = Not1, H = Hinf*I, and E = EcoRV. Hatched rectangles labeled represent deletions.*

available for human and mouse is *NotI* in our experience (we examined more than 10,000 gels). As for the second enzyme (in the standard method; *Eco*RV), *PstI* (C*TGCAG) and *Pvu*II (CAG*CTG) are both usable.

As for the third enzyme for *in situ* digestion, *Mbo*I (*GATC), *Bam*HI (G*GATCC), *PstI*, and PvuII are usable (although *Mbo*I is somewhat expensive). These enzymes are usable in any combination without problems. RLGS can be applicable to other species and it is easier if the genome size of the study species is smaller than human or mouse.

We have examined the following species: (1) *Medaka* (a fish; the genorne size is 1 ×10^9 base pairs [bp]): *NotI*, *Eco*RV, and *HinfI* combination , (2) *Sacchromyces* (yeast; 3 × 10^7 bp): *Hind*III (A*AGCTT; used as landmark enzyme, labeled with only [^{32}P]dATP) and *HinfI*, also *Hind*III and *PstI*, and (3) E. *coli* (4 ×10^6 bp): *Eco*RI (landmark enzyme, labeled with only [^{32}P]dATP) and *HinfI*.

For the preparation of RLGS for small-genome species, many other enzymes can be used as the landmark enzymes, but for *in situ* digestion in agarose only limited enzymes, the five aforementioned enzymes are practical.

POINTS TO REMEMBER

1. Intact high molecular weight DNA is very viscous and it is often difficult to measure its volume with a pipet. In such a case, digest the genomic DNA with *Eco*RV. Digest approx 5-10 Vg of DNA with 5 U of enzyme in 100 µL 1 × high buffer for 1 h at 37°C. Ethanol precipitate by adding 10 µL of 5 M NaCl and 250 µL of ethanol.

   Recover the partial digest by centrifugation and dissolve in 100 µL of TE. Estimate the concentration by measuring the OD_{260}. Ethanol precipitate the same again and dissolve in

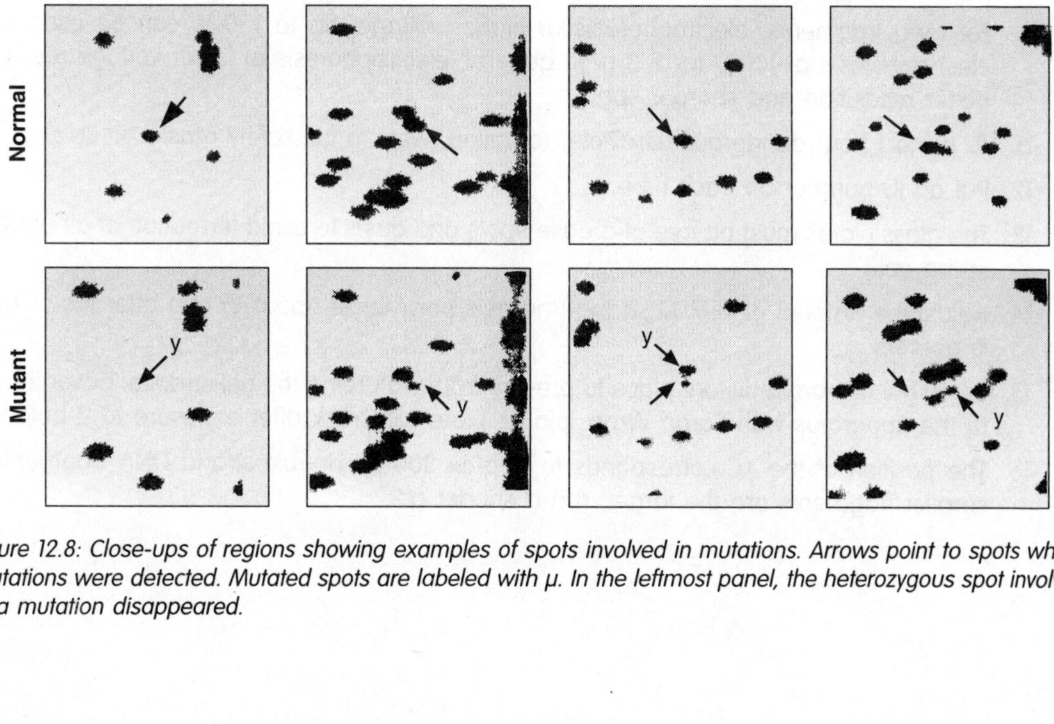

Figure 12.8: Close-ups of regions showing examples of spots involved in mutations. Arrows point to spots where mutations were detected. Mutated spots are labeled with µ. In the leftmost panel, the heterozygous spot involved in a mutation disappeared.

TE at a concentration of 300 μg/μL. Use this partially digested DNA instead of the high molecular weight DNA.

2. Unless stated otherwise, all buffers and solutions are prepared with Milli-Q H_2O that has a resistance value of 18 MΩ. Deionized and distilled H_2O is also usable.
3. Rstriction enzymes (*NotI*, EcoRV, and *HinfI*) manufactured by NEB are also usable. All enzymes should be stored in a −20°C freezer without defroster.
4. Measure the weight of the agarose solution before and after boiling and adjust the weight by adding H_2O after boiling.
5. Acrylamide and bis-acrylamide are often contaminated with metal ions. Stock solutions can be purified by stirring overnight with 1 g of a mixture of ionic exchange resins (Amberlite IRN-150L, Amersham) per liter of solution. Fresh solutions should be prepared every few months.
6. The digests can be stored at 4°C for several days before isotope labeling; do not freeze.
7. Carefully examine the gel for air bubbles. If any are present in the gel, it may be necessary to prepare another gel.
8. The final volume of the reaction is 21.9 μL in theory, but the volume is decreased to approx 21 μL or less in practice.
9. Examine any air bubbles in the buffer on the gel; if present, carefully remove by drawing up with a 90°-cut long needle fitted to 1-mL syringe.
10. The higher the temperature, the faster the dye and DNA fragments will move, and thus there are variations in the duration of electrophoresis. In a rush experiment, electrophoresis for short fragments can be done at 200-250 V after running at 130 V for 2 h.

    For long fragments, electrophoresis at higher voltage, up to 150 V, can be used after electrophoresis at 100 V for 2-3 h. In general, electrophoresis at lower voltage results in better resolution and sharper spots.
11. To protect from dangerous ultraviolet radiation, wear a full safety mask.
12. Put an ID number on each tube.
13. The glass plates must be free of grease spots and dusts to avoid formation of air bubbles in the gel.
14. Adjust the amount of TEMED so that the gels polymerize about 10 min after the solution is poured.
15. Spray gently from a distant place to prevent from disturbing the gel surface. Cover the top of the apparatus with Saran Wrap: plastic plate will crack after exposure to 2-butanol.
16. The position of the XC corresponds to approx 300-bp double-strand DNA fragments. If smaller fragments are the target, run a shorter gel.

Chapter 13

Transcription Factors

One of the major cell mechanisms for gene expression control is at the level of transcription. Transcription factor (TF) DNA-binding proteins recognize relatively short DNA "signals" (typically 6-12 bp) in the vicinity of transcription start sites (TSS) and initiate (activators) or repress (repressors) gene transcription. Although each TF has a preferred set of transcription factor binding sites (TFBS), this set is usually degenerate. Figure 13.1 A shows an example of 30 aligned rodent binding sites of a TF.

Once such a set of sites has been identified (usually with some biochemical method like DNA footprinting, SELEX, ChIP, and so on) then the information of their alignment can be organized and subsequently used to search for more TFBS in the promoters of other genes. The most widely used method for information encoding of an alignment of TFBS is the position-specific scoring matrix (PSSM).

A PSSM model is a $4 \times L$ weight matrix *(L is* the length of the target site of the particular TF) in which each column reflects the observed frequencies of each of the four bases in the particular target position. Figure 13.1 B shows the count matrix of the alignment and Figure 13.1 C the corresponding log-transformed matrix. For practical reasons, this is the most commonly used form of PSSM models, sometimes also corrected for the background frequencies.

The log-frequency values have been shown to correspond to the binding energies for some TFs. The binding preferences of a TF can be graphically represented by a LOGO. Each position is represented by a stack of symbols. The total height of the stack represents the information content at this position (i.e., how conserved this position is); and the height of each symbol in the stack corresponds to the relative frequency the corresponding base appears in this position in the set of known sites. Figure 13.1 D shows a LOGO example for the 20 aligned sites.

Even with the knowledge of the binding preferences of a TF (e.g., in the form of a PSSM model) the computational identification of new DNA targets becomes difficult because of the short length of the "signals," the degeneracy of the patterns and the quality of the PSSM models. This is especially true for searches on mammalian promoters, where regulatory elements can be observed many kilobases away from the TSS of the gene, thus increasing the search space.

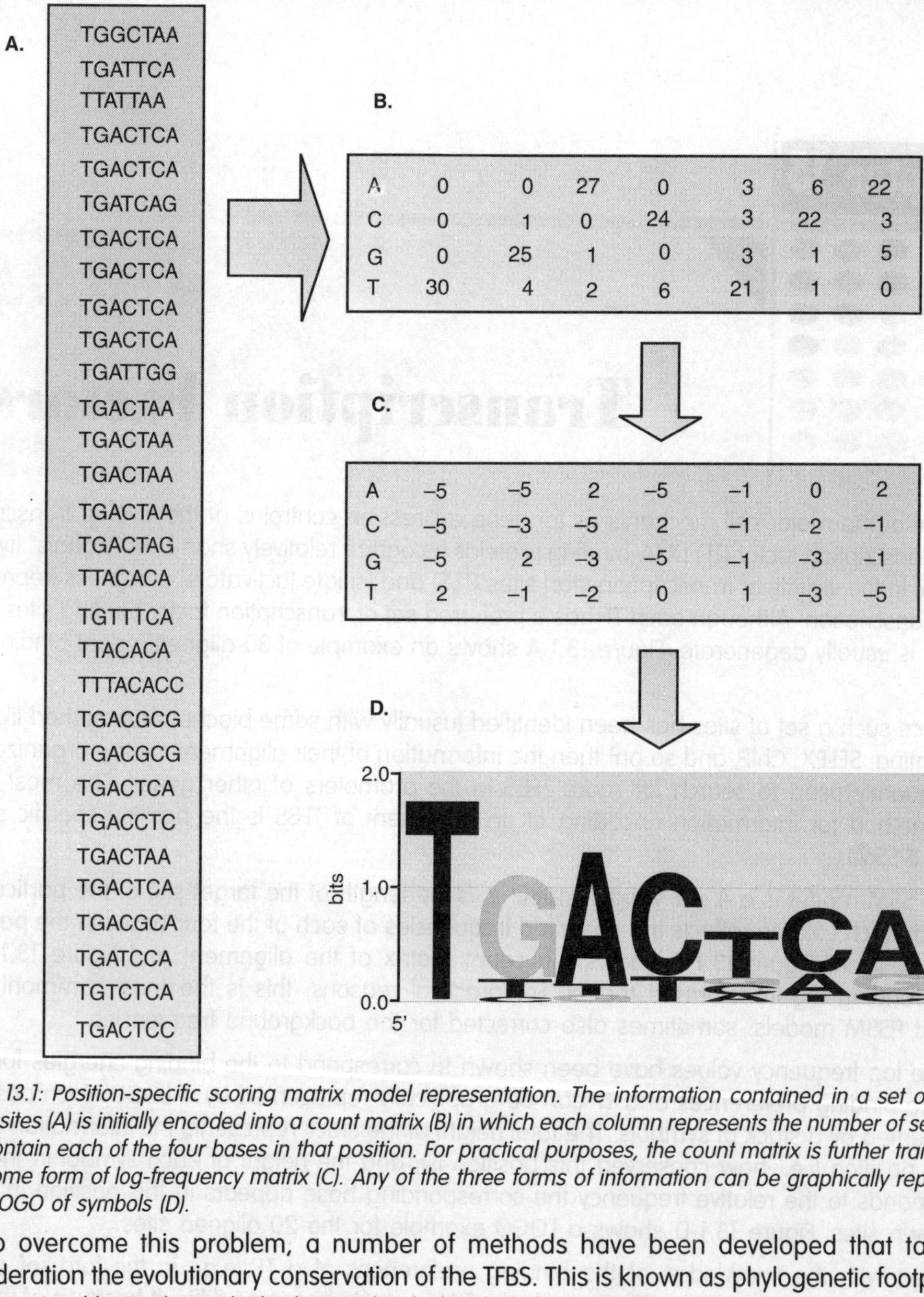

Figure 13.1: Position-specific scoring matrix model representation. The information contained in a set of aligned target sites (A) is initially encoded into a count matrix (B) in which each column represents the number of sequences that contain each of the four bases in that position. For practical purposes, the count matrix is further transformed into some form of log-frequency matrix (C). Any of the three forms of information can be graphically represented as a LOGO of symbols (D).

To overcome this problem, a number of methods have been developed that take into consideration the evolutionary conservation of the TFBS. This is known as phylogenetic footprinting, a term coined by Tagle et al. The basic idea behind phylogenetic footprinting is that the biologically important DNA cis-regulatory elements will be conserved throughout evolution. In other words, they should occur in roughly the same location in the promoters of two (or more) orthologous genes.

Clearly, this method is going to miss TFBS that have been acquired by an organism in a more recent time, but it is expected that the number of false-positive predictions will be significantly reduced.

This is important for the biologists when they design experiments to test putative binding sites. These tests are usually time-consuming and the focus in a smaller set of high-probability candidate sites is desired.

In this chapter we will focus on the practical aspects of FOOTER, the phylogenetic footprinting algorithm designed for searching for binding sites of known TFs in the promoters of mammalian genes. FOOTER is not designed for the identification of novel (unknown) DNA patterns in the promoters of multiple coexpressed genes. The detailed description of the algorithm and a brief introduction to the web tool has been done elsewhere.

FOOTER differs from other similar methods in two ways: one is that it concurrently evaluates both the location of a putative TFBS in the promoter of the gene as well as its PSSM score conservation. In other words, a site is considered to be "true" if the location *and* the PSSM score are conserved between species. Second, the evaluation of these two criteria is quantitative and a weighted average *p*-value (WAP) is calculated.

REQUIREMENTS

1. System configuration. The FOOTER web tool and necessary databases are stored on our Dell PowerEdge 2650 server machine consisting of dual 2.8 GHz Xeon processors with HT technology and 2 GB of RAM. The algorithm is written in Perl (v5.8.0) using the PG package with a web interface written in PHP (v5.0.3) and associated with MySQL 4.1 database. Alignment is currently performed with program DBA.
2. Web-tool accessibility. The FOOTER web tool is available over the web at http:/ /biodev.hgen.pitt.edu/Footer/. Although it works with virtually any web browser, the application has been optimized for Firefox v1.0.7. Internet browsers are freely available for download from http://www.mozilla.com/firefox/ (Firefox), http://browser.netscape.com (Netscape), and http://www.microsoft.com/windows/ie/ (Microsoft Internet Explorer).

PROCEDURES

The FOOTER algorithm has been developed for the efficient identification of evolutionary conserved DNA cis-regulatory elements. Currently, it runs on human and rodent promoters only. Each of the human and rodent promoters is scanned independently for candidate TFBS from a set of 95 mammalian TFs with known binding preferences.

The PSSM models we use are derived from selected target sequences deposited in TRANSFAC database. Whenever enough human and rodent sequences are available in TRANSFAC, species-specific PSSM models are constructed; otherwise, a mammalian-specific PSSM model is used.

By species- or mammalian-specific model, we mean that the target sequences are in fact biochemically verified targets in human, mouse, or rat genes. We decided to follow this approach instead of general all-species matrix that other programs use because we have found species-

or class-specific differences in the binding preferences of some TFs. The top 10 scoring sites for each 3 kb of analyzed sequence in each promoter are retained for further analysis.

An alignment between the two promoters provides some guidance as to where the conserved sections are located and helps determine the distance between putative sites, while correcting for local insertions/deletions. The idea behind that is that the "location conservation" is biologically important because TFs usually act in concert with other TFs and other proteins. Thus, the local conservation is more important than their absolute distance from the TSS.

This is more easily understood if one thinks about the TFBS, which are located few kilobases away from the TSS. It is important to note that the promoter alignment is only used to point to regions of conservation. However, because each program has its limitations, so does DBA (the alignment program we use). Thus, we allow putative sites to be subsequently analyzed without the restriction of being in a DBA-"conserved" region.

The subsequent analysis consists of comparing pairwise all putative sites in the two promoters and scoring their similarity according to the two criteria: (1) their relative distance in the two promoters (determined locally by the conserved regions' boundaries) and (2) their relative PSSM scores according to the species- (ideally) or mammalian-specific matrices. For each of the two criteria, a *p*-value score is assigned that reflects the probability of observing the corresponding distance and PSSM scores merely by chance.

The two *p*-values are weighted and combined in a single WAP score. The pairs with the best (lowest) WAP scores are reported as true binding sites. An outline of the work-flow of the web tool is presented in Figure 13.2.

Methods of Feeding

1. Promoter sequences. FOOTER requires the input of two DNA promoter sequences in FastA format. According to the FastA format the sequence is preceded by a single line starting with ">" followed by a sequence name or identifier. The rest of the lines contain the raw DNA sequence. One of the input sequences should be human and the other should belong to a rodent; each being placed into the proper field for that species.

   The sequences can either be copy-pasted (directly) or uploaded as a text file. FOOTER has been shown to be very successful in detecting cis-regulatory signals in large promoter sequences.

2. Selecting transcription factors. FOOTER has PSSM models for 219 TFs, including most of the well-known TFs such as Sp1, EGR-1, and NF-κB. The TFs have been classified into 15 families, based on their structural properties. The individual factors can be found within a folding tree consisting of the protein families. Factors can be selected individually by their checkboxes, or as a family (i.e., all TFs belonging to that family) by the family checkboxes.

   The users also can select to search for all available TFs by the "check all" function. If a user wants to select a particular factor but does not know its exact name or which family it belongs, there is a string search function, which returns the TF names that contain the searching string in their name or in one of the synonyms (a TF can have multiple synonyms). Each factor name is hyperlinked so that the users can obtain information about the TFs synonyms as well as a graphical representation of the available PSSM model(s).

A. Promoter Alignment

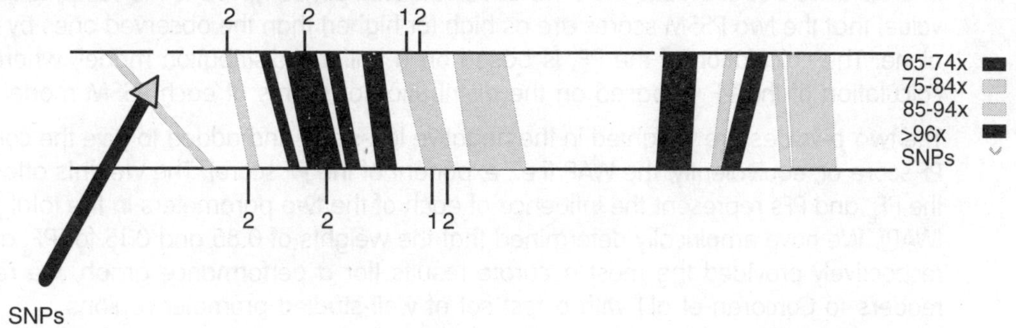

B. Table of Results

Binding Sites SNPs	All Found Human SNPs	All Found Rodent SNPs				
Factor	Human Site	PosH (orient,)	Rodent Site	PosR (orient.)	Score	WAP
(1)HNF-14	CTTCTGCAGGGCCGAGC AAAAATATTAAATGGAAG CAAATGAAGCATCG	1755(-)	CTTTTGCAAAGCACAGA AAAAATATTAAATGGGAA TAAATGAAGCGTCG	1886 (-)	9.5	0.00007
(2)HNF-3B	TAATTGGTTTACAGCCTT TTTTGTTTATTTATCCAT AAGAGCTGCTGTTAA	1225(-)	TAATTGGTTTACAGCCGT TTTTGTTTATTTATCCAT AAGAGCTGCTGTTAA	1343 (-)	9.9	0.00005
(2)HNF-3B	TTCTCATTCCACACATCA CATCGTTTGCTTTACAGA ATTTTCCAACTCAAG	825(-)	GTACTCTCATGAAATATA CATGGTTTGCTTTCTTGG GTTTTCTCAACTCAG	935 (-)	9.5	0.00007
(2)HNF-3B	TGGTTTACAGCCTTTTTT GTTATTTATCCATAAGA GCTGCTGTTAAATGG	1229(-)	TGGTTTACAGCCGTTTTT GTTTATTTATCCATAAGA GCTGCTGTTAAATGG	1347 (-)	9.5	0.00008
(2)HNF-3B	TGCAATTAGTTGAGAAGT GCAAAGTAAACACTCCGG GGGTGGGCAGGCGAGG	1830(+)	TGCAATTAGACCAGAAGT GCTAAGCAAACATCCTGG GGTGTGGGTTAGGCA	1963 (+)	8.6	0.00019

Figure 13.2: Example of FOOTER results. FOOTER results page consists of primarily two sections. The upper section presents a color-coded alignment of the two promoters, with different colors denoting different conservation percentages. The lower part of the output presents the predicted sites in a tabular format. This table can be copy-pasted in and analyzed further with any spreadsheet program.

3. Single-nucleotide polymorphisms (SNP) presentation. FOOTER has recently made available a function that will allow the identification and presentation of known SNPs in the examined promoters. The SNPs are derived from the dbSNP database. Though the presence or absence of SNPs in the promoter region(s) does not weight into the FOOTER analysis, it can provide useful information to the user about possible sites that could be affected by a polymorphism.

Parameters

1. PF_d and PF_s weights. PF_d is the tail probability (*p*-value) that two sites are found by chance

in a distance equal or less than the observed one. Similarly, PFs is the tail probability (*p*-value) that the two PSSM scores are as high (or higher) than the observed ones by chance alone. The calculation of the PF_d is based on a uniform distribution model, whereas the calculation of the PF_s is based on the distribution of scores of each PSSM model.

The two *p*-values are weighted in the negative log-scale and added to give the combined PF score or, equivalently, the WAP (i.e., exponent of the PF score). The weights attached to the PF_d and PFs represent the influence of each of the two parameters in the total *p*-value (WAP). We have empirically determined that the weights of 0.85 and 0.15 for PF_d and PFs, respectively provided the most accurate results (for a performance graph, we refer the readers to Corcoran et al.) with a test set of well-studied promoter regions.

2. WAP. The WAP represents the probability that a pair of putative binding sites could have been found in the observed distance (or closer) with the observed PSSM scores (or better) merely by chance. The two distributions that are taken into account for this value are the PF_d and PF,.

   The weight given to each of these two distributions in the calculation of the WAP can be adjusted. Based on our previous tests we have empirically determined that a WAP cutoff of 5×10^{-4} provides the most accurate results (for a performance graph, we refer the readers to Corcoran et al.). This parameter can also be adjusted on the results page.

3. Number of seed sites. The seed parameter adjusts the number of potential binding sites of each TF that FOOTER will retain for subsequent analysis in each promoter. We have decided to retain "blindly" *a* particular number of sites per kilobase of DNA sequence rather than setting a strict threshold in the PSSM score for selecting sites. The reason for that is that we would like to allow more suboptimal sites to be subsequently compared to the other predictions.

   The idea is that if two sites are in the same location and similarly "suboptimal" (according to their PSSM scores), then these might be biologically relevant sites. By default, FOOTER retains one seed site in each promoter for every TF for every 300 bp. For a 3000-bp long promoter (the size we usually examine) this corresponds to 10 seed sites.

Interpretation

1. Interpreting the results. An example of the FOOTER results page is presented in Figure 13.3. The output is in both tabular and graphical format showing the reported TFBS predictions. The table lists the TF name, target sequence, flanking sequence, position, and WAP for each pair (human-rodent) of predicted sites that have met the user-specified WAP cutoff.

   The produced PNG image displays regions and percent of conservation between the human and rodent promoter sequences as calculated by the DBA, program and the reported binding sites. Also available on the results page are links to the following text files: (1) the input sequences, (2) the DBA alignment, (3) the complete FOOTER output (including sites that did not meet the reporting criteria).

2. Adjusting the WAP. On the results page the user can adjust the WAP value threshold, which will reproduce the table and PNG image showing the predicted binding sites that

meet the new cutoff. The recalculation of the reported sites is done without the need of rerunning the algorithm.

3. SNPs. If the user had selected to search the human and rodent sequences for SNPs, their location will be displayed in the PNG image. If any SNPs fall within 20 by of a reported predicted site, the SNP will be identified as a highlighted hyperlink in the tabular results. The link will redirect the user to the dbSNP information page for that particular SNP. Two HTML pages will also be produced that show every SNP found as a highlighted hyperlink within the individual promoter regions.

Web Tools

FOOTER web tool includes help pages that describe its main features, the sequence input, the parameters and the output of the program. Another page describes the FastA format (input sequences) in detail. In the main page, an *e*-mail address *is* also listed for help, comments, and *so* on. Finally, a "Load example" option provides the user a first experience with the program.

POINTS TO REMEMBER

1. The methods that are using prior information, like PSSM models, to search for new DNA motifs are limited by the quality of the PSSM models. There are several factors that affect the quality of a PSSM model. One is the number of sequences that are available. FOOTER reports the number of sequences in the graphical display of LOGOs (just click on the TF name.) Another factor is the degree of conservation of the pattern, or equivalently, its information content.

   As a general rule, the higher the information content, the better the model. The information content is measured in bits of information, which can be calculated from the LOGO graphical representation of the pattern. For example, EGRI pattern has almost 13 bits of information whereas GATA-3 has 9 bits. Finally, the quality of a PSSM model can be influenced by the number and diversity of species that contribute sequences to it.

   Because TFs are proteins that change through time, the same is true for their binding preferences. Thus, a PSSM model derived from sequences from a variety of evolutionary distant organisms (e.g., mammalian, avian, and amphibian) can obscure the mammalian motifs, resulting in an increased number of false-positive predictions. FOOTER uses species-specific (when possible) or mammalian-specific PSSM models.

2. Because phylogenetic footprinting methods depend on evolutionary conservation to distinguish between true, biologically important DNA "signals" and spurious events, they are going to miss functional *cis*-regulatory motifs that have been recently acquired by an organism. In our case, FOOTER is going to miss the motifs that human or rodents acquired after their last common ancesor.

3. FOOTER is following a single-gene multiple-species approach. Thus, it is not designed to find motifs in the promoters of multiple coexpressed genes, e.g., from microarray experiments (multiple-genes single-species approach). There are a number of other methods available for this purpose.

The difference between the two approaches is that the first aims to identify targets of known TF proteins, whereas the latter identifies DNA "signals" that are presumed to be targets of some (usually unknown) TF. Recently, a hybrid approach was published. The SOMBRERO algorithm, a self-organizing map method, used prior information (PSSM models of familial binding profiles) to search the promoters of coexpressed genes. Thus the predicted motifs were associated with the TF where the initial familial binding profile came from.

4. When one tries to predict DNA regulatory sites in the promoters of genes, it is always useful to run different computational tools and compare the results before one decides to embark in time-consuming biochemical experiments. Besides FOOTER, there are two other methods that offer web-accessible tools for searching mammalian promoters for binding sites of known TFs: ConSite (http://www.phylofoot.org/consite/) and rVista (http://rvista.dcode.org/).

   Both algorithms evaluate the candidate sites qualitatively, taking into consider-ation mainly the position of the site in the promoter (the PSSM score is used to identify the candidate sites based on a threshold). ConSite uses mammalian-specific matrices to search mammalian promoters, whereas rVista uses all-species matrices. In a recent comparative analysis, FOOTER outperformed both ConSite and rVista, but the latter algorithm has been updated since.

5. The correct identification of the promoter sequences is important for FOOTER. In the past, we had offered an option for the automatic identification of the promoter sequence of human and mouse genes from a single protein sequence (human or mouse). However, we found that this approach was problematic. Currently, we direct the FOOTER users to obtain the promoter sequences themselves.

   The recommended source of orthologous promoter sequences is the TRED database (http://rulai.cshl.edu/cgi-bin/TRED/tred.cgi?process=searchPromForm). If the promoter of interest is not in the TRED database, then we recommend to the users to obtain the promoter sequences from one of the publicly available genome browsers, like the UCSC Genome

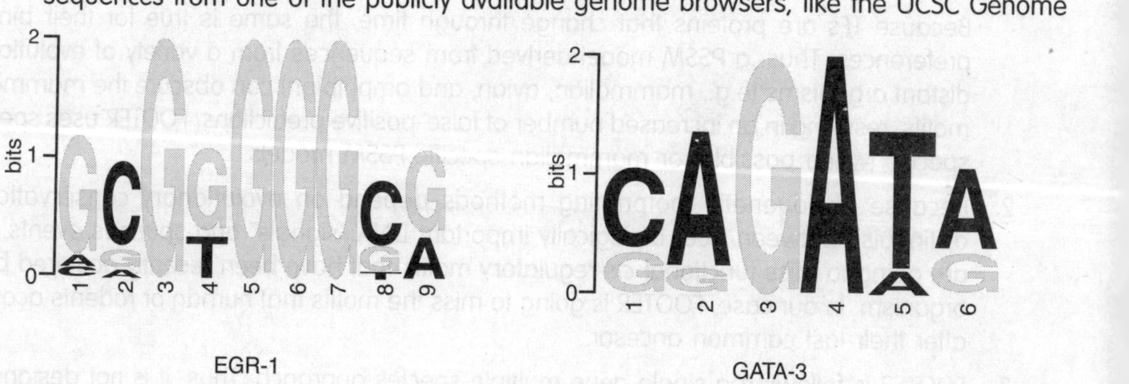

Figure 13.3: Examples of LOGOs. The information content of the various transcription factors is indicative of the expected frequency of its target sites in the genome. The examples of two well-known factors are presented here for comparison. Based on that, GATA-3 sites are expected to be more frequent (by chance) in the genome than the EGR-1 sites.

Browser (http://genome.ucsc.edu/cgibin/hgGateway) or the EnsEMBL (http://www.ensembl.org/).

6. Promoter length. FOOTER has been tested on various promoter lengths and it has been shown to increase its sensitivity (sensitivity is the percent of true [confirmed] sites that FOOTER identified) with the length up to 3 kb. This is partly because the number of seed target sites that are retained for further analysis is proportional to the analyzed promoter length.

   Thus, as the promoter length increases, more seed sites are retained. FOOTER has not been tested in promoters with length more than 3 kb, but longer promoters might cause problems related to the reliability of the alignment. Usually, the sequence conservation decreases when the distance from the TSS increases. Thus, for comparison of longer promoters (e.g., 5, 10, 20 kb), it might be useful if the promoters are "split" into smaller, overlapping pieces of 2-4 kb.

   Promoter pieces in one species that do not share any similarity to the promoter pieces in the other species (typically, long insertions) can thus be omitted from the analysis, resulting into more accurate predictions.

Chapter 14

Phylogeny of Chromosomes

During the last two decades, *comparative cytogenetics* and genomics has evolved from a specialized *discipline* to a highly *dynamic* field of research. This development was driven by major technological advancements as well as emergence of the deeper insight that many aspects of human genome function can be better understood when information about its *evolutionary* changes is taken into account.

Whole-genome *sequencing* projects of *biomedical* model species and domesticated animals provided important clues to the molecular mechanisms that shaped the human genome. These strategies were *complemented* by the launch of the *chimpanzee* genome project, leading to the recent publication of the first chimpanzee draft sequence and its alignment with the human reference sequence. The objective of this chapter is to:

(1) review recent technical developments in the field of comparative cytogenetics and genomics provide knowledge about ancestral primate chromosomal traits and on evolutionary landmark rearrangements;

(2) recapitulate the molecular cytogenetic evidence for the evolutionary history of human *chromosomes* with special *emphasis* on great apes; and

(3) summarize the present data about the genomic environment of evolutionary chromosomal breakpoints in human and great apes.

Rates and the direction of evolutionary rearrangements are discussed in the context of the *emerging* patterns of evolutionary genomic changes and future perspectives including whole genome sequencing and microarray based *approaches* are *addressed.* The introduction of fluorescence *in situ* hybridization (FISH) to *comparative karyotype* analysis marked a *paradigm* shift from the analysis of chromosome *morphology* toward *chromosomal* DNA content.

The use of human chromosome-specific probes in cross-species chromosome painting experiments for the first time allowed the secure identification of chromosomal *homologies* between species at a resolution of 3-5 Mb within *primates* and 5-10 Mb when comparing human with *nonprimate* mammals.

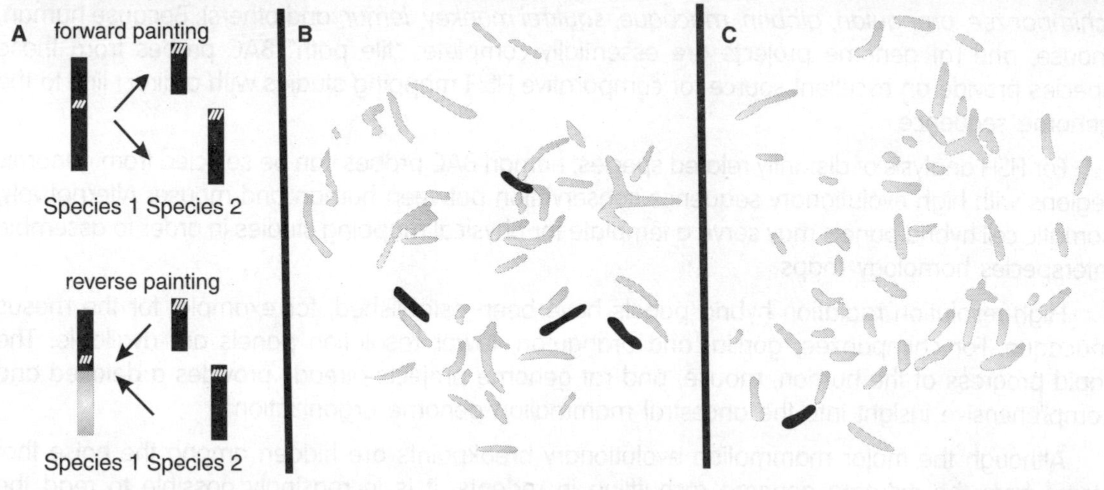

Figure 14.1: (A) The principle of reciprocal chromosome painting. (B,C) Reciprocal chromosome painting between human and orangutan: (B) in situ hybridization of human chromosome 2 probe to an orangutan metaphase delineates two homologs. (C) Reverse painting of the orangutan 2q homolog to human chromosomes hybridizes to the 2q 13-qter segment.

A second landmark was the establishment of a *reproducible* Zoo-FISH protocol for the analysis of any mammalian species with human probes. In addition, *chromosome-specific "painting"* probes were established from several nonhuman primates as well as from other mammalian and vertebrate species by *fluorescence-activated* chromosome *sorting* and subsequent degenerate oligonucleotide-primed polymerase chain reaction *amplification*, for example, *from lemurs, mouse,* and *chicken*.

The availability of chromosome paint probes from nonhuman species for ZooFISH experiments allowed the assembly of *comparative* chromosome maps in a reciprocal way. This approach is particularly helpful when *analyzing* distantly related species in which the hybridization efficiency is low. Moreover, when performing reverse painting to human chromosomes, evolutionary breakpoints can be localized on the human sequence map for a subsequent detailed characterization of evolutionary breakpoints.

Further development of this strategy led to the concept of multidirectional chromosome painting, where members of a species group of interest are *systematically* analyzed with human paint probes and in addition with paint *probes* of a species from the targeted species group. Cross-species chromosome painting certainly has a number of *limitations*: apart from the limited resolution, *intrachromosomal* rearrangements usually *escape* detection.

For the construction of more detailed *comparative* genome maps, several *complementary* methods have been established, among others Zoo-FISH employing chromosome bar *codes* and *subregional* DNA probes, *comparative* radiation hybrid mapping and, ultimately, whole genome sequencing. In order to obtain *high-resolution* comparative chromosome maps, Zoo-FISH with *vectorcloned* DNA probes can be performed.

Bacterial artificial chromosome (BAC) *libraries* from *numerous* species are *publicly* available, among them human, chicken, mouse, rat, cat, pig, and cow, and several *nonhuman primates* (*chimpanzee, orangutan, gibbon, macaque, squirrel monkey, lemur*, and others). Because human, mouse, and rat genome projects are essentially complete, "tile path" BAC probes from these species provide an excellent source for comparative FISH mapping studies with a direct link to the genome sequence.

For FISH analysis of distantly related species, human BAC probes can be selected from genomic regions with high evolutionary sequence conservation between human and mouse. Alternatively, somatic cell hybrid panels may serve a template for physical mapping studies in order to assemble interspecies homology maps.

High-resolution radiation hybrid panels have been established, for example, for the rhesus macaque. For chimpanzee, gorilla, and orangutan, lower resolution panels are available. The rapid progress of the human, mouse, and rat genome projects already provides a detailed and comprehensive insight into the ancestral mammalian genome organization.

Although the major mammalian evolutionary breakpoints are hidden among the noise that stems from the extreme genome reshuffling in rodents, it is increasingly possible to read the human/mouse sequence alignment like cross-species chromosome painting data.

With other more recently launched genome projects (i.e., chicken, dog, and chimpanzee [12]) advancing even faster, the view becomes increasingly clear. For example, the fusion point of human chromosome 2 in 2q 13 can readily be identified on the human/chimpanzee comparative sequence map at the junction of chimpanzee chromosome 12 and 13 homologous sequences.

PRIMATE KARYOLOGY

Because chromosomal homology maps between human and approx 30 nonprimate mammals and more than 50 primates have been established, the data set available provides a firm basis on which proposals for common ancestral mammalian chromosomal traits and shared derived primate-specific chromosome forms can be made.

For the sake of clarity and simplicity, in the following sections chromosomes are always numbered according to their human homologous counterparts. When comparing the karyotypes of species from different placental mammalian orders, a surprisingly high degree of conservation can be observed for the majority of species.

The homologs of human autosomes 1, 5, 6, 9, 11, 13, 17, 18, and 20 are found conserved as separate entities in several different orders and are therefore assumed to represent ancestral mammalian chromosome forms. In addition, human chromosome 3, 4, 14, 15, and 21 homologs are entirely conserved in other mammalian orders, however, translocated.

Syntenic associations of human homologous chromosomes 3/21, 4/8p, and 14/15 are, therefore, also considered to be ancestral for placental mammals. Human chromosome 2, 7, 8, 10, 16, and 19 homologs are found split in two separate syntenic segments, some of them are associated with further *chromosomal* material.

Human chromosome 2 is a fusion product of two separate ancestral homologs 2pter-q 13 (2a) and 2q 13-qter (2b). Human chromosome 7 is a complex fusion product of ancestral chromosome forms (7p21 -q 11.21, 7q 11.23-21.3, 7q22.1-qter) and (7pter-p22, 7q 11.21-11.23, 7q21.3-

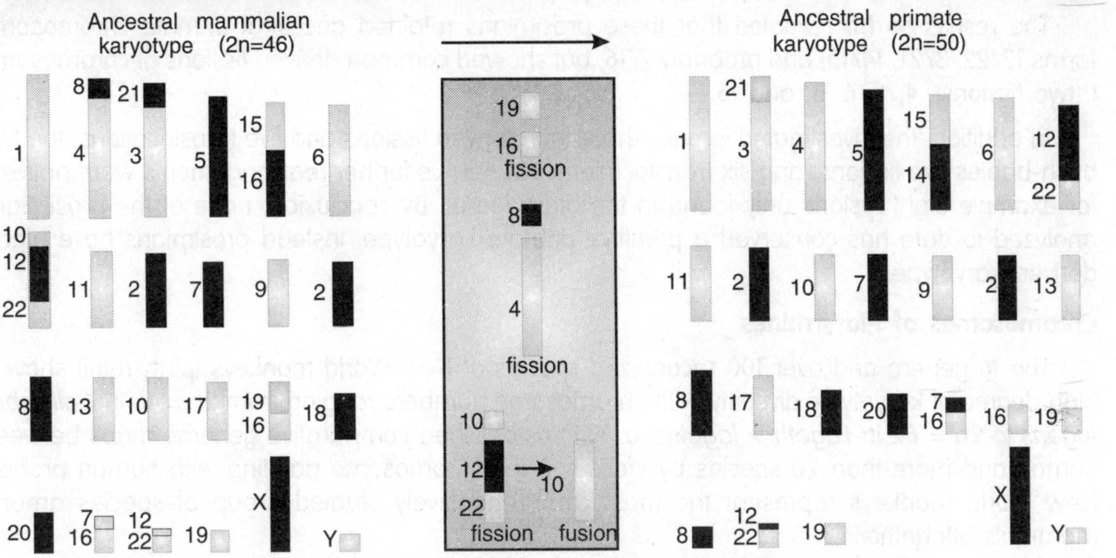

Figure 14.2: Inferred ancestral mammalian and ancestral primate karyotypes with the assignment of human homologies to the left, based on multidirectional chromosome painting data. The primate karyotype is derived by fissions of I Op and 12pter-q23.3, 16q and 19q, and 4 and 8p, followed by fusion of 10p/ 10q.

22.1)/I 6p. Human chromosomes 12 and 22 resulted from a reciprocal translocation t(12;22) (q23.3;q12.3) of the two ancestral homologs I Op/I 2pter-q23.3/22q 12.3-qter and 22q 11.2-22q 12.3/12823.3-qter.

Chromosome 16q and 19q homologs are associated on a single chromosome, whereas 8q, 10q and 19p are found as separate entities. The ancestral karyotype of placental mammals may therefore have comprised of 2n = 46 chromosomes. The *putative* ancestral primate *karyotype* shows only a very few differences when compared to the ancestral *mammalian karyotype.*

In the inferred primate ancestor *presumably* fissions of 10p and 12pter-q23.3, 16q and 19q, and 4 and 8p occurred, followed by fusion of 10p/ 10q, resulting in a karyotype of 2n = 50 *chromosomes* with conserved human homologous chromosome (*segments*) 1, 2pter-q1.3, 2q13-qter, 4, 5, 6, (7p21-811.21, 7q11.23-21.3, 7q22.1-qter), 8p, 8q, 9, 10, 11, 13, 16q, 17, 18, 19p, 19q, 20, X, Y, and the association of 3/21, (7pter-p22, 7q 11.21-11.23, 7q21.3-22.1)/16p, I 2pter-q23.3/22q 12.3-qter, 22q 11.2-22q 12.3/ 12q23.3-qter, 14/15.

The following section provides an overview of the current knowledge on karyotype evolution in prosimians and higher primates (anthropoidea), the latter being subdivided in New World monkeys (platyrrhini), Old World monkeys (cercopithecoidea), and apes (hominoidea, gibbons, and great apes).

Chromosomes of Prosimians

Prosimians are comprised of lorises, bush-babies, and lemurs. Only five species were fully analyzed using human chromosome painting probes: brown, black, and ring-tailed lemur (4, 9), and two bush-babies. In addition, black lemur chromosome-specific probes were characterized by reverse painting to human chromosomes.

The results so far revealed that these prosimians retained ancestral primate chromosome forms 12/22, 3/21, 14/15, and probably 7/16, but showed common derived fissions of chromosome I (two fissions), 4, 5, 6, 8, and 15.

In addition, the investigated lemurs share four derived fissions and five translocations, the two bush-babies five fissions, and six translocations. Numerous further rearrangements were noticed, for example eight fusions only found in the black lemur. By conclusion, none of the prosimians analyzed to date has conserved a primitive primate karyotype, instead prosimians have highly derived karyotypes.

Chromosomes of Platyrrhines

The 16 genera and over 100 recognized species of New World monkeys (platyrrhini) show a high degree of karyotypic diversity with chromosome numbers ranging from 2n = 16 in *Callicebus lugens* to 2n = 62 in *Lagothrix lagotricha.* With established comparative genome maps between human and more than 20 species by cross-species chromosome painting with human probes, New World monkeys represent the most comprehensively studied group of species among mammals altogether.

In addition, several species have been analyzed by multidirectional chromosome painting employing tamarin *(Saguinus oedipus)* and woolly monkey *(L. lagothricha)* chromosome-specific probes. In the inferred ancestral New World monkey karyotype of 2n = 54 chromosomes (17), human chromosomes 4, 6, 9, 11, 12, 13, 17, 19, 20, 22, X, and Y homologs are found entirely conserved as separate chromosomes: Chromosome 5, 14, 18, and 21 homologs show conserved synteny, are however associated with other homologs (5/7a, 14/15a, 8a/18, and 3a/21).

The remaining human homologs are fragmented: 1 a, 1b, 1c, 2a, 2b/16b, 3b, 3c, 7b, 8b, 10a/16a, 10b, and 15b. Among the family callitrichidae, two tamarins *(S. oedipus and Leontopithecus chrysomelas),* three marmoset species *(Callithrix jacchus, Cebuella pygmaea,* and *Callithrix argentata),* and *Callimico goeldii* were analyzed by multidirectional chromosome painting. The data showed that *C. goeldii* exclusively shares derived syntenic associations 13/9/22 and 13/17/20 with all callithrichids.

A study on an interspecies hybrid between a female Common marmoset *(C. jacchus,* 2n = 46) and a male Pygmy marmoset *(C. pygmaea,* 2n = 44) with a diploid chromosome number of 2n = 45 gave further support for the inclusion of *Cebuella* within genus *Callithrix.* Genomic imbalances between this interspecies hybrid and other callithrichidae, visualized by interspecies Comparative genomic hybridization (iCGH), were confined to centromeric and subtelomeric heterochromatin.

Cross-species FISH with a micro-dissection derived *C. pygmaea* repetitive probe revealed species-specificity of several 50 Mb and larger blocks of heterochromatin, thus providing a dramatic example for amplification of noncentromeric repetitive sequences within approximately 5 million years of evolution. Zoo-FISH data are also available from the squirrel monkey, capucin monkeys, and three species of titi monkey (family cebidae). The squirrel monkey shares the derived syntenic association 2/15 with callithrichidae.

Capuchin monkeys have conserved almost completely the ancestral New World monkey karyotype. Titi monkeys *Callicebus moloch, Callicebus donacophilus* (both 2n = 50), and *C. lugens (2n* = 16), are phylogenetically linked by common derived associations 10/11, 22/2/22, and 15/7.

The low chromosome number of C. hrgens is the result of at least 22 fusions and 6 fissions. The diploid chromosome numbers of atelidae (Howler monkeys, genus *Alouatta,* Spider monkeys, genus *Ateles,* Wooly monkeys, genus *Lagothrix,* and Wooly spider monkeys, genus *Brachyteles)* range from 2n = 32 in *Ateles* to 2n = 62 in *Brachvteles* and *Lagothrix.* Species from all four genera have been analyzed by multi-directional chromosome painting.

All atelidae analyzed share the derived fissions of human chromosome 1, 4, and 5 homologs, inversion of the 10/16 and 5/7 homologs, and a translocation 4/15. The ancestral atelidae karyotype is comprised of 2n = 62 chromosomes and is conserved *inLagothrix* and *Brachyteles.* Howler monkeys represent the genus with the most extensive karyotype diversity within platyrrhini with high levels of intra-specific chromosomal variability.

Molecular cytogenetic studies in Spider monkeys (genus *Ateles)* revealed that at least 17 fusions and three fissions are necessary to derive the putative ancestral *Ateles* karyotype conserved in *A. b. marginatus* (2n = 34).

Chromosomes of Catarrhines

Macaques and baboons have strongly conserved and uniform karyotypes with 2n = 42 chromosomes. Compared to the human karyotype all chromosomes show conserved synteny, except for two human chromosome 2 (two homologs). The reduced chromosome number is the result of two fusions, leading to syntenic association of chromosome 7/21 and 20/22 homologs. They further conserved the primate ancestral association of 14/ 15 homologs.

Among guenons, fissions are the main mechanism driving the evolutionary trend toward higher chromosome numbers of 2n = 60 in the African green monkey *(Chlorocebus aethiops)* and to 2n = 72 in *Cercopithecus wolfs.* Leaf-eating monkeys have fairly conserved karyotypes, compared with human with chromosome numbers of 2n = 44 in the black and white colobus *(Colobus guereza)* (34) and 2n = 48 in the proboscis monkey *(Nasalis larvatus).*

These Colobines share the derived association of human chromosomes 21/22. The Asian members further share a reciprocal translocation of the human chromosome 1 and 19 homolog that was followed by a pericentric inversion.

Chromosomes of Hominoids

The four gibbon subgenera show distinct karyomorphs with 2n = 38 in *Bunopithecus, 2n = 44* in *Hylobates,* 2n = 50 in *Symphalangus,* and 2n = 52 in *Nomascus.* During the last decade all four subgenera were studied by Zoo-FISH, which revealed extensive chromosome reshuffling in all gibbons.

First studies employed human chromosome paint probes (36-39). More recently, gibbons were reanalyzed by multidirectional painting. The inferred ancestral karyotype of all extant gibbons differed from the putative ancestral *hominoid karyotype* by at least five reciprocal *translocations,* eight inversions, 10 fissions, and one fusion. It included homologs to human chromosomes 7, 9, 13, 14, 15, 20, 21, 22, X, and Y with conserved synteny and homologous segments of human chromosome I (three segments), 2, 3, 4, 5, 6 (two segments each), 8, 11, 12, and 17, respectively.

Finally. syntenic associations of segments homologous to human chromosomes were included: 3/8, 3/12/19, 5/16, 5/16/5/16, 19/12/19, 10/4 (twice), 12/3/8, 17/2/17/2, and 18/11. The Whitecheeked

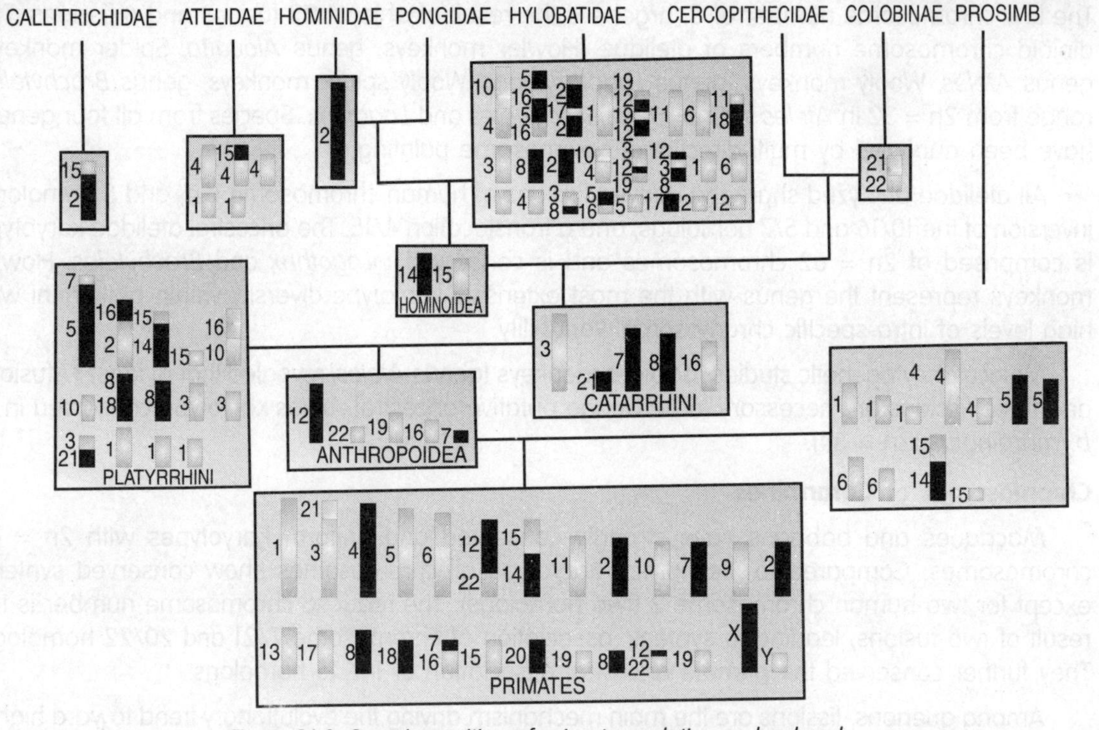

Figure 14.3: Superimposition of primate evolutionary landmark rearrangements onto the consensus phylogenetic tree.

gibbon, Siamang and White-handed gibbon further share at least five common derived chromosome forms: associations 2/7, 8/3/11/18, 4/5, 22/5/16, and an inversion.

In the last common ancestor of the White-cheeked gibbon and the Siamang, one fusion and four fissions may have occurred. In addition, each gibbon accumulated a large number of speciesspecific rearrangements (between 10 in the White-handed gibbon and 28 in the Hoolock).

Comparative high resolution G-banding analysis of human and great apes indicated, that the only interchromosomal changes are species-specific: a fusion of chromosome 2 in the human lineage, a reciprocal translocation t(5;17) in the gorilla and a band insertion ins(8;20) in the orangutan.

Chromosome painting of great ape genomes with human probes confirmed the fusion of human chromosome 2 and the translocation t(5; 17) in the gorilla, but not the insertion ins(8;20) in the orangutan.

Further, G-banding indicated that human and great ape genomes would differ by the presence and chromosomal location of constitutive heterochromatin, in the number of nucleolar organizer region-bearing chromosomes, and by several inversions. For example, chimpanzee and human chromosome 1, 4, 5, 9, 12, 15, 16, 17, and 18 homologs would differ by pericentric inversions and chromosome 7 homologs of chimpanzee and gorilla by a paracentric inversion. Human and gorilla chromosome 1.6 homologs as well as human and orangutan chromosome 3 and 17 homologs would have diverged by both peri- and paracentric inversions.

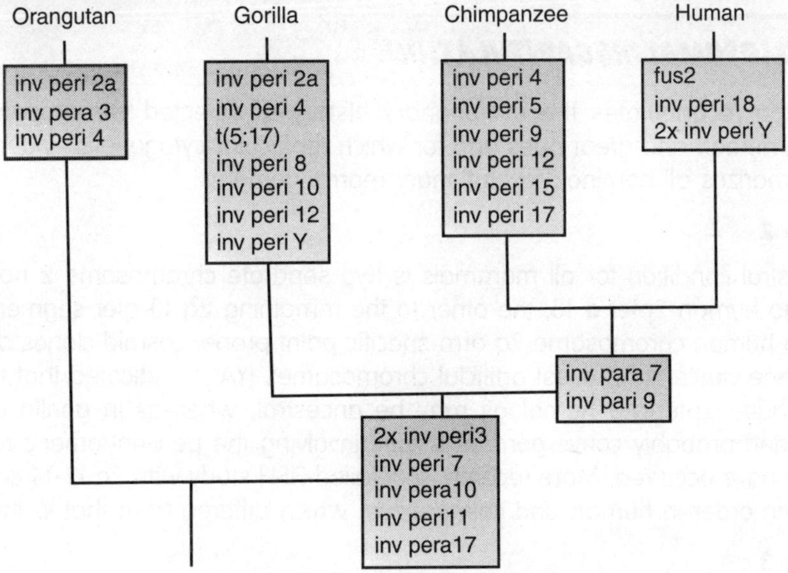

Figure 14.4: Overview on chromosome rearrangements that occurred in orangutan, gorilla, chimpanzee and human.

Phylogenetic Rearrangements in Primates

On the basis of molecular cytogenetic evidence it is possible to draw increasingly accurate conclusions on landmark rearrangements that occurred during primate evolution. These landmarks are shared by defined subgroups of primates and thus provide fundamental phylogenetic links between members of these species groups.

They can further be superimposed onto the consensus primate phylogenetic tree, which permits an estimate of the timing of these events. The putative *ancestor* of all primates most probably acquired separate chromosomes 4 and 10 homologs by fission/fusion *events* and lost the *association* of 16/19 homologs by fission about 60 million years ago.

The inferred ancestor of higher primates only retained *ancestral* primate syntenic associations 14/15 and 3/21, but acquired separate chromosome 12 and 22 homologs by reciprocal translocation, a single chromosome 19 homolog by a fusion, and lost the association of human 7/16 homologs by fission.

These events can be dated before the emergence of New and Old World monkeys (approx 30-40 million years ago). After the New/Old World monkey split, platyrrhini acquired common derived associations 5/7, 2/16, 10/16, 8/18, and several fissions, whereas in the catarrhini ancestor (about 25 million years ago) a fission led to separate chromosomes 3 and 21, whereas fusions resulted in single chromosome 7, 8, and 16 homologs.

Approximately 18 million years ago, in the common ancestor of hominoids fission of the association 14/15 took place, leading to separate chromosomes 14 and 15. Finally, less than 6 million years ago and after the divergence of human and chimpanzee, the fusion of human chromosome 2 occurred.

CHROMOSOMAL RECAPITULATION

This section recapitulates the evolutionary history of selected chromosomes, which are rearranged in human and great apes and for which molecular cytogenetic evidence is available. Figure 4 summarizes all hominoid evolutionary rearrangements.

Chromosome 2

The ancestral condition for all mammals is two separate chromosome 2 homologs. One is homologous to human 2pter-q 13, the other to the remaining 2q 13-qter segment. Comparative studies with a human chromosome 2q arm-specific paint probe, cosmid clones derived from the human V k gene cluster, and yeast artificial chromosomes (YACs) indicated that the chimpanzee and the macaque 2pter-q13 homologs may be ancestral, whereas in gorilla and orangutan, independent and probably convergent inversions involving the pericentromeric region of the 2p homolog may have occurred. More recently, a detailed FISH study with 2q 12-14 cosmids revealed the same clone order in human and chimpanzee, which differed from that in the macaque.

Chromosome 3

According to one study, the putative ancestral primate chromosome 3 homolog is conserved in the Brown lemur, from which a pericentric inversion led to the ancestral Old World primate homolog conserved in the Bornean orangutan. Human/African ape homologs would differ from this chromosome form by two inversions. An even more complex scenario involving several recurrent sites of new centromere seeding was proposed by Ventura et al., according to which human and Bornean orangutan would differ by three inversions.

A third hypothesis suggested that from the ancestral simian homolog a common derived and two independent inversions would lead to Bornean orangutan and human chromosome 3 homologs. In conclusion, the evolutionary history of human chromosome 3 is probably the most dynamic and complex of all human chromosomes studied in detail so far.

Chromosome 4

Detailed comparative studies on the evolution of human chromosome 4 homologs of great apes and the macaque indicated that the human homolog would represent the ancestral hominoid chromosome form. A minimum of seven different breakpoints were observed, some of them were located in the 4p pericentromeric region.

For three of these inversions breakpoint spanning YAC clones were identified, of which one was confirmative for a previous analysis. Interestingly, one clone showed a split signal in chimpanzee and macaque, indicating two independent evolutionary breakpoints in close proximity to each other.

Chromosome 5

Both chromosome bar codes and detailed Zoo-FISH studies with YACs and subregional paint probes identified the chromosome forms shared by macaque, orangutan, and human to be ancestral for hominoids. From this, the homolog of the chimpanzee is directly derived by a pericentric inversion, those of the gorilla by a reciprocal translocation t(5; 17).

Chromosome 6

In a recent FISH study, it was shown that the remarkable conservation of chromosome 6 is also present at the subchromosomal level. Despite this, evolutionary centromere relocation events were observed. One of these events may have occurred in the ancestor of great apes, where the centromere moved from 6p22.1 to the present day location.

This hypothesis gained support from the observation that in the assumed ancestral location in a cluster of intrachromosomal segmental duplications was found, which the authors explained as remnants of duplicons that flanked the ancestral inactivated centromere.

Chromosome 7

A FISH study on evolutionary changes of human chromosome 7 revealed that the ancestral mammalian homologs were comprised of two chromosomes (7a and 7b/16p) as observed in carnivores. The ancestral primate segment 7a shared by a lemur and higher Old World monkeys is the result of a paracentric inversion.

The ancestral higher primate chromosome form was derived by a fission of 7b and 16p, followed by a centric fusion of 7a/7b in higher Old World primates as observed in the orangutan.

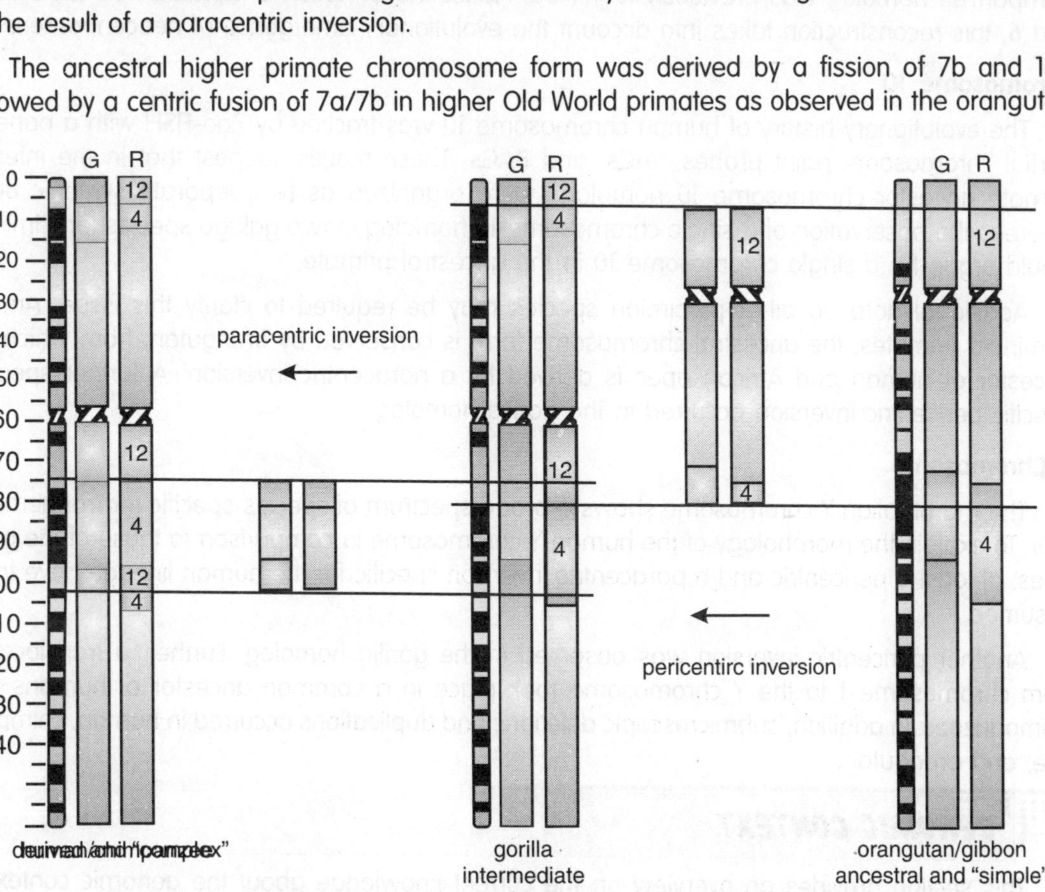

Figure 14.5: Hominoid chromosome 7 evolutionary rearrangements delineated by comparative sequence maps of human and rat (R) and cross-species fluorescence in situ *hybridization with gibbon painting probes (G). Both data sets visualize the inversion breakpoints and their evolutionary direction by an increasingly simple pattern when tracking the rearrangements in the evolutionary reverse direction. Horizontal bars indicate inversion breakpoints.*

In hominoids two further inversions with four distinct breakpoints occurred: the pericentric inversion in the human/African ape ancestor and the paracentric inversion in the common ancestor of human and chimpanzees.

Chromosome 9

Zoo-FISH with 12 evenly spaced YAC clones revealed an identical marker order in Old and New World monkeys, which may therefore represent the ancestral chromosome form for higher primates. A paracentric inversion would derive the ancestral hominoid chromosome form, which was conserved in orangutan and gorilla. A further pericentric inversion would lead to the chromosome form of the last common ancestor of human and chimpanzee.

Human conserved this evolutionary intermediate chromosome form, from which the homolog of chimpanzees differs by another pericentric inversion. One of the inversion breakpoints in the chimpanzee homolog was previously identified. As for the evolution of human chromosomes 3 and 6, this reconstruction takes into account the evolutionary emergence of neocentromeres.

Chromosome 10

The evolutionary history of human chromosome 10 was tracked by Zoo-FISH with a panel of partial chromosome paint probes, YACs, and BACs. These results suggest that in the inferred primate ancestor chromosome 10 homologs were organized as two separate syntenic units, whereas the observation of a single chromosome 10 homolog in two galago species (prosimians) would argue for a single chromosome 10 in the ancestral primate.

Additional data on other prosimian species may be required to clarify this issue. Among hominoid primates, the ancestral chromosome form is conserved by orangutan, from which the ancestor of human and African apes is derived by a paracentric inversion. A further species-specific pericentric inversion occurred in the gorilla homolog.

Y Chromosome

The mammalian Y chromosome shows a broad spectrum of species-specific rearrangements (60). To explain the morphology of the human Y chromosome in comparison to those of the great apes, at least a pericentric and a paracentric inversion specific for the human lineage have to be assumed.

Another pericentric inversion was observed in the gorilla homolog. Further, a translocation from chromosome 1 to the Y chromosome took place in a common ancestor of humans and chimpanzees. In addition, submicroscopic deletions and duplications occurred in human, chimpanzee, and orangutan.

GENOMIC CONTEXT

This section provides an overview on the current knowledge about the genomic context of evolutionary chromosomal breakpoints in human and great apes. By analogy to the functional importance of genomic alterations related with acquired chromosomal aberrations in cancers, it was anticipated that changes of the genomic environment caused by evolutionary chromosome rearrangements might hold the key for a better understanding of the origins of our own species.

Head to Head Arrays

Two head-to-head arrays of degenerate telomere repeats are found directly at the fusion site in 2q13. Their inverted orientation indicates a telorneric fusion. Subsequently, the inactivation of one of the two ancestral centromeres must have occurred.

Indeed, close to the fusion point in band 2q21 a degenerated alphoid domain was found by low stringency FISH of a satellite DNA probe. Like in other subtelomeric regions, large blocks surrounding the fusion point are comprised of duplicated sequences. Chromosome 2q 13 paralogs of 96-99% identity were identified on chromosome 9pter, 9p 11.2, 9q 13, 22qter, and 2q 11.2.

The emergence of some of these segmental duplications could be dated back prior to hominid divergence, whereas others appeared to be of more recent origin. It can be speculated that these duplicons may have been the cause for the fusion and that human chromosome 2 is the product of paralogous recombination between two different chromosomes using a duplicated segment as recombination substrate.

The fusion point is located in a gene-rich region, with at least 24 potentially functional genes and 16 pseudogenes located within less than 1 Mb distance or in paralogous regions elsewhere in the genome.

At least 18 of these genes are transcriptionally active, for example members of the cobalamin synthetase W domain *(CBWD)* and forkhead domain *FOXD4* gene families, thus providing an example of genomic innovation connected with duplication and evolutionary rearrangement of subtelomeric and pericentromeric regions.

Molecular Characteristics

A detailed molecular cytogenetic characterization of evolutionary inversion breakpoints in hominoid chromosome 3 homologs revealed that the ancestral pericentromeric region is associated with both large-scale and micro-rearrangements. Small segments homologous to human 3q 11.2 and 3q21.2 were repositioned intrachromosomally in the orangutan lineage.

The breakpoint in the human 3p12.3 homologous region of the orangutan is associated with extensive transchromosomal duplications observed in multiple subtelomeric regions. A second breakpoint in the same region, but with a distinctly different location, is flanked by sequences present in all subtelomeric regions of the Siamang (gibbon) genome.

Reciprocal Translocation t(4;19) in the Gorilla

The breakpoints of the reciprocal translocation t(4;19) in the gorilla are located in regions syntenic to human 5q14.1 between *HMGCR* and *RASA 1* genes, and in 17p 12 *with an approx* 383-kb region-specific low-copy repeat (LCR) 17pA. The 17p 12 region is also susceptible to constitutional rearrangements in human.

The authors proposed a series of consecutive evoloutionary segmental duplications involving LCR17pA and approx 191-kb LCR 17pB copies that resulted in complex genome architecture in the rearrangements. Detailed comparative analysis of the corresponding region identified remnants of DNA-transposable element MER 1Charlie3 and retroviral ERVL elements at the translocation breakpoint in a pre-gorilla individual. In addition, genomic rearrangements involving LCR17pA and LCR 17pB resulted in the creation of novel genes at the breakpoint junctions.

Inversions of Chromosome 7 in Great Apes

Comparative FISH analysis with BAC clones that were derived from the Williams-Beuren syndrome region in 7q11.23 and which contained LCRs including *NCF1(p47-phox)* sequences revealed duplicated segments in the 7q11.23 homologous region of chimpanzee, gorilla, orangutan, and a gibbon.

As in human, cross-hybridization was observed in the inversion breakpoint regions at 7q22 and 7p22 in African apes, but not in the homologous chromosome regions in orangutan and gibbon. Zoo-FISH analysis employing BAC probes confined the 7p22.1 breakpoint of the pericentric inversion in the human/African ape ancestor to 6,8 Mb on the human reference sequence map and the 7822.1 breakpoint to 97,1 Mb in regions with predominantly inter-chromosomal duplications with paralogs on human chromosomes 2-4 and 8-15.

These duplicons were already present in the orangutan, but spread to a variety of additional chromosomes in the gorilla. The paracentric inversion breakpoints in the common ancestor of human and chimpanzees were found in 7q11.23 between 76,1 Mb and 76,3 Mb and in 7q22.1 at 10 1,9 Mb, respectively.

Intrachromosomal duplicons mark an at least 110-kb stretch of nearly identical DNA sequence, which is most probably directly flanking both breakpoints. The 7q11.23 breakpoint is further located in close proximity to a 200-kb or larger insertion of chromosome 1 material, which could be dated back to the African ape ancestor.

Considering that the duplicated sequences flanking the four inversion breakpoints as well as the chromosome I transposition were already present in the evolutionary ancestral state prior to the inversions, segmental duplications may have been the cause rather than the result of both rearrangements.

Inversions Pericentric in Chimpanzee and Gorilla

Both chimpanzee and gorilla show derived pericentric inversions of the chromosome 12 homologs. In a first comparative FISH study, the 12p12 breakpoints in both species were mapped to the same YAC clone, whereas the 12q15 breakpoints were shown to be located in distinctly different regions.

In addition, a chimpanzee BAC clone was identified, which also spanned the 12p12 breakpoint. Recently it could be demonstrated that this clone did not span the 12p12 breakpoint in the gorilla, thus, demonstrating that both the 12q and I 2p inversion breakpoints differ in chimpanzee and gorilla. Sequence analysis of the breakpoints in the chimpanzee genome revealed two large duplications in both breakpoint regions, which probably emerged in concert with the inversion because they were shown to be chimpanzee-specific.

Centromere in the Great Ape Ancestor

The fission that separated human chromosome 14 and 15 homologs led to the inactivation of the ancestral centromere in 15q25 and to the formation of two new centromeres in the locations where they can be found in human.

Detailed comparative molecular cytogenetic analysis of the region 15q24-26 revealed 500 kb of duplicons, which flank both sides of the single ancestral centromere in Old World monkeys.

Notably, the same duplicons are associated with neocentromeres in 15q24-26 in two clinical cases. This suggests that neocentromere formation in human pathology in a region of an evolutionary inactivated centromere may be triggered by the persistence of recombinogenic pericentromeric duplications.

Inversion Pericentric in Chimpanzee

Chimpanzee and bonobo share a derived pericentric inversion of their chromosome 15 homologs. A comparative FISH and *in silico* approach was used to narrow down the breakpoint interval of this pericentric inversion to the 15q11-q13 homologous region.

The breakpoint mapped to a 600-kb segmental duplication cluster. Sequence analysis indicated that this region comprises a duplication of the *CHRNA 7gene* and several Golgin-linked-to-PML duplications. Notably, this evolutionary breakpoint did not colocalize with one of the three major common disease rearrangement breakpoints in 15811-q13.

Inversion Pericentric in Chimpanzee

FISH was used to investigate the derived pericentric inversion with breakpoints in l 7p 13 and 17g21.3 homologous regions, by which chimpanzee chromosome 19 differs from human chromosome 17. Breakpoint-spanning BACs were subsequently used to clone and sequence the junction fragments.

Both breakpoints were localized in intergenic regions rich in *Alu* and LTR elements, but were not associated with LCRs, duplicated regions, or deletions. The findings suggest that repetitive sequence mediated nonhomologous recombination has facilitated inversion formation.

In close proximity of the breakpoints, *NGFR* and *NXPH3* genes are located in 17g21.3 and *GUC2D andALOX15B in* 17p13. Most likely, the genomic structure, the expression level or the replication timing of these genes was not affected by the inversion.

Inversion Pericentric in Human

Human chromosome 18 differs from its great ape homologues by a pericentric inversion, which can be assigned to the human lineage. Recently, chimpanzee BAC clones that span one of the breakpoint regions where were identified by FISH, because in human split signals were observed on 18p11.3 and 18q11. Interspecies sequence comparisons indicated that the ancestral break occurred between the *ROCK1* and *USP14* gene.

In human, the inversion translocated *ROCKJ* near centromeric heterochromatin and *USP14* into proximity of 18p subtelomeric repeats. *USP14* is differentially expressed in human and chimpanzee cortex as well as fibroblast cell lines.

Further, a 19-kb segmental duplication with paralogs in pericentrorneric regions of chromosomes 1, 9 and the acrocentric chromosomes is also flanking both 18p1 1.3 and 18g11 regions in inverted orientation.

The authors propose a model, according to which the 19-kb 18q segment containing part of the *ROCK1* gene was first locally duplicated and then transposed to the pericentromeric region of the ancestral 18p. Subsequently, the segmental duplication may have mediated an intrachromosomal crossover, resulting in modern human chromosome 18.

Diversified Chromosomal Changes

Primates, like mammals in general, show strikingly divergent diploid chromosome numbers ranging from 2n = 16 in the New World monkey C. *lugens* to 2n = 72 in the Old World monkey *C.* woo. Interestingly, several examples provide evidence against a certain trend toward higher or lower chromosome numbers.

In New World monkeys, for example, the ancestor of the monophyletic group atelidae has 2n = 62 chromosomes, compared to the ancestor of all platyrrhini with 2n = 54 chromosomes. More recently in atelidae evolution, the chromosome number in the common ancestor of genus *Ateles* was reduced to 2n = 34. On average, extant species differ from the inferred ancestral eutherian karyotypc by 32 chromosome breaks.

The rate of chromosomal exchange, however, varies greatly. The cat and the mink conserved the ancestral mammalian chromosome forms almost unchanged (11-16 breaks), whereas in the dog genome at least 66 breaks occurred. Rodents show even greater differences between evolutionary conserved and extremely reshuffled genomes.

A recent summary on the current state of the rat genome project listed 278 disrupted chromosomal syntenies between human and rat and 280 between human and mouse as a consequence of 353 rearrangements. The vast majority of these rearrangements can be assigned to the muridae lineage because squirrels, which also belong to the order rodentia, show a highly conserved genome organization (21 breaks).

A similar situation is found in primates: human chromosomes differ by only 21 breaks from the eutherian ancestor, whereas gibbon genomes accumulated at least 90 breaks. In summary, a "chromosomal evolutionary clock" does certainly not exist. Instead, the rate of chromosomal evolution may change over time in a phylogenetic lineage and could be interrelated with changing environmental impacts and population dynamics.

Phylogeny of Genomic Changes

Together with retroposon integrations, indels (insertions and deletions), changes of the mitochondrial gene order, gene duplications, and genetic code changes, evolutionary chromosome rearrangements can be classified as rare genomic changes. A bioinformatics approach to reconstruct the succession human/mouse chromosome rearrangements revealed a surprisingly large number of breakpoint hotspots: 190 of the 245 breaks analyzed reused a genomic region, although the authors emphasized that they were not necessarily located in exactly the same nucleotide position.

The existence of breakpoint clusters would argue against random breakage and in favour of a fragile breakage model. Some other reports support this nonrandom breakage model. On human chromosome I and its homologs in placental mammals, evolutionary fissions cluster in two breakpoint hotspots on the long arm. Further, at the resolution of chromosome banding, in the macaque 9 out of 17 evolutionary breakpoints correspond to evolutionary conserved fragile sites.

Because the vast majority of the rearrangements analyzed are confined to the muridae lineage, a careful inspection of breakpoints in other mammals is required in order to determine whether this model applies to mammalian genome evolution in general. In the majority of mammalian species, Robertsonian (centromeric) and tandem fusions or fissions appear to be the predominant type of evolutionary rearrangements. For example, in all New World primates that were analyzed by Zoo-FISH, 64% of the 149 recorded rearrangements were fissions, 24% fusions, and 10% intra-chromosomal rearrangements.

Only two rearrangements were identified as reciprocal translocations. Notably, some of the whole chromosome fusions were not simple centromeric head-to-head fusions, but head-to-tail fusions between centromeric and telomeric regions.

The majority of the evolutionary rearrangements appear to be mediated by highly repetitive regions of the genome and are accompanied by gain or loss of centromeric or telomeric function. A second group of repetitive sequences, which came into the focus of attention as cause for genomic disorders and evolutionary genome instability, are segmental duplications or LCRs. So far, they have been found to be associated with almost every evolutionary chromosomal rearrangement in hominoid primates that was characterized in detail.

Among these were the fusion of human chromosome 2, the reciprocal translocation t(5;17) in the gorilla, several inversions and the formation of neocentromeres. Within the primate lineage, the explosive expansion of the majority of these elements can be correlated with the burst of primate *Alu* retroposition activity approx 35 million years ago.

At least for some of the rearrangements in hominoid primates it was demonstrated that the associated segmental duplications were present prior to the rearrangement and may have been a cause for it rather than a consequence. In addition, segmental duplications are statistically significantly enriched in genomic regions, which are differentially expressed in the human and chimpanzee brain.

It would certainly be interesting to know to which degree the chromosomal distribution of duplicated segments in rat and mouse genomes corresponds with sites of genome reshuffling in these two species.

CONCLUDING REMARKS

Sequencing whole genomes of a large number of mammalian species would obviously provide the most comprehensive information on the molecular mechanisms of genome organization and evolution. Recently eight nonprimate mammals (among them the African elephant, the European common shrew, the European hedgehog, guinea pig, the nine-banded armadillo, the rabbit, and the domestic cat) and the orangutan were added to the National Human Genome Research Institute sequencing pipeline.

These species will complement sequence data from other species, which are already available *(Drosophila, Fugu, Caenorhabditis elegans,* mouse, and rat) or rapidly emerging (cow, dog, chicken, zebrafish, chimpanzee, and rhesus macaque). Despite these efforts, whole genome sequencing projects will always be limited to key species from different phylogenetic lineages. For an in depth study of primate genome evolution, genomic sequence from at least two species from each major evolutionary branching point would be desirable to reduce species-specific noise.

An alternative scenario could be that comparative molecular cytogenetic- and array-based screening technologies will guide the way to relevant genomic regions, which will then be targeted by genome sequencing projects. A key technology with great promise for evolutionary studies employs genomic clonebased microarrays. In the past, iCGH to metaphase chromosomes provided only limited information about genomic imbalances between species owing to a low resolution of approx 10 Mb.

The introduction of comparative genomic hybridization to microarrays with almost 2500 human BAC clones evenly spaced in approx 1-Mb distance (array CGH) lead to a 10-fold increase in resolution. A first evolutionary array CGH study (933) identified 63 sites of DNA copy-number variation between human and the great apes.

Most of these copy number changes affected interstitial euchromatic chromosome regions suggesting that such large-scale events are not restricted to subtelomeric or pericentromeric regions. Another recently introduced genomic microarray based technique was termed "array painting". It involves a BAC microarray as the target for reverse painting of aberrant chromosomes for the analysis of chromosomal breakpoints.

The complete composition and the breakpoints of aberrant chromosomes can be analyzed at the same resolution as mentioned above for array CGH. Initial experiments with gibbon chromosome paint probes successfully demonstrated that this technology has the potential to significantly speed up the molecular characterization of evolutionary breakpoints.

The impact of the higher order nuclear architecture on the complex epigenetic mechanisms responsible for cell type specific gene expression patterns in multi-cellular organisms has become an important field of genome research. Initial studies on the gene-density correlated radial arrangement of chromatin revealed a remarkable evolutionary conservation over a period of at least 30 million years.

As in certain human cell types, in a variety of higher primate species gene-dense chromatin is preferentially located in the nuclear center, whereas genepoor chromatin shows an orientation toward the nuclear periphery. This probabilistic but highly nonrandom arrangement is still strictly maintained in species like gibbons with extremely reshuffled genomes. This evolutionary conservation argues for a still unknown functional significance of distinct radial higher-order chromatin arrangements.

It may also be hypothesized that *evolutionary* chromosome *rearrangements* are triggered by certain *nonrandom* chromatin arrangements. *Alternatively*, such events may lead to the dislocation of chromosomal material to a different nuclear environment, which in turn may have consequences for the transcriptional activity of the affected loci.

Chapter 15

Detecting Mutations

Since the completion of the human genome sequencing project in 2001, more and more scientists are sequencing samples to identify the mutations responsible for disease. Most common diseases are complex *genetic traits*, with multiple genetic and environmental components contributing to *susceptibility*. It has been proposed that common genetic variants, including single nucleotide *polymorphisms* (SNPs), influence susceptibility to common diseases.

This proposal has begun to be tested in numerous studies of association between genetic variation at these common DNA *polymorphisms* and variation in disease *susceptibility*. A polymorphism is a variation in DNA sequence that has an allele frequency of at least 1% in the population. Approximately 1 in 1,000 base pairs of the human genome differs between any two *chromosomal homologues*.

Studies comparing individuals in a population have shown that, on average, polymorphisms in occur every 200-300 hundred base pairs. There are several types of polymorphisms, the most common, SNPs, and repeat polymorphisms (*microsatellites*), insertions or deletions, which range *extensively* in size. It is estimated that 93% of genes contain a SNP. A variation (or SNP) is termed *a mutation* when it leads to a change in amino acid sequence and consequently an alteration to the protein and its function.

High-throughput technology is now used to genotype SNPs and genetic variants, although it is still not practical to genotype all SNPs in the genome for association with disease phenotypes. Therefore, SNPs and mutations are selected on their ability to affect the function of a protein or its expression, as those polymorphisms with molecular consequences are more likely to be involved in influencing disease risk. However, many SNPs have no direct effect on cell function, but it is believed that they could predispose people to diseases or influence their response to a drug.

The 5′ flanking regions upstream of genes are known to harbor transcriptional control elements and therefore are *routinely* screened for sequence variants that may help define a clinical phenotype. Recently, attention has focused on the effects of *polymorphisms* in noncoding regions, such as 3′ and 5′ UTR (untranslated region), *regulatory* regions, and *intronic* regions, rather than the traditional *promoter* and coding region *mutations*, as these regions also can contribute to the control of gene expression. The role of UTRs and introns in gene regulation may have been underestimated. Scientists believe SNP maps will help them identify the multiple genes associated with such complex diseases such as cancer, diabetes, vascular disease, and some forms of mental illness. These associations are difficult to establish with conventional gene-hunting methods because a single altered gene may make only a small contribution to the disease.

By comparing patterns and frequencies of SNPs in patients and controls, researchers hope to identify which SNPs are associated with which diseases. A number of methods are available for SNP or mutation identification, including DNA sequencing; enzymatic digestion combined with gel-electrophoresis (SNP-RFLP, restriction fragment length polymorphism); pyrosequencing; ARMS-PCR (amplification refractory mutation system), and conformational analysis (WAVE); all of which require an initial target sequence PCR amplification.

Another, more recent SNP detection method is by hybridization to DNA or oligonucleotide microarrays. This method does not require prior PCR amplification, as total genomic DNA is used directly in labeling and hybridization. Additionally, there are several high-throughput methodologies for SNP detection, including ABI TaqMan assays and mass spectrometry. The current methods have been comprehensively reviewed by Wang et al..

This chapter describes the laboratory and computational methods for identifying SNPs and mutations employing fluorescent-based sequencing protocol using ABI's 3700 DNA analyzer. DNA sequencing is the preferred method for SNP and mutation detection, as it provides information on the type and location of the variants at very high accuracy, whereas other methods require sequencing as an additional step to confirm results due to higher false positive rates. However, sequencing can be a costly method and is less adaptable to multiplex and automation than other methods.

REQUIREMENTS

Step Wise Methodology

1. The DNA template can be either purified plasmid DNA or PCR product. Total plasmid DNA is isolated from bacterial recombinant clones, whereas the PCR technique is used for amplifying the required target sequences using specific flanking primers and polymerase.
2. Qiagen Plasmid Mini Kit.
3. LB broth medium: 1% tryptone, 0.5% yeast extract, 0.5% NaCl (Sigma Aldrich, L3022-1 kg).
4. Nuclease-free water (Sigma Aldrich W4502).
5. PCR primers for amplifying a specific region of the genome for SNP analysis, (e.g. vector primers T7 and SP6):
6. T7-5′ AATACGACTCACTATAG 3′.
7. SP6-5′ ATTTAGGTGACACTATAG 3′.
8. PCR buffer and reagents (Promega PCR Core System 1, M7660).
9. Gel electrophoresis equipment.
10. 1 X TBE: 54g Tris-base, 27.5g boric acid, 20 mL of 0.5M EDTA, pH 8.0. Make up the vol to

1 L with water (Sigma Aldrich, 93290-1 L).

11. Electrophoresis-grade agarose (BioRad, 161-3101 [125 g]).
12. Ethidium bromide (Sigma Aldrich, E8751-5G).
13. Agilent bioanalyser and DNA chips (Agilent, 5067-1506).
14. Gel Extraction Kit (Qiagen, 28704).
15. QIAquick PCR Purification Kit (Qiagen, 28104).
16. Incubating shaker.
17. Thermal cycler for PCR amplification.
18. 15-mL Falcon tubes (Fisher, CFT-420-079D).

Sequencing Reactions

19. PCR primers for sequencing reaction (e.g. T7 Universal primer): T7- 5′ TAA TAC GAC TCA CTA TAG GG 3′
20. ABI Prism Big Dye Terminator Cycle Sequencing Kit (ABI, 4337455).
21. SeqSaver Sequencing Premix Dilution Buffer (Sigma Aldrich, S3938) or BigDye Terminator v3.1 5X Sequencing Buffer (ABI, 4336699).

Isolating Reactions

22. 0.5 M EDTA (Sigma Aldrich, E-7889).
23. 70% and 95% ethanol.
24. Vacuum centrifuge.
25. Hi Di Formamide (ABI, 431132p).

Sequencing of DNA

26. ABI 3700 DNA analyzer.

Use of Computer

27. Computer, minimum requirement is Windows 98.
28. Chromas or similar sequencing software. Chromas is available for free download (noncommercial use).
29. Internet access.

PROCEDURES

The following description covers both PCR and plasmid sequencing methods and analysis of data for the detection of SNPs or mutations.

Qiagen Kit

Several methods of DNA preparation and purification are available, all of which work reasonably well. However, the most commonly used kit, with a high rate of success, is the Qiagen kit for plasmid purification. The following protocol is suitable when small quantities of DNA is required:

30. PCR product preparation. Pick a colony and transfer into 5 mL of LB-broth medium.
31. Grow overnight at 37°C with shaking at 220-250 rpm.
32. Centrifuge the overnight culture in 15-mL Falcon tube at 600g for 10 min at room temperature.
33. Discard the medium and resuspend the pellet in (250μL) Cell Resuspension Solution, which is provided in the Qiagen kit.
34. Follow the Qiagen Plasmid Miniprep Kit protocol, in the booklet provided.
35. Elute the DNA using 100 μL of nuclease-free water.

PCR Reaction

Before sequencing PCR products, it is important to run them on an agarose gel to ascertain the success of the PCR reaction.

36. Amplify the DNA samples using preoptimized PCR primers and PCR cycling conditions to amplify a region of the genome for SNP analysis.
37. Run samples on an agarose gel or Agilent Bioanalyser DNA chip to determine whether the PCR amplification has generated a single band.
38. Prepare 1% TBE agarose gel with ethidium bromide (1μL of 1mg/mL EthBr in 100mL gel). Load 5μL of the PCR product on the gel for each sample. Alternatively, load 1μL of each PCR product into the wells of a DNA 7500 chip and run on the Agilent Bioanalyser following the protocol provided with the instrument.
39. If a single band is present, purify the PCR product using Qiagen PCR Purification columns, following the protocol provided. If more than one band is seen, load the remaining PCR sample onto a 1% TBE agarose gel, containing ethidium bromide. Purify the correct band by cutting out of the gel with a scalpel blade and use a Qiagen Gel Extraction Kit to isolate DNA from agarose slices.

Cycle Sequencing Reactions

The following table indicates the quantities of starting DNA for sequencing reactions as recommended by Applied Biosystems (ABI), using its Big Dye Terminator sequencing reaction kit.

Template	Quantity
PCR product size:	
100-200bp	1-3 ng
200-500 by	3-long
500-1,000 by	5-20 ng
1,000-2,000 by	10-40 ng
>2,000 by	40-100 ng
Plasmid (single stranded)	50-100ng
Plasmid (double stranded)	200-500ng
Cosmid or BAC DNA	0.5-1.0μg
Bacterial genomic DNA	2-3 μg

These template quantities are recommended for all types of primers. The amount of PCR product to use will be dependant on the length and purity of the products.

Big Dye Cycle Sequencing Kit

This section describes how to prepare sequencing reactions with templates such as plasmids and PCR products. The first step in the sequencing of PCR products or plasmids is to incorporate fluorescent dNTPs. The automated sequencer detects fluorescence from four dyes that are used to identify A, G, C, and T extension reactions.

Each dye emits light at a different wavelength when excited by laser energy, thus all four dyes and reactions can be detected and distinguished in a single gel lane. All the sequencing reactions described here are carried out using dideoxy terminator chemistry. Several versions of ABI sequencing kits are available, and all have easy to follow protocol manuals. The reaction described here uses the ABI Prism Big Dye Terminator v3.1 cycle sequencing kit (Applied Biosystems, UK).

Dye terminator chemistry uses four 3¢-dye labeled dideoxynucleotide triphosphates (dye terminators) each labeled with a different dye. The exact base sequence of the primers used for the sequencing of a specific gene region are not relevant, but all sequencing primers should be designed with an approximate GC content of 50% and an annealing temperature of 50°C. Sequencing reactions are carried out using a single primer.

The reagents needed for sequencing PCR products are provided together as a "ready reaction" mix supplied by ABI, except for the primers and milliQ water, as shown in the following table. The dNTP mix in the ready reaction contains dITP in place of dGTP to minimize band compressions, and dUTP instead of dTTP, as dUTP improves the incorporation of the T terminators and results in a better T pattern.

1. For each sample add the following reagents in order:

Reagent	*Quantity*
Terminator Ready Reaction Mix	4.0 μL
SeqSaver dilution buffer	4.0μL
Template	See the table elsewhere in this chapter
Primer (forward/reverse)	3.2 pmol
Sterile water	Make up to 20 μL

2. Keep the reaction mix on ice.
3. Vortex and briefly centrifuge.
4. Place the reaction tubes in a PCR machine.
5. Typical run parameters using a Thermal Cycler follow:0

   Lid preheat, 110°C.
   Temperature gradient to 96°C at 1 °C/s.
   Hold at 96°C for 10s.

Temperature gradient to 50°C at 1 °C/s.

Hold at 50°C for 5 s.

Temperature gradient to 60°C at 1°C/s.

Hold at 60°C for 4 min.

Repeat cycle loop (25 cycles).

Store at 8°C.

Extracting Samples

After the extracted samples have undergone the sequencing PCR to incorporate fluorescent dNTPs, the PCR product must be purified to remove any remaining dye terminators, as these interfere with the data analysis. Several methods are available for postreaction cleanup. However, a widely used small-scale method is described in detail next:

1. Pipette entire contents of PCR reaction into a 1.5 mL microfuge tube.
2. Add 2µL of 250 mM EDTA.
3. Add 64µL of nondenatured 95% ethanol.
4. Close the tubes and vortex briefly.
5. Leave at room temperature for 15 min.
6. Place the tubes in a microcentrifuge and mark their orientations. Centrifuge the tubes for 20 min at 20,000g.
7. Carefully aspirate the supernatants with a separate tip for each sample and discard the tip. Pellets may or may not be visible.
8. Add 250µL of 70% ethanol and vortex briefly.
9. Centrifuge the tubes for 20 min at 20,000g.
10. Carefully aspirate the supernatants.
11. Repeat steps 8-10.
12. Place the tubes in the microcentrifuge in the same orientation as step 6 and spin for 10min at 20,000g.
13. Aspirate the supernatant carefully as in step 7.
14. Dry the samples in a vacuum centrifuge for 10-15 min or to dryness.

ABI DNA Analyzer

The samples are now ready to load on an ABI DNA analyzer and can be resuspended to the desired volume for the sequencer. The detailed protocol for setting up and running the ABI 3700 DNA analyzer is beyond the scope of this chapter. Full instruction and training on the DNA analyzer would be provided by ABI on installation of the analyzer.

In many cases, DNA analyzers are maintained and run by core facilities and specialized technical staff; therefore, the individual researcher need not have full knowledge of the machine parameters and programs.

Analyzing Results

Once the sequencing run is complete, the data are assimilated by computer using the software collection supplied with the DNA sequencer. The latest software upgrade is the ABI Sequencing Analysis Software v5.2 with KB Basecaller Software v1.2. The basic feature of the software is that it interprets the four fluorescent dyes (base calling), and produces the raw data nucleotide sequence as a display of peaks: an electropherogram.

An electropherogram is a graph measuring the intensity of the wavelengths corresponding to the four fluorescent colors at every nucleotide position. The intensity is registered as a colored peak on the electropherogram. The highest peak at every specific nucleotide position thus is allocated as the nucleotide corresponding to the color of that peak. The colors allocated for each nucleotide are

Red = T (thymine).

Blue = C (cytosine).

Green = A (adenine).

Black = G (guanine).

The sequence can be read from the electropherogram as the definitive nucleotide order. If the software were unable to interpret a particular fluorescent peak a letter *N* is assigned for that base position. This could be caused by two bases being present in one position (heterozygosity) or weak peak height.

In addition to the chromatogram trace, automated DNA sequencers generate a text file of sequence data by calling the bases associated with each peak (base calling). It is important to manually check the electropherogram against the predicted base, as errors are reasonably common. Commonly, errors occur near the beginning and the end of any sequencing run.

Sequence Analysis Programs

A number of sequence analysis programs are available. The most popular free program that displays the electropherogram is Chromas Lite. The program opens *chromatogram* files from Applied Biosystems and Amersham MegaBace DNA sequencers, in addition to opening SCF format *chromatogram* files created by ALF, Li-Cor, Visible Genetics OpenGene, Beckman CEQ 2000XL and CEQ 8000, and other sequencers.

Analyzing Sequence Data

The following protocol is recommended for analyzing the sequence data using Chromas.

1. Load Chromas.
2. Open the sequencing file (.abl). Click on File, Open.
3. Scan the chromatogram for sequence quality, that is, noise, check base calling and resolution by scrolling through the sequence.
4. Check the signal levels.
5. Search for Ns. Click on Edit, Search next N.
6. Search for heterozygotes and mutations by scanning the electropherogram for double

peaks or anomalies.

7. Compare the sequence to a known sequence (from database or previous sequencing) and search for sequence differences.
8. Save sequence text file as FASTA format.
9. The saved text file can be used in a number of sequence alignment programs to identify novel variants. Often, a sequence generated from a clinical diseased sample is compared against the sequence of a normal sample. Any sequence differences then are investigated to see if they play a role in the etiology of the disease.

BLAST Analysis

The Basic Local Alignment Search Tool (BLAST) finds regions of local similarity between sequences. The program compares nucleotide or protein sequences to sequence databases and calculates the statistical significance of matches.

1. Load the search engine (http://www.ncbi.nlm.nih.gov/BLAST/).
2. Select Nucleotide-nucleotide BLAST (blastn).
3. Enter the sequence (cut and paste the FASTA sequence into search box).
4. Select the genome of interest to search sequence against (i.e., Human).
5. To search, press the BLAST button.

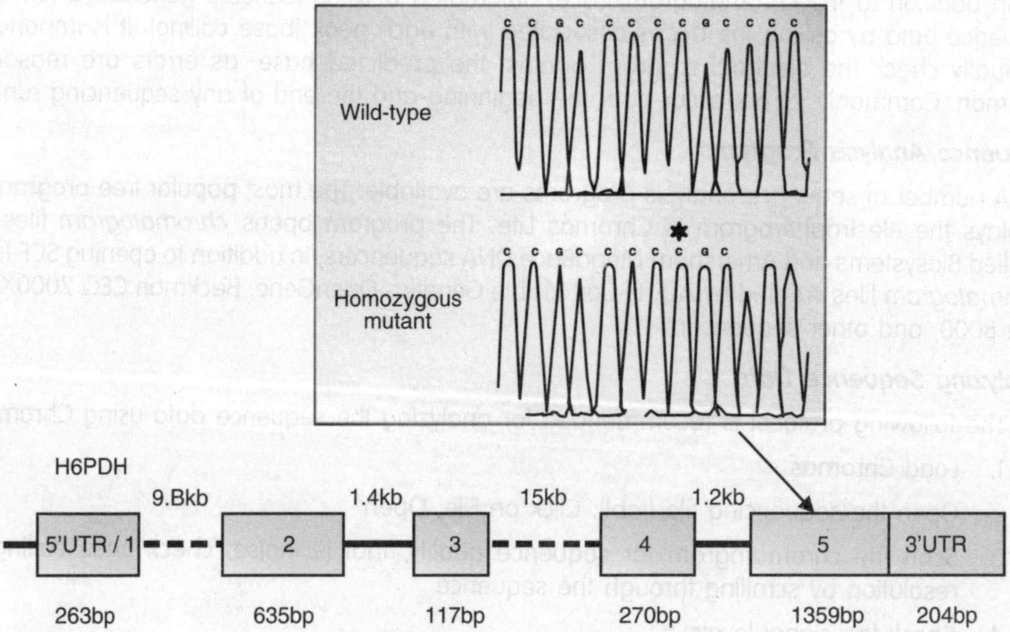

Figure 15.1: Diagrammatic representation of H6PDH gene showing position of mutation. Chromatogram showing single C-T base pair mutation in exon 5 of H6PDH, which causes the change R453Q: (top) wild type; (bottom)-mutant; indicates base change.

6. Press BLAST again on Format page. Wait for alignment results.
7. Search the alignment result for highlighted differences, that is, Ns or changed bases. These are polymorphism or mutations in the sequence.

BLAST analyses highlight regions in the newly generated sequence that do not match with the known sequence for that region in the GenBank database. These variations in the sequence can be termed an *SNP* if it is a single base change, or they may be other types of variants such as microsatellites, insertions, or deletions.

A single base change may be classified as a common polymorphism or SNP, however; if it changes an amino acid, it would be termed *a mutation,* as it may lead to a change in protein function. Figure elsewhere in this chapter shows that a difference of one base between sequences caused a homozygous mutation in exon 5 of the gene H6PDH, resulting in an amino acid change (R to Q), identified by BLAST analysis.

Individual BLAST analyses are not practical for analyzing multiple sequence results. For multiple sequence analyses, a number of programs are available whereby the known sequence can be uploaded and batches of unknown sequences can be aligned and compared to the known sequence, for example, DNAstar.

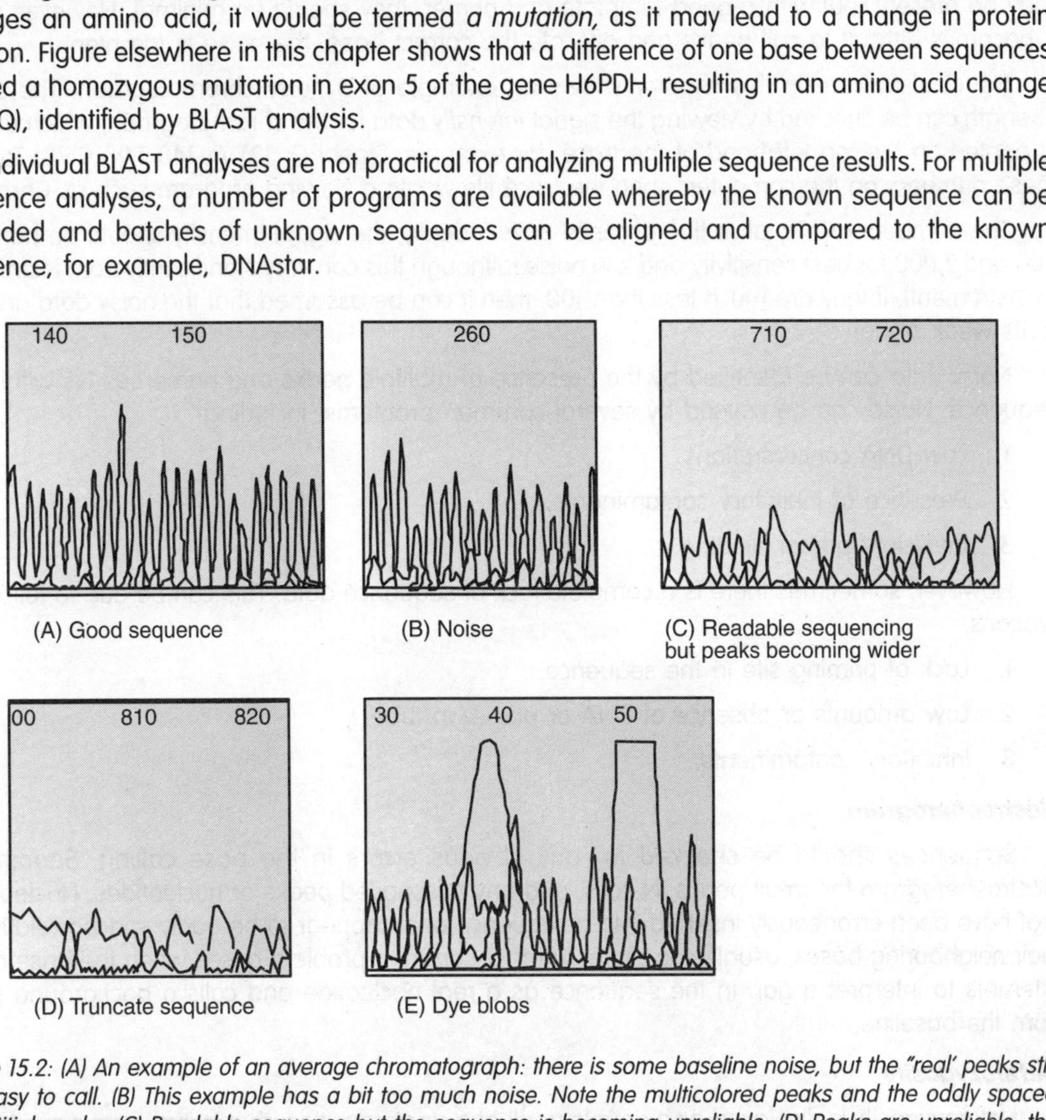

Figure 15.2: (A) An example of an average chromatograph: there is some baseline noise, but the "real' peaks still are easy to call. (B) This example has a bit too much noise. Note the multicolored peaks and the oddly spaced interstitial peaks. (C) Readable sequence but the sequence is becoming unreliable. (D) Peaks are unreliable, the sequence should be truncated. (E) Artifact, dye blobs.

Interpretation of DNA Sequencing

This section explains how to examine and interpret a DNA sequencing chromatogram. First, when examining an electropherogram, it is best to get an overview of how clean the sequence is. Second, the sequence should be checked for miscalls and truncated when errors become frequent.

The sequence chromatograph should have well-defined, evenly spaced peaks, each with only one color. Figure elsewhere in this chapter shows an example of a good sequence chromatograph. The height if each peak may vary up to threefold, and baseline (noise) peaks may be present, but with a good template and primer, they should be minimal. However, when it becomes difficult to call peaks and allocate the correct base, the noise is too high.

Figure elsewhere in this chapter shows an example of a noisy chromatograph. The signal strength can be checked by viewing the signal *intensity* data for the chromatograph, which usually is printed on the top left hand of the trace, for example, Signal G: 131 A: 140 T:98 C:78. To view these numbers on the computer, open the trace file within a viewing program such as Chromas.

The numbers are "relative fluorescence units." Ideally, the signal intensity should be between 500 and 2,000 for best sensitivity and low noise (although this can vary significantly from instrument to instrument). If they are much less than 100, then it can be assumed that the noisy data are due to its weak signal.

Noisy data can be identified by the presence of multiple peaks and numerous Ns within the sequence. Noise can be caused by several common problems, including

1. Low DNA concentrations.
2. Presence of inhibitory contaminants.
3. Inefficient primer binding.

However, sometimes there is a complete lack of sequence data. This can be due to following reasons:

1. Lack of priming site in the sequence.
2. Low amounts or absence of DNA or primer in tube.
3. Inhibitory contaminants.

Electropherogram

Sequences should be checked for any obvious errors in the base calling. Search the electropherogram for small peaks, N calls, and any misspaced peaks or nucleotides. Nucleotides that have been erroneously inserted into a sequence often appear to be oddly spaced relative to their neighboring bases, usually too close. Another common problem arises when the base caller attempts to interpret a gap in the sequence as a real nucleotide and calls a background peak from the baseline.

Heterozygosity

Heterozygosity is displayed as a double peak at a single position, that is, two peaks overlaying the same position. This is common when sequencing a PCR product derived from diploid genomic

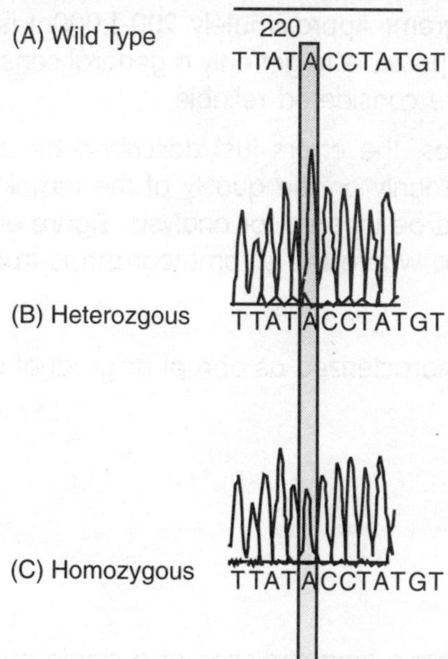

Figure 15.3: (A) Wild-type sequence; the sequence is a C. (B) Heterozgous sequence, two peaks showing GT. (C) Homozygous mutant sequence; the sequence is a T.

DNA, where polymorphic positions show both nucleotides simultaneously. Usually the two peaks show 50% intensity of a single neighboring peak (homozygous). Figure elsewhere in this chapter shows an example of a PCR amplicon from genomic DNA, with a clear heterozygous single-nucleotide polymorphism.

In this case, one allele carries a C, while the other has a T. However, a heterozgote can be missed by the base caller as the text sequence may show only a C. If all your other sequences also had a C at that location, the heterozgous SNP would be missed. The presence of multiple peaks within the sequence can be caused by numerous other factors that are not due to sequence polymorphisms. To help determine the cause, it can be useful to look at two aspects, where the multiple peaks begin and the overall signal strength of the sample.

Base Caller Program

Chromatograms stop giving accurate-data toward the end of the trace, because as the gel progresses, it loses resolution. This is normal as peaks broaden and shift, making it difficult for the base caller program. The sequencer will continue attempting to "read" the data, but errors become more and more frequent. Generally, up to approximately 800 nucleotides, the sequence is quite reliable.

The peaks will be broader and less well-resolved, but there still will be evident separation between them and usually few miscalls. Figure elsewhere in this chapter shows data from later regions in a normal chromatogram. Approximately 900-1,000nt is the very limit of resolution on an ABI 3700 DNA analyzer. It is possible to get only a general sense of the sequence at this point, as only a few base calls can be considered reliable.

As the sequence progresses, the errors just described become more and more frequent. Where this occurs depends primarily on the quality of the template. When the error rate is too high, the remaining data should be ignored for analysis. Figure elsewhere in this chapter shows an example of sequencing data where the chromatograph is truncated.

Truncated Sequences

Truncated sequences are characterized as abrupt or gradual and can be caused due to any of the following reasons:

1. Secondary structure.
2. Too much DNA.
3. Salts.

Homopolymeric Regions

Homopolymeric regions contain long stretches of a single nucleotide and can be difficult to sequence through accurately. Sequence data up to and including the polynucleotide region may be fine, but the last base of the poly region and all peaks following it may show a wavelike, stuttering pattern of double peaks that cannot be interpreted.

Sequencing Artifacts

The capillary-based system appears to be more susceptible to sequencing artifacts than the original slab-based gel sequencers, mostly due to the increased sensitivity of the instrument. There is a link between template preparation and the degree to which artifacts can be a problem.

Dye blobs are a common sequencing artifact, caused by unincorporated dye terminator molecules that passed through the cleanup columns and remain in solution with the purified DNA loaded onto the sequencers. They are seen most often with samples that have low signal strength. In general, dye blobs appear as broad, undefined peaks with the true DNA peaks underneath and tend to occur relatively early in, the data. An example of a dye blob is shown in Figure elsewhere in this chapter.

POINTS TO REMEMBER

1. Always start using a freshly grown culture (inoculate the previous day on an LB agar plate containing the appropriate antibiotic).
2. Elute DNA in water. Do not use a Tris-EDTA buffer, as EDTA inhibits the polymerase.
3. Cycle-sequencing reactions are made up in a final volume of 20μL. This volume allows for up to 8μL of DNA template and 4μL of primer (0.8 pmol/μL). If the sample DNA is not concentrated enough and you need to add more than 8 μL of DNA template, then compensate for the additional volume by using a more concentrated solution of primer.
4. For each DNA sample, it is advisable to prepare a forward and reverse reaction covering

the same region. This enables each sequence to be checked by comparing the forward- and reverse-complementary sequences to ensure 100% complementarity.

5. The Terminator reaction mix supplied with the Cycle Sequencing kit can be diluted to increase the number of samples that can be labeled. Dilution can provide for more than 200+ reactions from a 100 reaction kit (typical dilution: 4μL of Reaction Mix + 4μL of dilution buffer = 8μL).
6. Due to variation in thermal cyclers it is recommended to test a couple of samples using the new program before preparing a large number of samples.
7. After cycling is completed, samples can be stored at –20°C to prevent excitement of the fluorescent dyes.
8. To purify the sequencing samples, a number of high-throughput 96-well kits are available, such as UltraClear Sequencing Reaction Clean-Up Kit (Sigma, UC9601) or Centri-Sep 96-well plates (ABI, 4367819).
9. Precipitation times of <15 min result in a loss of very short extension products. Precipitation times >24h increase the precipitation of unincorporated dye terminators.
10. The supernatant must be removed completely, as unincorporated dye terminators are dissolved in them.
11. Do not overdry.
12. After drying, it is recommended that 10μL of Hi Di Formamide is added to each sample and left for 30 min to allow the DNA to resuspend before loading onto the 3700 DNA Analyzer.
13. A number of other sequence analysis programs are available that allow chromatograph visualization, such as DNAstar. This has a sequence analysis program, called *Lasergene,* which is a comprehensive software for DNA and protein sequence analysis, contig assembly, and sequence project management.

    BioEdit is another biological sequence alignment editor written for Windows 95/98/NT with many convenient features, making alignment, manipulation, and viewing of sequences relatively quick and easy.
14. Check the stock concentrations, calculations and dilutions.
15. The sample must be repurified.
16. In the cycle sequencing reaction, the primer and template annealing step occurs at 50°C. Therefore, increasing the primer Tm by adding additional bases to the 5¢ or 3¢ end to raise the Tm to be within the range of 52-58°C provides better sequence quality.
17. Check the plasmid maps and sequences to make sure the priming site is present in the vector.
18. Check the stock concentrations and dilutions.
19. The sequencing reaction is very sensitive to the presence of certain contaminants, some of which completely inhibit the sequencing enzyme. The sample may need to be repurified

to sufficiently remove the inhibitory components. The following table is a list of potential inhibitors and the amounts that are tolerable:

Contaminant	Amount tolerated in sequencing reaction
RNA	1µg
PEG	0.3%
NaOAc	5-10 mM
Ethanol	1.25
Phenol	0%
CsCl	5 mM
EDTA	0.25 mM

20. Note that the basecaller may mark the base position as an N, or it may call only the larger of the two peaks.
21. To detect all of the polymorphic positions in a large SNP-detection project, a computer program should be used to scan the chromatograms; for example, the phred/phrap/polyphred/consed suite of programs.
22. To obtain good sequencing results, it is important to download and examine the chromatogram. By using the text data alone, many anomalies maybe missed and the data may be invalid.
    (a) In multiple priming sites involving vectors, the primer may have a secondary hybridization site that may be identical or closely related, with different nucleotide sequences following each site, giving superimposed bands within the sequence. In this case, a new primer design is required.
    (b) In mixed plasmid prep, a plasmid prep that is contaminated by more than one product, such as two vectors, generally shows an early section of clean sequence data (common vector multiple cloning site sequence) followed by double peaks. Check that only a single colony is picked.
    (c) A frame shift mutationcan occur when one or more bases are inserted or deleted into the template DNA and if multiple products are present in the sample. There is clean sequence up to the point of the mutation, followed by double peaks caused by the shift in the nucleotide sequence.
23. For SNP detection, only high-quality sequence data can be used.
24. GC-rich DNA is predisposed to secondary structure formation, causing hairpins that can restrict the passage of the sequencing polymerase and thus be very difficult to sequence through reliably. A secondary structure may appear as a sharp termination of signal with no sequence data after, or if the loop has been relaxed slightly, there may be some weaker peaks following that are still quite accurate.

Adding a DNA denaturant such as DMSO to the sequencing reaction can help melt the duplex formation and allow the polymerase to pass through. Changing the cycle sequencing parameters to include a higher denaturation temperature (98°C vs. 96°C) and elimination of the 50°C annealing step sometimes is useful.

25. Overloading of DNA exhibits early topheavy peaks followed by rapidly weakening peak height and strength, leading to premature termination of signal.

26. Excessive amounts of salts give rise to premature termination with strong signal followed by progressively weakening signal. Salts have an inhibitory effect on the sequencing Taq polymerise, which can lead to an overabundance of short fragments. If salts are potentially a problem, an ethanol precipitation could be performed.

27. In repetitive regions, the nucleotide composition as well as the size of a repetitive region can play a large role in the success of sequencing through the area. In general, G-C and G-T (often seen in bisulfite-treated DNA) repeats are the most troublesome. Usually, it is possible to sequence partially through the repetitive region but the signal then begins to fade and eventually becomes unreadable.

    This may be due to premature dNTP depletion, secondary structure formation, or enzyme slippage. When sequencing cloned DNA with a homopolymer region, several options can be tried. In the BigDye Terminator sequencing chemistry, dTTP has been replaced with dUTP, which lowers the melting temperature of the DNA.

    An alternative sequencing chemistry can be used: dRhodamine chemistry, where dTTP is still in the reaction and generally gives better results through polyA regions. The addition of DMSO or alterations in cycle sequencing parameters can help. If the repeat region is not excessively large, sequencing from the opposite strand to complete the region can be successful.

Chapter 16

Evolutionary Footprinting

The identification of transcription factor binding sites (TFBSs) is an important part of defining the regulatory network of an organism. TFBSs exert significant control over gene transcription through the binding of their cognate transcription factors (TFs). The promoters of genes regulated by a common TF, either within a species or across species, can be analyzed to predict potential regulatory sites.

Experimental methods exist to detect coexpressed genes within a species, and can be used to identify a subset of sequences for analysis. Phylogenetic footprinting, on the other hand, does not require experimental data, and instead uses orthologous sequence data from multiple species (genomes). This method relies on the assumption that among closely related species, orthologous genes are likely to be regulated by a common TF; thus, the TFBSs will be conserved, whereas the nonregulatory portion of the promoter sequences will be less conserved.

Phylogenetic footprinting has been successfully applied to both prokaryotic species and eukaryotic species to locate putative TFBSs. Although there have been a number of computational tools developed for detecting conserved sequence elements, the Gibbs Motif Sampler is one of the most mature. Gibbs sampling was first applied in the field of bioinformatics in 1993, and there have been numerous enhancements to, and applications of, the original Gibbs sampling technique since its introduction.

Here, the method is described only conceptually; mathematical details of the procedure as applied to the search for motifs in biopolymer sequences are described elsewhere. A key feature of the Gibbs sampling approach is the use of motif models to capture sequence patterns shared by multiple sequences. A motif is a common model of a collection of binding sites. The motif is usually represented as a position-weight matrix in which each row represents a position in the conserved pattern and each column represents a nucleotide.

The elements of the matrix are typically probabilities or counts of nucleotides. The advantage of using motif models of multiple sequence alignments lies in the increase in the signal-to-noise ratio that results from averaging over the individual sequences. However, this advantage decreases when the sequences are correlated (possess a high degree of identity), as is the case for orthologous data from evolutionarily close species, in which insufficient time has elapsed for mutations to accumulate in nonsite regions.

The Gibbs Motif Sampler also uses a principled Bayesian method for determining the number and positions of the TFBSs, and provides a number of options designed to model biological features of TFBSs, such as palindromic and directly repeating models. Although Gibbs employs a relatively mature algorithm, it is important to bear in mind that experimental validation is ultimately necessary to substantiate the biological function of predicted motifs.

REQUIREMENTS

Accessing the Gibbs

There are two ways to access Gibbs for analyzing data:

1. Web interface: Gibbs may be accessed on the internet at http://bayesweb.wadsworth.org/gibbs/gibbs.html and at http://www.bioinfo.rpi.edu/applications/bayesian/gibbs/gibbs.html. These websites also provide access to a user manual, various data files, and online descriptions of phylogenetic footprinting and the analysis of prokaryotic coexpression data. Furthermore, each option and data input field is linked to context-sensitive help describing its use.
2. Command line program: Gibbs is available as a command-line tool for the Solaris, Solaris.x86, and Linux operating systems. In addition, a version is available for Microsoft Windows, which requires Cygwin (http://cygwin.com/).

   Gibbs is distributed free of charge to the academic research community for noncommercial, nonprofit internal research use. A license request may be obtained at http://bayesweb.wadsworth.org/GIBBS-SAMPLER-ACADEMIC.htm. A license for commercial use may be obtained at http://bayesweb.wadsworth.org/GIBBS-SAMPLER-COMMERCIAL.htm.

   A version for parallel computer systems supporting the Message Passing Interface is also available from the authors for computing clusters using the Linux, Solaris, or Solaris.x86 operating systems. This chapter will focus on the use of the command-line tool, but most of the comments will also apply to the web-based version.

Materials

1. Hardware: a Sun workstation running Solaris or an x86-based PC running Solaris.x86, Linux, or Microsoft Windows are required.
2. Software: the current command-line version of Gibbs is 3.00 or later. The package as distributed contains the appropriate Gibbs binaries, the binaries to perform a Wilcoxon signed-rank test of statistical significance, and the unifiedcpp binary, which produces the background composition files for use as input to Gibbs.

   The Cygwin package is needed for the Microsoft Windows version. X-Windows support is required for the Solaris version. Memory and processor speed requirements vary depending on the size of the data sets being analyzed. The examples in this article can be run quite comfortably on a laptop computer with 512 Mb of memory and an Intel Pentium processor.

Methods

Gibbs is distributed as a gzipped tar file. To install Gibbs, copy the Gibbs.tar.gz file to a directory. In the next example, we use the directory Gibbs, but any appropriate directory name may be used. At a command-line, type (user entry is in bold):

```
~/  Gibbs>gunzip-Gibbs.tar.gz
~/  Gibbs>tar-xvf  Gibbs.tar
```

The "tar" command will create the appropriate subdirectories for the various versions. The top level directory will contain a README file and a sample data file, crp.dat. The subdirectories will contain the binary files for the appropriate operating systems.

Sequence Data

Gibbs accepts sequence data in FASTA format. The Gibbs distribution contains a sample data file, crp.dat. This file, along with all data files used in this article, is available for download at http:/ /bayesweb.wadsworth.org/ gibbs/module.

PROCEDURES

Gibbs has a large number of options and modes of operation . In this chapter we will concentrate on the subset of these options that are used for predicting TFBSs in cross-species data. The following examples illustrate the principles involved in computational detection of TFBS in prokaryotes, however, the principles are similar for the analysis of eukaryotic sequences. We will point out differences in the analysis of eukaryotic sequence data in the Note section.

Sequence Phylogenetic

In phylogenetic footprinting, the sequences being analyzed represent orthologous promoter sequences, and as such are assumed to contain binding sites for a common TF. We typically chose a target species of interest and additional species based on their phylogenetic relationship to the target species. Orthologous promoters are identified by first identifying orthologous gene sequences in these species.

If the species' genomes have been annotated, a pairwise reciprocal BLAST methodology such as INPARANOID will efficiently predict orthologous genes from their protein translations. Alternatively, using the protein sequences from only the target species, a TBLASTN procedure may be used to identify orthologous genes in the raw genome sequences of the related species.

Briefly, the protein sequences from the target species' genome (which must be annotated) are used as TBLASTN queries to a nucleotide database composed of the target species' genome and the genomes of the related species. A number of heuristics are then applied to ensure that BLAST hits are likely true orthologs, rather than paralogs or domain-level matches:

1. The expectation value must be less than 10^{-20}.
2. The expectation value must be less than the second best hit in the target species.
3. The BLAST hit must start within 20 amino acids of the target query sequence.

Once orthologous genes are identified by either of the previously listed approaches, the orthologous upstream intergenic sequences are extracted. If orthologs have been identified by INPARANIOD from fully annotated genomes, the annotations are used to define intergenic sequence boundaries. Using the TBLASTN procedure, the boundaries of an intergenic region can be delineated using the BLAST results for the bounding orthologous gene, if the BLAST results indicate that the gene order between species is conserved.

If conserved gene order is not detected, a maximum of m bases (= 500 by by default), upstream of the orthologous gene is used. With this approach, the target species data will include only intergenic sequence, whereas the sequences for the additional species may or may not be trimmed to exclude upstream coding region. For eukaryotes, similar techniques may be used; however, the intergenic regions extracted are typically much larger. In these cases, we limit the sequence length, m, to a maximum of 3000-5000 bp.

Parsing INPARANOID or BLAST output and generating orthologous intergenic sequence files on a whole-genome scale is generally accomplished using *ad hoc* Perl scripts. The two examples shown next describe the use of Gibbs to predict TFBSs upstream of orthologous *lexA* genes. The file, gamma lexA.fa, contains the *lexA* promoter sequences for *Escherichia coli K12* and six additional y*-proteobacterial species: *Salmonella enterica* serovar Typhi CT18, *Yersinia pestis CO-92, Vibrio cholerae El* Tor, *Haemophilus influenzae* Rd, *Pseudomonas aeruginosa PAO1*, and *Shewanella oneidensis.*

The file, alpha lexA.fa, contains the *Rhodopseudomonas palustris lexA* promoter sequence plus seven additional orthologous sequences from a-proteobacterial species: *Bradyrhizobium japonicum, Brucella suis* 1330, *Caulobacter crescentus, Rhodobacter sphaeroides, Rhodospirillum rubrum, Novosphingobium aromaticivorans,* and *Hyphomonas neptunium.* Both these example files were generated using the TBLASTN procedure previously described, using the LexA protein sequences from *E. coli* and *R. palustris,* respectively, as the queries.

Sampling

Gibbs sampling is a Markov-chain Monte Carlo technique and site positions are sampled into the motif model based upon a probability model. The probability that a particular position in the sequence is sampled as a site is calculated as the ratio of the probability of the site under the motif model to the probability under a background model. The background model describes the sequence in the absence of TFBS.

Using a background model that is homogeneous in composition is problematic, because noncoding sequence is often heterogeneous in composition, particularly in eukaryotes. Local variations in nucleotide composition have been shown to adversely affect sequence alignments. To address this, the Bayesian segmentation algorithm is used to produce a position-specific background model.

The algorithm calculates the probabilities of observing each of the four bases at each position in a sequence, based on the sequence's compositional heterogeneity and on the uncertainty in the heterogeneity. These probabilities are then used as position-specific background models in the Gibbs sampling procedure. To create a background composition file, use the unifiedcpp program, supplied with the Gibbs package:

```
~/ Gibbs>unifiedcpp gamma_lexA.fa
```

Unifiedcpp will produce a number of files in the directory containing the FASTA file. These files will begin with the same name as the FASTA formatted sequence file, but will have an additional extension. The file ending with info-det, in this example gamma lexA.fa info-det, contains the background probabilities.

This file will be used as an input parameter to Gibbs. The background composition file does not have to be recreated unless the number of sequences or length of the sequences in the FASTA file changes.

Building Options

We have attempted to build a number of options into the Gibbs software to model the biological properties of TF-TFBS interactions, because the more accurately a computational analysis reflects the underlying biology of the system under study, the more likely it is to be successful. Therefore, it is important to consider the biological context of the data being analyzed, and use this as a guide when choosing Gibbs parameters.

For example, many prokaryotic TFs bind as symmetric, homodimeric protein complexes that have corresponding palindromic DNA binding motifs (e.g., *E. coli* Crp), whereas others bind as directly repeating multimers and therefore have direct-repeat binding patterns (e.g., E. *coli* PhoB). These binding site structures (palindromes and direct repeats) can be specified using options in Gibbs.

In addition, the exact width of the binding pattern may-not be known. Gibbs can infer this width using a modified version of the fragmentation algorithm. The command-line parameters control how the program makes inferences about these features. Gibbs sampling is a stochastic process and thus also provides several parameters to control program runtime. Gibbs uses the posterior probability of the alignment, the MAP (maximum *a posteriori* probability), as a measure of the quality of the alignment.

The MAP is calculated as the log of the alignment probability minus the log of the empty or "null" alignment. Thus, it is a measure of the extent to which a particular alignment is better than background. Gibbs sampling begins by generating a random alignment, then iterating through the sequences, examining each in turn and sampling motif sites.

During each sampling iteration, the program updates the background and motif counts by adding the positions occurring in each motif element to motif counts and deleting them from the background counts.

Sampling proceeds until a plateau is reached in the MAP values (i.e., the MAP does not improve for some specified number of iterations). For details of the sampling process, see refs. elsewhere in this chapter.

Because it is possible for the Gibbs sampling procedure to become stuck in local optima, the program performs a number of random restarts called seeds. Command-line parameters control the number of seeds, the maximum number of iterations, and the plateau period. The user can obtain a complete list of options as follows :

```
~| Gibbs> Gibbs -h
```

The user can also obtain more extensive information about these options by consulting the website (http://bayesweb.wadsworth.org/gibbs/bernoulli.html).

Experimental Evidence

Using informed priors with Gibbs is another way in which information on the biological system under study can be leveraged; two commonly used types of prior information are described here. In some cases, there may he experimental evidence about which TF regulates a given gene, and it may be possible to obtain position weight matrix (PWM) for the specific TF. For example, the PRODORIC, TRANSFAC, and JASPAR databases contain PWMs for a number of common TFs.

These matrices can be used to provide clues about the expected binding patterns of TFs, but do not control the inference of sites and motifs by Gibbs. Also, an investigator may be interested in a TF that is a member of a family of TFs, but for which there is no available PWM. In these circumstances, an informed prior from the family can enhance the sensitivity of this algorithm.

Details of using PWMs as informed priors can be found at http://bayesweb.wadsworth.org/gibbs/prior.html. In the absence of prior motif information, Gibbs will calculate uninformed priors based on the average background composition. In our examples, we use uninformed prior models. In addition, it is known that many TFs bind to more than one site in a promoter and often bind cooperatively, particularly in eukaryotes.

Thus, we allow Gibbs to predict more than one site per sequence. In our example sequences, we have reasonably high confidence that among closely related species, an orthologous gene will be regulated in a similar way. Therefore, we allow Gibbs to search for a motif with up to two sites per input sequence and provide prior information that there is relatively low prior probability of finding no sites (p = 0.05), and equal probabilities of finding one or two sites (p = 0.475) in each sequence. To do this, we create a text file (called a prior file) containing the following statements:

```
>BLOCKS
0.05  0.475  0.475
>
```

In our examples, this text file is named lexA.pr and was provided as an argument at the Gibbs command-line. Further details on prior information may be found at the URL previously mentioned. As discussed next, the Gibbs sampler also permits searching for motifs of multiple TF simultaneously.

Binding Sites

This first example searches for the LexA binding site in the *lexA* promoters from seven -y- proteobacterial species. Enter the following command to analyze the _*lexA* sequences in gamma lexA.fa.

```
~/Gibbs> Gibbs gamma_lexA.fa 16 7 -n -E 2 -r -R 1, 1, 8 -M 1,
24-B gamma_lexA.fa_info-det-PlexA.pr-ogamma_lexA.out
```

The parameters are:

1. gamma _lexA.fa: the file name of the FASTA format sequence file.
2. 16: the initial motif width. This is the number of conserved positions in the TFBS. Fragmentation will allow the site to expand.
3. 7: the initial estimate of the total number of sites. For cross-species studies, we typically

set this estimate equal to one site per input sequence.

4. -n: indicates that the FASTA format file contains nucleotide data and a DNA alphabet should be used.
5. -E 2: the maximum number of sites allowed in each sequence. In this case, each sequence may contain zero, one, or two sites.
6. -r: disables searching of the reverse complement of the input sequence data.
7. -R1, 1, 8: specifies that motif number one is a palindrome. The first eight conserved positions will be combined with the reverse complements of the last eight positions to form the palindromic model. When using a palindromic model, searching for sites on the reverse complement of the input sequence data should be disabled using -r option.

   Not doing so when using a palindromic model generally results in an artificially asymmetrical motif model. When searching for non-palindromic models, it is often useful to search for TFBSs in both the forward and reverse complement directions.

   Even with non-palindromic models, searching both strands of the DNA may not be desired, however, if the input sequence data have been oriented with respect to the gene, as in the direct-repeat example below.
8. -M 1,24: set the maximum fragmentation width for motif model number one. This setting allows Gibbs to search for TFBS with widths between 16 (previously specified) and 24 bp.
9. -B gamma_lexA.fa info-det: specifies the background composition file created by unifiedcpp.
10. -P lexA.pr: specifies the prior file containing the probabilities of the number of sites per sequence.
11. -o gamma_lexA.out: specifies the file that will be created by Gibbs to contain the results. If this option is omitted, the Gibbs output is written to the terminal window (STDOUT).

Except for the FASTA file name, the motif width, and "-n," which indicates that the FASTA file contains nucleotide data, the other parameters are strictlyn optional. However, note that if an estimate of the total number of sites and the maximum number of sites allowed per sequence are not specified, Gibbs will default to a site sampling mode in which exactly one site will be identified in each sequence, a behavior that is generally not desirable for phylogenetic footprinting.

There are a number of technical command-line options that control behavior of the program. These are set to default values that we have found useful in our research, but may require adjustment depending on the type and amount of sequence data to studied be . These parameters include:

1. -s: -s followed by an integer indicates the number of random restarts (seeds) of the Gibbs sampling process. The default is 20 seeds.
2. -p: -p followed by an integer sets the plateau period. For example, -p 50 sets a plateau period of 50 iterations; if there is no improvement in the alignment after 50 sampling iterations, then Gibbs will initiate the next random restart. The default value for this parameter is 50 iterations.

3. -i: -i followed by an integer sets the maximum number of iterations for each seed; the default is 500. When a large number of sequences are being examined, the default number of seeds and plateau period may be too low to effectively search the alignment space. Thus, as the amount of sequence data increases, it is a good idea to increase the number of seeds and the plateau period.

   If the plateau period is increased,_ the maximum number of iterations may be reached before a plateau in the MAP is reached. To avoid this problem, the maximum number of iterations may be increased by including the option, -i, followed by an integer. A rule of thumb that has proven useful is that the maximum number of iterations should be at least five times the plateau period. For example, if the option, -p 200 were included, including the option, -i 1000 would be appropriate.

As it executes, Gibbs will write some temporary progress information to the user's console. Running a file such as the one in the example (seven sequences, each less than 500 bp) should take from 1 to 2 s to slightly less than 1 min on most computers. The amount of time required to process a set of sequences will depend on the length of the individual sequences, the number of sequences, the number of seeds, and the plateau period.

Results

Figure 16.2 hows a portion of the output file that results from running the example previously described. The complete output file from this example can be downloaded from http://bayesweb.wadsworth.org/gbbs/module/ and a complete description of the Gibbs output format can be found in, as well as on the Gibbs website. At each sampling iteration, the posterior probability (the MAP) of the sampled alignment is calculated and saved.

When all the seeds (random starts) have been run, Gibbs saves the alignment with the maximum MAP. Then, for this alignment solution, the algorithm continues to sample sites for a fixed number of iterations, to explore variations in the models. Gibbs counts the frequency at which each site is sampled to assess its reproducibility; these frequencies are sampling estimates of the probability that each site belongs to the common motif model. Sites selected with a frequency greater than 50 % are also displayed in the output.

These sites represent a frequency solution. The frequency solution is often the same as the optimal MAP solution, but because it represents the reproducible samples from among the possible alignments, it is less likely to contain false-positives than the MAP solution.

Figure 16.2 A shows the motif model detected in the 'y-proteobacterial *lexA* example. This model consists of a table with a column for each of the nucleotides and a row for each conserved position in the motif.

The numbers in the table indicate the frequency of occurrence of each nucleotide at each position within the motif. The last column is an information parameter, expressed in bits, that indicates how much the column adds to the model. Using the information value, it is possible to determine which positions are most conserved. Figure 16.2 B shows the TFBSs predicted in the frequency solution in each species.

The first column identifies the sequence number, followed by the motif element number for that sequence. The next column indicates where the motif element starts within' the sequence.

The fourth column contains the predicted TFBS in upper case; flanking sequences are shown in lower case. The motif element is followed by the ending site position within the sequence.

Column six shows the sampling estimates of the probability of each of the sites in the solution. Finally, a portion of the FASTA header is displayed. The row of asterisks below the binding site indicates the fragmentation pattern. In this case the motif model was not fragmented. Figure 16.2 C shows the MAP solution. LexA is the TF that regulates the DNA damage response (SOS response) and is widely distributed across prokaryotes.

In *E. coli,* more than 30 genes are likely regulated by LexA as part of the SOS response regulon, and LexA regulates its own expression via auto-regulatory binding sites upstream of the *lexA* gene. In *E. coli* and related γ-proteobacteria, LexA is known to bind to a palindromic DNA sequence with the consensus, CTG-N_{10}-CAG. This is the pattern of the TFBSs shown in Figure 16.2.

The probability values in column six of Figure 16.2 C indicate that most of the sites have high reproducibility (0.9 or greater in this case) and thus we have high confidence that they match the predicted motif. However, several sites have low sampling probabilities and should be regarded with suspicion. These sites are not included in the frequency solution.

Orthologous Promoter Sequence

Because it is usually not known what types of TFBSs to expect in a given set of orthologous promoter sequences during phylogenetic footprinting, we typically do not use a motif PWM in our prior information, and we run Gibbs on the data several times using different parameters, for example, specifying a palindromic, non-palindromic, or direct repeat model. Gibbs motif predictions are then collected and considered potentially significant if they have a positive MAP.

For example, running Gibbs with the α-proteobacterial promoter sequences in the file alpha lexA.fa and using the same parameters as previously described, with the exception of varying the option for model type, produced the distinct motif alignments shown in Figure 16.3.

Specifying a palindromic model resulted in a motif consisting of only three sites. Running the data with either a non-palindromic or direct repeat model produced similar alignments: both alignments are composed of sites from seven of the species, however the sites are shifted slightly and fragmented differently in the two alignments.

To specify a direct repeat, a -I parameter was used in place of the -R parameter previously used for a palindromic model, and a non-palindromic model was specified by not using either of these parameters. In addition, the -r option can be omitted when using direct repeat or non-palindromic models, thereby allowing Gibbs to search for sites on the reverse complement DNA strand, as well as the forward 'strand.

However, we included the -r option in this example, because the sequences in alpha_lexA.fa are oriented 5' to 3', with respect to the promoter of interest; thus searching both DNA strands is not required.

The motifs in Figure 16.3 B, C exhibit a direct repeat pattern, GAAC-N_7-GAAC, that has been shown to bind the *R. palustris* LexA protein in gel-shift experiments. Furthermore, the alignment in Figure 16.3 C had the highest MAP, suggesting that of the three models tested, a direct repeat model is the most appropriate model for the a*-proteobacterial LexA binding sites.

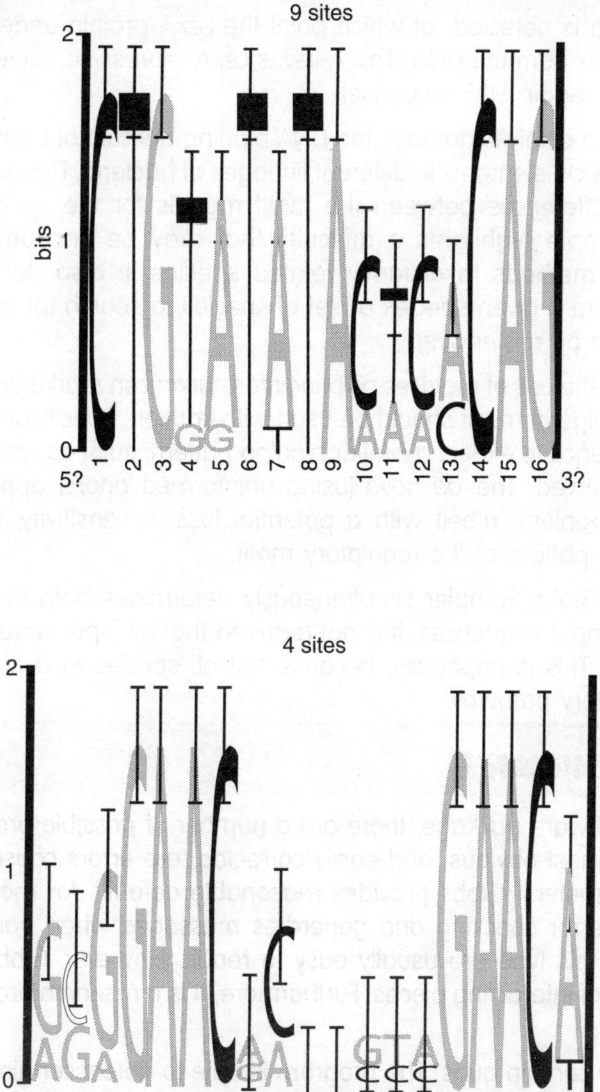

Figure 16.1: Sequence logos for the -γ-proteobacterial LexA sites (top) compared to the a-proteobacterial LexA sites (lower).

In each of these cases, the MAP and frequency solutions were identical. Note that, regardless of the Gibbs parameters used, none of the motif patterns in Figure 16.3 match the LexA motif in Figure 16.2 from the γ-proteobacteria. The observation that the LexA protein is conserved across bacterial lineages, could lead one to naively assume that the LexA binding motif would be conserved across a broad spectrum of species.

However, this is not the case, because the conserved region of LexA is involved in catalysis rather than DNA binding. Specifically, the LexA repressor remains bound to its cognate binding sites until DNA damage is detected, at which point the LexA protein undergoes auto-proteolytic cleavage and dissociates from the DNA. This relieves LexA repression, allowing the expression of genes involved in DNA repair (SOS response).

Unlike the conserved catalytic domain, the DNA binding domain of LexA is not well conserved and recognizes different cis-elements in different lineages of bacteria. Figure 16.4 shows sequence logos illustrating the differences between the motif models for the -γ- and α-proteobacteria, respectively. This example highlights a difficulty that may be encountered when applying comparative genomic methods to distantly related species. It also demonstrates that using precomputed PWMs from a given species or set of species to search for cis-regulatory elements in another genome can be misleading.

Therefore, although the use of matrices as prior information can markedly improve the sensitivity of the Gibbs sampling algorithms, it should be used with caution. Specifically, they should be used only when there is evidence to expect a similar binding pattern, such as when DNA binding motifs are known to be conserved. The *de novo* (using uninformed priors) approach described here avoids this potential problem, albeit with a potential loss in sensitivity, because it makes no assumptions about the pattern of the regulatory motif.

Also, because the Gibbs Sampler simultaneously determines both the motif model and its distribution across the input sequences, it is not required that all input sequences contain a given cis-regulatory element. This is important, because not all species in a comparative study may employ a given regulatory pathway.

POINTS TO REMEMBER

1. As with any software package, there are a number of possible problems that may arise in its use. The most obvious, and easily corrected, are errors caused by incorrect syntax or use of parameters. Gibbs provides reasonable defaults for most parameters. It also has extensive error checking and generates messages when parameter values are in error. Errors of this type are usually easy to repair. However, Gibbs is a large program with a number of interacting pieces. Furthermore, it is a research program under continual modification.

   As such, it may contain bugs. The program is able to detect certain conditions that might lead to a fatal error. When such a condition is encountered an error message beginning with the following phrase is output:

   ```
   FATAL INTERNAL ERROR : :
   ```

   If the user receives such a message, please contact the authors for assistance. Other problems are more subtle. They are typically caused by naive choices of the data sequences or inappropriate parameter choices.

2. We have tried to show how one method of analysis, phylogenetic footprinting, can be used to predict TFBS, particularly in bacterial species. Analysis of regulatory regions from higher organisms (e.g., vertebrates) has additional complications relative to studies in prokaryotes. A number of repetitive elements have been identified in eukaryotic genomes, and it is useful to mask these repeat sequences using RepeatMasker prior to attempting

motif prediction.

Furthermore, although prokaryotic promoter regions are typically short (less than 500 bp), in eukaryotes the upstream regulatory region may be quite a bit larger, extending to thousands or tens of thousands of bases. Also, eukaryotic TFBS tend to be somewhat shorter (10-14 bases on average) than those in bacteria, and are less likely to be palindromic.

The sites also typically occur in clusters, called regulatory modules, that consist of sites for multiple different TFs. Therefore, it is useful to search eukaryotic promoter data for multiple different motif models simultaneously. This can be done with Gibbs as follows:

~/Gibbs> **Gibbs file. fa 10,10,10 5, 5, 5 -n -E 5 -P file.pr -B file. fa_info-det -p 200 -i 1000 -o file. out**

In this example command-line, three motifs are specified, each with an initial motif width of 10, and an initial estimate of number of sites of 5. We will allow from zero to five sites in each sequence. Even though eukaryotic TFBSs tend to occur in modules, in a collection of relatively long orthologous eukaryotic promoter regions, it is quite possible that the short regulatory motifs (TFBSs), will be lost in the background noise.

It has been shown, however, that in a collection of orthologous human and rodent promoters from skeletal muscle genes, 98 % of experimentally defined sequence-specific binding sites of skeletal-muscle-specific TFs are found in the 19 % of human sequences that are most conserved in the orthologous rodent sequences. In practice, this means that by concentrating on the highly conserved regions of the sequence, we can increase the signal to noise ratio in the data and more effectively predict TFBS.

This can be accomplished by prealigning the promoter sequences from orthologous genes and sampling simultaneously from the aligned groups of sequences. Despite this, it is difficult to analyze promoter regions from single genes successfully using mammalian cross-species data, because of the relatively recent divergence of mammals. As additional mammalian sequences are added to an orthologous data set, the correlations in the sequence data tend to limit the marginal contribution from the sequences of multiple, related mammalian species.

This version of the Gibbs sampler does not account for the phylogenetic relatedness of input species, but a version that takes into account the phylogenetic relationships among the input species is now under development, and will be included in updated releases.

3. Although this chapter has focused on phylogenetic footprinting, Gibbs is also commonly used to predict regulatory elements in coexpression data for a single species. Advances in technology have made feasible, in select organisms, the detection of coregulated genes, as well as protein-DNA interactions under a variety of physiological conditions. Traditional microarray techniques and promoter fusions are commonly used to identify coregulated genes. Protein-DNA interactions are readily detected using gel-mobility shift assays or ChIP-chip assays. In addition, cutting edge technologies like ChIP-PET have potential for whole-genome analysis of cis-regulatory elements in eukaryotes.

   These types of experiments are used to identify genes whose coexpression is owing, at least in part, to regulation by a common TF, and Gibbs sampling can be used to identify the putative binding motif for the factor. We expect high-throughput experimental data, especially expression array data, to contain more noise, in the form of false-positives and secondary effects, than other more targeted techniques; therefore, additional optimization may be needed when using Gibbs to search for conserved sequences.

For example, when analyzing microarray gene expression data, which provides evidence of coexpression, but not direct evidence of a commonality in the regulatory mechanisms, the prior probabilities on the number of sites per sequence can be changed to reflect the fact that only a subset of coexpressed genes may be regulated by a particular TF or module of TFs, by setting a higher expectation of zero sites.

This setting reflects our expectation that more of the input promoter sequences are not regulated by a single regulatory mechanism, and thus are more likely to contain zero sites from the predicted module. Examples of the analysis of coexpression data are provided online at http://bayesweb.wadsworth.org/web help text.CE.apr232007.html. The examples presented here, and in the supplementary online resources, demonstrate how the Gibbs Motif Sampler can be used to detect conserved regulatory motifs.

The convergence of high-throughput sequencing initiatives and parallel experimental methods is providing the data necessary to delineate the complex regulatory networks of many organisms. The Gibbs Motif Sampler is under continuous development and future versions will allow us to better utilize this wealth of genomic and experimental data.

Chapter 17

DNA Libraries and Mutagenesis

Random mutagenesis of DNA is a powerful tool to study gene function and protein structure and is a prerequisite for preparation of randomized DNA libraries, required for many genetic selections. Currently used methods for preparing DNA libraries fall into a few broad categories, each with its own benefits and disadvantages. Therefore, choosing a suitable mutagenesis strategy is a key step for obtaining libraries with the desired complexity for successful selections.

Conventional in vivo random mutagenesis involves the treatment of cells containing a gene of interest, usually in suitable expression plasmids, with chemical mutagens such as 2-aminopurine, 5-bromouracil, ethylmethane sulfonate, or *N*-methyl-*N*¢-nitro-*N*-nitrosoguanidine (MNNG). These chemicals are DNA-modifying agents, which are highly toxic and require special care in handling.

Moreover, these chemical mutagenesis methods produce extremely low frequencies of mutation, and only a certain type of mutation can be generated. For example, MNNG preferentially produces GC-to-AT transitions at only 1 mutation per 10,000 by. Recently, the use of toxic chemicals for mutagenesis has been superseded by the use of bacterial strains carrying single or multiple mutations in the DNA repair pathways, such as the recently developed XL1-red mutator strain (Stratagene).

This mutator *Escherichia coli* carries a combination of a DNA repair deficiency *(mutT, mutS)* and a defect in the c-subunit of DNA polymerase III *(mutD),* which produces mutations during DNA replication. To generate mutations in the gene of interest, the plasmid DNA is simply transformed into these bacterial cells and random mutations are introduced during each cycle of DNA synthesis.

The use of such mutator strains is by far one of the simplest mutagenesis strategies, and a very large library of randomly mutated target DNA cloned into a plasmid vector can be obtained. However, this strategy can provide only relatively low mutation frequencies. For example, we found that XL1-red triple mutator strain produced approx 1 mutation per 2000 by of the target DNA, although the mutation frequency was slightly increased by expanding the culture period to longer than the overnight period suggested by the manufacturer.

Among the most popular in vitro mutagenesis strategies is the error-prone polymerase chain reaction (PCR), which is based on the creation of mistakes in the incorporation of nucleotides into newly synthesized PCR products. A commonly used thermostable DNA polymerase in error-prone PCR, *Taq* DNA polymerase, already possesses high intrinsic error rates ($1/10^4$-10^5 bps) because it lacks detectable 3¢-5¢ proofreading exonuclease activity (for the error rate of commercially available thermostable DNA polymerases, see http://micro. nwfsc.noaa.gov/protocols/taq-errors.html).

Several conditions are known to enhance the misincotporation by *Tag* DNA polymerase. These include raising the Mg^{2+} concentration to stabilize noncomplementary pairs, the addition of Mn^{2+} to diminish template specificity, use of dNTP analogs or alteration of the nucleotide ratio to promote misincorporation, and use of a high amount of *Taq* DNA polymerase to overcome the mismatch positions.

Other factors that influence misincorporation events during PCR include the number of amplification cycles and the initial number of the template copies. Basically, in a reaction containing a lower concentration of template, more rounds of amplification can be achieved and more mutations can be accumulated in the PCR product.

Although error-prone PCR offers the full spectrum of mutations and allows experimental control of the misincorporation level, the PCR products must pass through a series of additional steps. These usually include PCR, restriction digestion, gel purification, ligation, and transformation. In addition to mutations generated during error-prone PCR, these laborious subcloning steps are key steps determining the size and complexity of the mutant library.

In contrast, when the target DNA is quite small (e.g., in the production of short peptide libraries), commonly used methods are based on the integration of oligonucleotide primers carrying random mutations into the gene of interest.

During the chemical synthesis cycles used to produce the oligonucleotides, the desired position and rate of mutation can be controlled. The oligonucleotide pools are subsequently incorporated into wild-type genes using a standard subcloning procedure.

However, if the integration is accomplished through a more efficient, ligation-free, DNA extension procedure, as described here, a significantly larger pool size can be obtained. The strategies described here have been used successfully to construct DNA libraries of the *Plasmodium falciparum* dihydrofolate reductase (pfDHFR) gene for various genetic selection studies. To search for novel drug-resistant mutants of pfDHFR other than naturally existing mutations, random mutagenesis of the entire pfDHFR gene was performed.

In nature, single or combinations of four mutations in pfDHFR, including N5 1I, C59R, S 108N, and 1164L, are associated with antimalarial antifolate resistance. To develop a system comparable to the natural selection process, libraries of pfDHFR mutants containing approx two to three changes in the entire 0.7-kb gene were constructed using error-prone PCR. Libraries were transformed into bacterial, host cells and drugresistant mutants were selected by bacterial complementation.

From our selection, both natural and novel drug-resistant mutants of the pfDHFR were obtained. Characterization of those mutants provided useful information for the design and development of antimalarial compounds. This chapter describes three different mutagenesis strategies. XL1-red. errorprone PCR, and degenerate oligonueleotides-Pfu (DOP) mutagenesis. Techniques and tips for the control of mutation frequency and producing large and diverse libraries of cloned, mutated target DNA are included.

REQUIREMENTS

XL 1-Red Mutator in E. coli

1. XL1-red mutator strain (Stratagene, cat. no. 200129). Keep frozen at -80°C until use.
2. Standard plasmid DNA extraction kit.
3. DNA of interest in a plasmid vector
4. Ice.
5. Falcon tubes: 15 mL.
6. β-Mercaptoethanol.
7. Water bath at 42°C; incubator at 37°C.
8. Luria Bertani (LB) broth medium: 10-g/L tryptone 140, 5-g/L yeast extract, 10-g/L NaCl in distilled H_2O.
9. LB-agar: LB + 15-g/L agar. Add antibiotic appropriate for plasmid vector in use.
10. Sterile toothpicks.
11. Spectrophotometer.
12. Microcentrifuge.

Error-Prone PCR Technique

1. PCR primers flanking the target DNA and containing appropriate restriction sites for cloning.
2. 10X mutagenesis buffer: 500 *mM KCl, 100 mM* Tris-HCl, pH 8.5, 1 % Triton-X100.
3. 10 mM dATP.
4. 10 mM dCTP.
5. 10 mM dTTP.
6. 10 mM dGTP.
7. DNA template encoding the gene of interest (e.g., genomic DNA or a large clone containing entire gene).
8. *Taq* DNA polymerase.
9. A standard thermocycler and mineral oil, if required.
10. Standard agarose, buffer and electrophoresis equipment, and 10-mg/mL ethidium bromide (EtBr) stock.
11. DNA markers of appropriate size and known concentration.

Mechanism of Desalting

1. Nitrocellulose (0.025 µm, White VSWP, 25 mm in diameter (Millipore cat. no. VSWP02500).
2. Sterile Petri dishes, sterile H_2O, and clean forceps.
3. Appropriate restriction endonucleases and buffers.

4. Standard agarose, buffers, 10-mg/mL EtBr, agarose gel electrophoresis equipment.
5. T4 DNA ligase and appropriate 10X buffer.
6. Clean scalpel blade.

Extraction of DNA

1. QIAquick gel purification kit (Qiagen cat. no. 28706 or equivalent kit).

Ligation

1. Plasmid vector, linearized with appropriate restriction enzymes.
2. Speed vacuum drier, if required.
3. T4 DNA ligase and appropriate lox buffer.
4. Sterile H_2O.
5. Water bath or incubator at 16 or 25°C.

Electroporation Technique

1. Electrocompetent *E. coli* cells, e.g., DH5α.
2. An electroporator and electroporation cuvets.
3. High-quality water, e.g., Milli-Q.
4. Standard plasmid DNA extraction kit.
5. LB medium and LB-agar plates with appropriate antibiotic.
6. Standard plasmid DNA extraction kit.

DOP Mutagenesis

1. Template plasmid DNA carrying gene or DNA segment to be mutated (this should be prepared from an *E. coli* strain that has an intact DNA methyl-transferase system).
2. Top- and bottom-strand degenerate oligonucleotide primers designed for this experiment (OPC-purified).
3. 10 mM dNTPs (as above).
4. *Pfu* DNA polymerase or high-fidelity thermostable DNA polymerise and the corresponding X buffer.
5. Standard thermocycler and mineral oil, if required.
6. *DpnI* restriction enzyme.
7. Standard agarose, buffers, 10-mg/mL EtBr, agarose gel electrophoresis equipment.
8. Nitrocellulose (0.025 pm, White VSWP, 25 mm in diameter) (Millipore, cat. no. VSWP02500), sterile Petri dishes, water, and forceps.
9. Electrocompetent *E. coli* cells.
10. LB medium and LB-agar plates with appropriate anitbiotic.
11. Standard plasmid DNA extraction kit.

PROCEDURES

In Vivo Mutagenesis Method

The use of XL1-red mutator cells is one of the simplest and most effective in vivo mutagenesis methods and requires no subcloning step. However, it is important to note that the rapid mutation rate also affects the bacterial chromosome. As a result, the cells grow extremely slowly and a reversal or alteration of mutator phenotype can occur. Hence, during the use of XL1-red cells, growth rate and doubling time must be monitored to ensure that the mutator phenotype is not undergoing change.

The mutation generated from this modified protocol is 1/1400 bp, compared to 1/2600 by obtained from growing XL1-red cells for a single overnight period only. Therefore, this mutagenesis strategy is suitable for low-frequency mutation of a target gene of 1 kb or larger.

1. Thaw the XLI-red-competent cells on ice.
2. Aliquot 100µL of the competent cells into a prechilled 15-mL Falcon tube.
3. Add 1.7 µL of β-mercaptoethanol provided with the kit to 100µL of cells (to get a final concentration of 25 mM).
4. Add 50 ng of the template DNA (e.g., a cloned gene of interest in an *E. coli* plasmid vector) to the tube and swirl gently. Incubate the tube for 30 min on ice.
5. Place the tubes containing the cells in a 42°C water bath for 45 s. This subjects the cells to heat shock, which causes transient permeabilization. Immediately return the tube to the ice bucket and incubate the cells on ice for 2 min.
6. Add 0.4 mL LB and incubate the tube at 37°C for 1h with vigorous shaking at 225-250 rpm.
7. Plate the entire reaction onto appropriate LB-agar plates containing the appropriate antibiotic, and incubate at 37°C for 24-48 *h*. Approximately 100-200 colonies in various sizes should be obtained from one transformation.
8. Use a sterile toothpick to select 200 individual colonies at random from the plate and inoculate each into 5 mL of LB broth containing appropriate antibiotics.
9. Incubate the cells at 37°C with shaking at 225-250 rpm.
10. To ensure the mutator phenotype of the XL1-red cells is retained, do not allow cell density to increase beyond desired point. Cell growth should be monitored every 4-6 h by measuring OD_{600}. The doubling time of the XLI-red cell in LB medium should be 4-6 h at 37°C.
11. When the OD_{600} reaches 0.6, dilute the cells by 1:100 with fresh LB medium (supplemented with appropriate antibiotics).
12. Culture the cells at 37°C with vigorous shaking.
13. Continue monitoring the OD_{600} of the culture and when OD_{600} is approx 0.6, again perform a 1:100 dilution with fresh LB medium containing appropriate antibiotics. Such cycles can

be repeated until an abrupt decrease in doubling time from 4–6 h to 1–2 h occurs. This indicates that the mutation efficiencyy declining (usually after 6–7 d).

14. Pellet the cells and purify the mutagenized plasmid DNA pool using a plasmid extraction kit or a standard phenol-chloroform extraction and precipitation with ethanol.
15. To assess the mutation frequency, transform the library into DH5a* (or any compatible E. *coli* strain) and randomly select 5-10 colonies for DNA sequencing analysis.

Error-Prone PCR

We present here a "mild" error-prone PCR protocol. A sufficient yield of PCR products should be obtained (at least 1 µg/ PCR reaction) at approx 1-2% mutation (I mutation in 100 by of DNA).

1. Prepare 10X mutagenesis buffer Make ali-quots and keep at –20°C until use.
2. Prepare 50-µL reactions as follows: 34 µL sterile H_2O; 1µL template DNA (e.g., a cloned gene of interest in an *E. coli* plasmid vector) (20 fmol/µL); 1µL PCR primer 1; 1µL PCR primer 2; 0.5 µL each of dATP and DTP; 2.5µL each of dTTP and dCTP; 5µL 10 X mutagenesis buffer.
3. Vortex the reaction mixture and spin down briefly.
4. If the thermocycler does not have a heated lid, the reaction should be overlaid with two to three drops of mineral oil.
5. Place each reaction you have prepared in the thermocycler and perform the PCR program as follows: 95°C for 3 min; pause the thermocycler and add 2µL (5-6 U) of *Taq* polymerase; mix thoroughly by pipetting. Spin the reaction down briefly. Continue 30 cycles of 95°C, 1 min; primer T_m–5°C, 1 min; 72°C, 1 min/kb of target DNA; 72°C for 10 min.
6. Electrophorese 5µL of the reaction in an agarose gel and stain with 0.5-µg/mL ethidium bromide to verify that the PCR products are the expected size. The amount of PCR product in 5µL can be estimated by comparing the fluorescence intensity under UV radiation of the PCR product in the gel to that of the a DNA marker containing a known amount of DNA in each band.

Method of Desalting

This simple desalting step is necessary when the salt conditions required in the subsequent enzymatic reactions are not compatible with the error-prone PCR buffer.

1. Pour 15-20 mL sterile H_2O into a sterile Petri dish.
2. Use sterile forceps to carefully place VSWP02500 nitrocellulose membrane (white membrane) on the water surface, glossy side up.
3. Pipet the PCR product on the membrane, while the membrane is floating on the water surface. Do not allow the membrane to sink at any time. One membrane can hold up to 100µL PCR product.
4. Allow to float and desalt for 20–30 min without any disturbance.
5. Use a pipet to gently transfer the desalted PCR product from the membrane to a sterile microfuge tube. Take care not to submerge any part of the membrane during this process.

Up to 90% of the PCR product should be recovered from the membrane.

6. Digest the PCR product and plasmid DNA with appropriate restriction endonucleases, using the condition for the best restriction efficiency according to the manufacturer's protocols.
7. Run the digested product on an agarose gel of appropriate percentage for the expected size of the product.
8. After ethidium bromide staining and visualization under UV light, excise the desired DNA bands from the gel using a sharp blade (precleaned with 70% ethanol).

Extraction of DNA

The desired DNA band from the gel slice may be purified using the QIAquick gel extraction kit (see the manufacturer's handbook for detailed principle and protocol). The equivalent protocols or kits for DNA extraction can also be used.

1. Weigh the gel slice and add 3 vol of buffer QG to 1 vol of gel (µL/µg).
2. Incubate at 50-55°C until the gel slice has completely dissolved. Mix occasionally.
3. Add an equal volume of isopropanol to the dissolved gel solution.
4. Apply the sample to the QIAquick column, and centrifuge for 1 min. Discard the flow-through (FT).
5. Wash the column with 0.5µL Buffer QG and centrifuge for 1 min. Discard the FT.
6. Wash the column with 0.75µL Buffer PE (in ethanol) and centrifuge for 1 min. Discard the FT.
7. Centrifuge for 2 min to ensure the complete removal of the ethanol in the PE buffer. 8. Place a column into a sterile microfuge tube.
9. Elute the DNA from the column by adding 30 µL of H_20 into the center of the column, incubating for 1 min, and centrifuging for 1 min.
10. Repeat step 9 to achieve maximum efficiency of the elution. Combine the eluates.
11. The purity and concentration of the eluted DNA fragments can be estimated by agarose gel electrophoresis.

Ligation

1. Add an appropriate amount of plasmid DNA, linearized with the appropriate restriction enzyme(s), and mutated PCR products in a single microfuge tube.
2. Set up a negative control ligation in a separate tube by adding an equal amount of plasmid DNA as in *step* 1 but no PCR product.
3. If the total volume of the DNA exceeds 7µL, the DNA should be dried under speed vacuum. This drying step also helps to get rid of ethanol carried over from the previous reaction, which greatly inhibits ligation.
4. To each tube, add 2 µL of 5X ligase buffer; 1 µL of T4 DNA ligase; and up to 10µL of sterile H_2O.

5. Incubate the mixture at 16°C for 16 h. This is particularly important if at least one of the sticky ends is a two-base overhang (for example, AT overhang produced by *NdeI*). However, if both overhangs are four bases or more, the ligation can be performed at 25°C.

Electroporation Technique

To obtain a large plasmid library of cloned, mutated PCR products from a ligation, a robust transformation method such as electroporation is required. It is recommended that the construction of libraries be performed using highly competent *E. coli* cells (for example, DH5a*). The library, in the form of intact plasmid DNA, can be harvested and subsequently transformed into cells required for selection of the desired function. To obtain the best result, media and solutions used during preparation of cells and electroporation should be prepared with Milli-Q water.

1. Thaw the electrocompetent cells on ice for about 5-10 min.
2. Add 1-2µL of ligation mixture into 50-µL electrocompetent cells. Incubate on ice for 2 min.
3. Transfer the cell-DNA mixture into a prechilled cuvet. Make sure that the cell-DNA mixture covers the bottom of the cuvet evenly and is free of large bubbles.
4. Place the cuvet in the chamber in the lock position.
5. If a 1-mm cuvet is used, subject the mixture to electroporation at 25 µf, 200 W, and 1.6 kV. For a 2-mm cuvet, the power should be increased to 2.5 kV. A time constant of 3.8-4.5 ms is obtained for high-efficiency electroporation.
6. Immediately take the cuvet out of the chamber and suspend the cell-DNA mixture in 1 ml-of LB.
7. Transfer the cells into a sterile tube using a sterile Pasteur pipet.
8. Incubate cells at 37°C for 1 h before plating the reaction on selective medium containing agar and an appropriate antibiotic.
9. Incubate the plates at 37°C overnight. A wide range of colony sizes might be observed from randomly mutagenized pools, especially if the product of the mutated gene is being expressed and is involved in one or more cellular processes.
10. Count or estimate the number of colonies to determine the pool size of the library. Up to several thousand colonies should be obtained from a single transformation, and approx 20,000-30,000 clones can be obtained from one ligation. If a larger pool size is desired, additional ligations and transformations can be performed.

    The number of colonies on negative control plates is an indication of the quality of the libraries. The more colonies on the negative control plates, the more wildtype background colonies are present in the libraries.
11. The colonies of the DNA library can be harvested from plates by adding 3 mL of LB onto the plates and washing off the colonies. Transfer the cells from plate to a new sterile tube.
12. Pellet the cells and purify the plasmid DNA, containing multiple randomly mutated copies of the gene of interest, using the plasmid DNA extraction kit or phenol–chloroform extraction.

13. To examine the type and frequency of mutations, the plasmid in the DNA library should be transformed into new DH5a* (or other standard) E. *coli* cells for DNA manipulation.

14. Select 10 colonies at random for DNA sequencing, to assess the type and rate of mutations by comparison of the sequence to that of the wild-type gene.

DOP Mutagenesis

DOP mutagenesis is modified from Quikchange site-directed mutagenesis (Stratagene). DOP mutagenesis utilizes a pair of oligonucleotide primers, degenerate in the region of interest but flanked by wild-type sequence at both ends. The target DNA templates are denatured at high temperature, then the primers are allowed to anneal to the target region in the plasmid.

In this reaction, thermostable DNA polymerase is added to start primer extension, and the DOP primers carrying random mutations are subsequently incorporated into newly synthesized plasmid DNA. Generally, high-fidelity DNA polymerase, for example, *Pfu* polymerase, is used in the primer extension reaction, to minimize undesired mutations outside the target region.

After successful extension cycles, parental plasmid DNA carrying wild-type sequence can be eliminated from the reaction by digestion with *DpnI,* which recognizes and digests only the methylated DNA that has been produced by replication in E. *coli.* The PCR products are not methylated.

The pools of mutant PCR products can be transformed directly into host cells, without additional cloning steps: they are introduced as nicked circular plasmid DNA, and the nicks are fixed inside the cells after transformation.

Complementary Primers

DOP mutagenesis exploits two complementary primers (top and bottom strands) of similar length. The randomized region (or the mutated region) containing degenerate nucleotides (up to 45 nucleotides) is at the center of the primers, flanked by wild-type sequence at both ends. These wild-type ends are essential for the annealing of the degenerate primers to the target region in the plasmid DNA template.

For a successful extension, each of these annealed sections must survive through an extension step (68-72°C). Hence, each end should have a minimum T_m of 70°C. The length of each entire primer should not exceed 80 nucleotides, which is a limit for the chemical synthesis of high-quality oligonucleotides.

Within the mutated region of the primers, both location and composition of the mutations can easily be controlled during primer synthesis. This unique feature allows the construction of ingenious libraries to maximize the possibility of obtaining mutants with desired phenotype(s). For example, in a selection for zinc finger mutants with altered specificity, four cysteine residues critical for the structural formation of the zinc finger should be intact. This can eliminate 160,000 (20^4) misfolded zinc fingers from the libraries prior to the selection process.

DOP Reaction

1. Prepare the reaction as follows:

   a. 39 µL sterile H_2O.

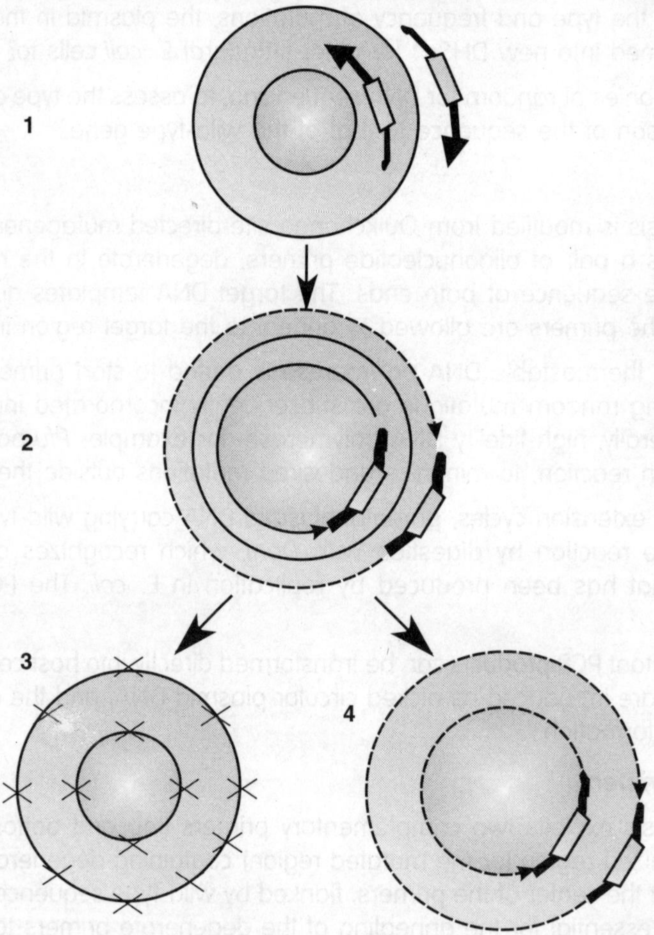

Figure 17.1: Primer design and schematic diagram of DOP mutagenesis. DOP primers are a set of complementary primers carrying degenerate oligonucleotides (*) in the central region. These degenerate nucleotides are surrounded by wild-type sequence (*) on both ends. After the plasmid DNA templates are denatured at high temperature, DOP primers are annealed to template DNA. Random mutagenized sequences in the primers are incorporated into newly synthesized DNA during primer extension reaction. The DNA templates carrying wild-type sequence are eliminated from the reaction by DpnI *digestion. Lastly, DNA libraries can be transformed directly into host cells.*

b. 1 µL Plasmid DNA template (50 ng/µL), prepared from standard E. *coli* cells that have an intact DNA methyltransferase system.

c. 2 µL dNTP mixture (10 mM).

d. 1 µL Top-strand degenerate primer (125 ng/µL).

e. 1 µL Bottom-strand degenerate primer (125 ng/µL). f. 5 µL 10X *Pfu* buffer.

2. Vortex and spin the reaction briefly.
3. If the thermocycler does not have a hot lid, the reaction should be overlaid with two to three drops of mineral oil.

4. Place the reaction in the thermocycler and start the reaction as follows:
   a. 95°C for 3 min.
   b. Pause and add 1 μL (3-5 U) of DNA polymerase and mix thoroughly.
   c. Continue 30 cycles of 95°C for 1 min T_m–5°C for 1 min 68-72°C for 2 min/kb (or according to manufacturer's recommendation).
5. Add 1 μL of *DpnI* restriction enzyme, mix thoroughly and incubate at 37°C for 1-2 h. This should be sufficient to completely digest 50 ng of the wild-type plasmid DNA templates. If a higher amount of DNA template is used, then add more *DpnI* and incubate the reaction for 10-16 h.
6. Verify that the amplified product is of the expected size by agarose gel electrophoresis. DOP amplification is essentially a linear amplification (i.e., not exponential). One DOP amplification yields approx 300-500 ng of amplified product of similar size to the linear plasmid DNA template.
7. Desalt the reaction using VSWP nitrocellulose membrane.
8. Transform 1-2 μL of the product into 50 μL electrocompetent cells.
9. Plate 50 μL of the transformed cells on a plate containing appropriate antibiotics. Incubate the plate at 37°C for 16 h.
10. Count the colony on plates to estimate transformation efficiency and pool size. Several thousand colonies can he obtained from a single plate, and up to 100,000 colonies should easily be obtained from a single transformation of 1-2 μL of the DOP product.
11. Harvest the cells and extract plasmid carrying mutagenized gene using a plasmid extraction kit or standard phenol-chloroform extraction methods.

POINTS TO REMEMBER

1. Generally, tiny colonies are visible after 24-30 h and the plate can be incubated up to 48 h for larger colonies. If the colonies are visible within 16-20 h, it is a sign of reversal phenotype and a very low mutation frequency might be obtained.
2. The error rate produced by the error-prone PCR can be increased by lowering the amount of DNA template. Using 2- and 0.2-fmol DNA template (e.g., a cloned gene of interest in an E. *coli* plasmid vector) should generate approx 5 and 11% mutation frequency, respectively, without any effect on the yield of PCR product.
3. In several error-prone PCR protocols, concentration of dATP was altered by 5-10-fold to promote misincorporation. We found that AT-to-GC transition was preferentially observed when a ratio 1:1:1:0.2 of C:T:G:A was used. However, unbiased mutation was obtained from a reaction containing 1:1:0.2:0.2 of C:T:G:A ratio.
4. The addition of Mn^{2+} is known to enhance the mutation frequency. However, it also interferes with the PCR reaction, leading to low yield of PCR product. If Mn^{2+} is to be used, the optimum concentration for desired mutation frequency with sufficient PCR yield should be determined experimentally.

5. It is suggested that the mutation rate can also be increased by consecutive rounds of PCR, using product of a previous reaction as a template in a subsequent error-prone PCR, and in the instruction manual for Diversify PCR Random Mutagenesis kit, Clontech). In practice, only a small fraction (one-tenth or one-twentieth) of the first round of error-prone PCR is transferred and amplified in the next PCR.

   This could lead to a clonal bias problem and could hamper the selection of mutants of interest if the transferred fraction contains mutations that are deleterious to the structure or function of the resulting proteins. If this is to be done, the total volume of the second PCR should be increased to ensure that all clones generated from the first error-prone PCR are further amplified. For example, if one-tenth of the first PCR is transferred to the next PCR, 10 reactions of the second PCR should be performed and combined.
6. To maximize the cloning efficiency, plan to use only robust restriction endonucleases known to work well with PCR products (for example, *EcoRI).* Try to avoid end-sensitive enzymes (such as *HindIII).* Those enzymes are less effective at recognizing and digesting the sites near the end of the PCR products, resulting in a low cloning efficiency.
7. According to the manufacturer's instruction, elution buffer (EB) is suggested for the elution of the DNA from the column. However, if the EB pH 8.5 is not compatible with subsequent ligation reaction, sterile H_20 is preferred for the elution of DNA.
8. In a single 10-µL ligation reaction, approx 200 ng of total DNA should be used. The optimum ratio between plasmid DNA and PCR can be varied, depending on the size of the plasmid DNA backbone and PCR insert. For 4-kb plasmid and 0.7-kb PCR product, we used 1:5 of plasmid:PCR product.
9. To minimize instability during freezing and thawing cycles, 5X ligase buffer should be divided into aliquots sufficient for several uses and stored at –20°C. If the buffer has been subjected to many cycles of freezing and thawing, 1 µL of 10 mM ATP should be added to the reaction (unless it is a blunt-end ligation).
10. For preparation of electrocompetent cells, *see* http://www.fhcrc.org/labs/hahn/methods/ mol bio_meth/electrocompetent or consult a standard laboratory manual.
11. If a high frequency of background colonies (wild-type) is obtained, the plasmid DNA backbone should be redigested and purified before carrying out another ligation. Additionally, the backbone can be treated with alkaline phosphatase to reduce the background.
12. Do not add an additional extension step, which is usually performed in regular PCR. Once the extension of the entire plasmid is completed, prolonged extension can promote overextension, and duplication of the primer region can occur.

Chapter 18

Evolutionary Hybridization

Metagenomics addresses the collective genetic structure and functional composition of a microbial environmental sample without the bias or necessity for culturing the microorganisms from the community in question. An extensive suite of molecular based approaches developed over the past decade has enabled the study of these uncultured microorganisms and is the direct outcome of the use of small subunit ribosomal RNA (SSU rRNA) targets which have been widely utilized to study the bacterial diversity, community structure, and microbial interactions within these microbial ecosystems.

Although these methods provide a wealth of information about microbial. community structure, they are limited to a subset of identifying sequences. Thus, metage-nomic studies are now beginning to take advantage of the plethora of complete genome sequences and the associated tools to discover novel genes and survey the structure and function of microbial communities. Community genome complexity and size have been estimated by analyzing the reassociation of genomic DNA from environmental samples.

However, construction of metagenomic libraries using bacterial artificial chromosome (BAC) and fosmid vectors is a more direct route to gaining insight into the functional nature of mixed microbial samples. The sequencing and assembly of these large insert libraries have also been hypothesized to lead to recon-struction of a nearly complete microbial genome. Complementary and less expensive methods to compare genomes from individual microbes have been utilized in comparative genomic studies.

Suppressive subtractive hybridization (SSH) is one such approach, which has been utilized to compare the genomic content of closely related species of bacteria. Recently, SSH has also been used as a comparative method to examine the microbial diversity (i.e., species composition) and functional differences (i.e., gene composition) in the genomic content of two different rumen environmental communities.

The core concept of subtractive hybridization was first documented in a publication by Bautz and Reilly (1966). Here, the authors demonstrated that a bacteriophage T4 deletion mutant identified by phenotypic methods could be utilized to isolate mRNA corresponding to the specific deleted gene from the wild-type counterpart without extensive knowledge of the gene and surrounding sequences.

Denatured (single-stranded) DNA from the deletion mutant was bound to a nitrocellulose column and used as a template to bind mRNA from the-wild type. The mRNA in common with the wild-type and deletion mutant bound its DNA-based template and adhered to the column while the mRNA species corresponding to the deleted (missing) DNA target of interest emerged from the column, resulting in an effective subtraction.

Approximately 20 years later this basic idea was systematically applied to isolating differences in genomic DNA from eukaryotes, whose genomes are several orders of magnitude larger than that of a bacteriophage. A series of three papers documented the identification of specific genomic DNA regions using an approach modified from that of Bautz and Reilly, and introduced the term "genomic subtraction" (as well as "difference cloning") to describe the process.

By 1990, the technique was formalized by the incorporation of three crucial concepts: (1) the introduction of the "tester" and "driver" nomenclature; (2) the use of an excess amount of "driver" DNA to "drive" the subtraction; and (3) the incorporation of polymerase chain reaction (PCR) to amplify the remaining "tester"-specific fragments following the subtraction process. Two papers demonstrated the utility of this streamlined approach.

Straus and Ausubel (1990) applied "genomic subtraction" to the isolation of wild-type DNA that was absent in a yeast deletion mutant, while Wieland et al. (1990) utilized a similar "subtractive hybridization" procedure to compare A phage DNA and placental DNA, as well as placental DNA compared with DNA from Epstein-Barr virus-immortalized lymphocytes. Both publications involved the use of an avidin-biotin-mediated approach to isolating DNA fractions.

Southern blot analyses illustrated the success of both approaches, and indicated the power of the technique relative to the investment of time and finances. However, the widest use for subtractive hybridization at the time was in the normalization of cDNA libraries. Because cDNA libraries are based on extracted mRNA from a specific eukaryotic tissue, certain mRNA species may be under-represented in one tissue type versus another.

Subtractive hybridization would allow the recovery of these low-expression mRNA species to produce a full representation of the possible mRNA species. An earnest description of the subtractive hybridization process utilizing computer-based modeling was soon developed. This line of research produced not only a series of computer utilities to predict the ideal parameters for a successful experiment but also a further refinement of the subtractive hybridization procedure-which came to be called suppressive subtractive hybridization.

Suppressive subtractive hybridization (SSH; other descrip-tions in the literature also refer to it as "suppression subtractive hybridization") introduced the use of specially designed adaptors (short double-stranded DNA segments). If these adaptors are ligated to themselves or to small DNA fragments then the secondary structure (a "panhandle"-like structure) intro-duced by their sequence content would effectively inhibit the PCR amplifi-cation in the final preparation.

This would efficiently decrease the detection of background fragments and allow the identification of larger. differ-ential DNA of sufficient "informational" length. The amended version of subtractive hybridization was demonstrated in a comparison of the genomes of two strains of

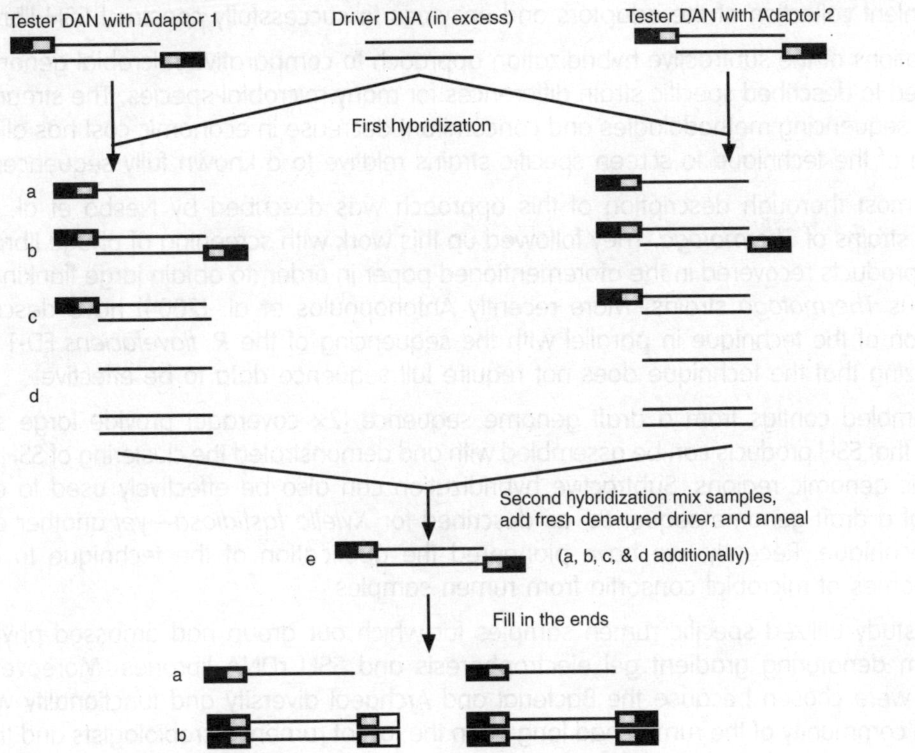

Figure 18.1: An overview of the suppressive subtractive hybridization procedure. Note the inhibitory "pan-handle" secondary structure ("b -> b'" bottom panel) formed in the final PCR amplification; this greatly decreases the recovery of "background noise" (i.e., short non-specific sequences).

Helicobacter pylori. A patented kit, produced by Clontech, is based on this publication and provides a convenient collection of the adaptors and enzymes to successfully construct SSH libraries.

Variations of the subtractive hybridization approach to comparative microbial genomics have been used to described specific strain differences for many microbial species. The streamlining of genome sequencing methodologies and concomitant decrease in economic cost has allowed the adoption of the technique to screen specific strains relative to a known fully sequenced one.

The most thorough description of this approach was described by Nesbo et al. (2002) in studying strains of *Thermotoga.* They followed up this work with screening of phage libraries with the SSH products recovered in the aforementioned paper in order .to obtain large flanking regions for various *Thermotoga* strains. More recently Antonopoulos et al. (2004) have described the application of the technique in parallel with the sequencing of the *R. flavefaciens* FD-1 genome, emphasizing that the technique does not require full sequence data to be effective.

Assembled contigs from a draft genome sequence (2× coverage) provide large sequence scaffolds that SSH products can be assembled with and demonstrated the clustering of SSH products to specific genomic regions. Subtractive hybridization can also be effectively used to direct the closure of a draft genome sequence, as described for *Xylella fastidiosa—yet* another extension of the technique. Recently, we have pioneered the application of the technique to compare metagenomes of microbial consortia from rumen samples.

This study utilized specific rumen samples for which our group had amassed phylogenetic data from denaturing gradient gel electrophoresis and SSU rDNA libraries. Moreover, rumen samples were chosen because the Bacterial and Archaeal diversity and functionality within the microbial community of the rumen had long been the foci of rumen microbiologists and there was a considerable knowledge base from both classical culture-based methods as well as current molecular approaches.

Therefore, for this study, SSH was employed to isolate DNA fragments present in one rumen sample, but absent from another. DNA of these subtracted samples was cloned, and the DNA sequences determined and analyzed via computer-assisted DNA and amino acid comparisons. Subtractive hybridization fragments of both Bacterial and Archaeal origin were identified, demonstrating that SSH provided a powerful complementary approach to current methods for the metagenomic analysis of complex microbial environments.

Based on simple quality control assessments of the SSH libraries constructed, the technique appears capable of discerning and identifying DNA fragments unique to one complex microbial community relative to another. In this chapter, detailed methods for performing SSH with environmental microbial consortia are presented, including quality control experiments to determine subtraction efficiency.

MATERIALS

Extraction of Total Genomic DNA from Environmental Samples

1. 0.1 M sodium phosphate buffer, pH 8.0: to make 200 mL *mix* 5.3 mL of solution A (27.6 g of $NaHPO_4.H_2O$/L [0.2 M]) with 94.7 mL of solution B (53.65 g of $Na_2HPO_4.7H_2O$/L [0.2 M]) and then dilute with H_2O to *200 mL.*

2. Lysis solution: 0.15 *M* NaCl, 0.1 *M* EDTA, pH 8.0. Autoclave and then add 15 mg/mL of lysozyme. Store at 4°C.
3. SDS solution: 0.1 *M* NaC1, 0.5 *M* Tris-HCl, pH 8.0, 10% sodium dodecyl sulfate.
4. Proteinase K (Sigma-Aldrich).
5. Buffered phenol, pH 8.0. Store at 4°C in a dark glass bottle.
6. Phenol-chloroform-isoamyl alcohol (25:24:l).
7. Chloroform- isoamyl alcohol (24:1).
8. 10.5 *M* ammonium acetate: dissolve 80.93 g of ammonium acetate per 100 mL of ddH2O. Filter sterilize using a 0.22-pm filter.
9. Cold isopropanol (4°C).
10. 70% Ethanol in water.
11. TE buffer, pH 8.0: 10 *mM* Tris-HCl, pH 8.0, *1 mM* EDTA, pH 8.0.
12. RNase OneTM Ribonuclease (Promega). Store at –20°C.

Suppressive Subtractive Hybridization (SSH)

The subtractive hybridization process utilizes the CLONTECH PCR-Select™ Bacterial Genome Subtraction Kit (Clontech Laboratories Inc. cat. no. 637404). The following listed items are supplied in the kit. Store 4× hybridization buffer at room temperature. All other reagents are to be stored at –20°C.

Endonuclease Digestion

1. 10× *Rsa*l restriction buffer: 100 mM bis-Iris-propane-HC1, pH 7.0, 100 mM $MgCl_2$, 10 mM dithiothreitol (DTT).
2. 12 µL of *Rsa*l (10 U/µL).

Adaptor Ligation

1. 21 µL of T4 DNA Ligase (400 U/µL; contains 3 mM ATP).
2. 5× DNA ligation buffer: 250 mM Tris-HCl, pH 7.8, 50 mM MgCl2, 10 mM DTT, 0.25 mg/mL of bovine serum albumin (BSA).
3. 30µL of adaptor 1 (10 mM).
4. 30 µL of adaptor 2R *(10 mM).*

Hybridization

1. 4× Hybridization buffer: 200 mM N-2-hydroxyethylpiperazine-N¢-2-ethanesulfonic acid (HEPES-HCl), pH 8.0, 2 M NaCl, 0.8 mM EDTA, pH 8.0. Store at room temperature.
2. Dilution buffer: 20 mM HEPES-HCl, pH 8.3, 50 mM NaCl, 0.2 mM EDTA, pH 8.0.

PCR Amplification

1. 50 µL of PCR primer 1 (10 mM).

2. 100 µL of nested primer 1 (10 mM).
3. 100 µL of nested primer 2R (10 mM).
4. 10µL pf PCR control subtracted cDNA.

Control Reagents

1. 5 µL of E. *coli* genomic DNA (1 mg/mL).
2. 10µL of control DNA (3 ng/µL; f*X174/HaeIII digest).
3. 50µL of 10 µM 23S RNA forward primer*.
4. 50 µL of 10 µM 23S RNA reverse primer*.

Cloning and Sequencing of Library

This process utilizes a TOPO TA Cloning Kit for Sequencing (with pCR®4-TOPO® vector) with One Shot® TOP 10 Electrocomp™ *E. coli* (Invitrogen; cat. no. K4580-01). The following listed items are supplied in the kit. Store all reagents at –20°C unless noted otherwise.

TOPO TA Cloning Kit for Sequencing (with pCR®4-TOPO® Vector)

1. 25µL of pCR®4-TOPOTM (10 ng/µL of plasmid DNA in 50% glycerol, 50 mM Tris-HCl, pH 7.4 [at 25°C], 1 mM EDTA, 2 mill DTT, 0.1 % Triton X- 100, 100 µg/mL of BSA, 30 M phenol red).
2. 100µL of 10x PCR buffer (100 mM Tris-HCl, pH 8.3 [at 42°C], 500 mM KCl, 25 mM MgCl2, 0.01% gelatin).
3. 10µL of dNTP mix (12.5 mM dATP, 12.5 mM dCTP, 12.5 mM dGTP, 12.5 mM dTTP neutralized at pH 8.0 in water).
4. 50µL of salt solution (1.2 M NaCl, 0.06 M $MgCl_2$).
5. 1 mL of sterile water.
6. 20 µL of M13 forward (-20) primer (0.1 µg/µL in TE buffer, pH 8 [407 pmol]).
7. 20µL of M13 reverse primer (0.1 µg/µL in TE buffer, pH 8 [385 pmol]).
8. 20µL of T3 primer (0.1 µg/µL in TE buffer, pH 8 [329 pmol]).
9. 20µL of T7 primer (0.1 µg/µL in TE buffer, pH 8 [328 pmol]).
10. 10µL of control PCR template (0.1 µg/µL in TE buffer, pH 8).

    *These primers will amplify a 374-bp fragment of the *E. coli* 23S rRNA gene.

11. 10 µL of control PCR primers (0.1 µg/µL each in TE buffer, pH 8).

M13 forward (-20)	5′-GTAAAACGACGGCCAG-3′
M13 reverse	5′-CAGGAAACAGCTATGAC-3′
T3	5′-ATTAACCCTCACTAAAGGGA-3′
T7	5′-TAATACGACTCACTATAGGG-3′

One Shot′ TOP 10 Electrocomp™ *E. coli* and Reagents

1. S.O.C. medium (2% tryptone, 0.5% yeast extract, 10 mM NaCl, 2.5 mM KCl, 10 *mM* $MgCl_2$,

10 mM $MgSO_4$, 20 mM glucose [6 mL]). Store at +4°C or room temperature.

2. 21× 50μL of TOP10 cells ElectrocompTM. Store at –80°C.
3. 50μL of pUC 19 control DNA (10 pg/μL in 5 mM Tris-HCl, 0.5 mM EDTA, pH 8). Store at -80°C.

Inventory of Other Supplies and Reagents

1. *Alu*I restriction enzyme (includes NEBuffer2 from New England Biolabs).
2. 1 kb DNA Ladder (Invitrogen).
3. 1.5-mL tubes (Fisher Scientific siliconized/ low-retention microcentrifuge tubes).
4. 0.2-mL thin-walled PCR reaction tubes (Fisher Scientific Costar PCR tubes).
5. Sterile, deionized H_2O
6. 5× TBE electrophoresis buffer: dissolve 54 g of Tris base, 27.5 g of boric acid, 20 mL of 0.5 M EDTA, pH 8.0, in 1 L of ddH_2O. Working solution for agarose gel electrophoresis is 0.5× (45 mM Tris-borate, I mM EDTA).
7. 0.2 M EDTA, pH 8.0.
8. Phenol-chloroform-isoamyl alcohol (25:24:1) (Sigma-Aldrich).
9. Chloroform-isoamyl alcohol (24:1) (Fluka via Sigma-Aldrich).
10. 3 M sodium acetate, pH 5.2.
11. 20 mg/mL of glycogen (Sigma-Aldrich).
12. 100% Ice-cold ethanol.
13. 80% Ice-cold ethanol.
14. TaKaRa Ex *Tay* DNA polymerase (includes dNTPs and buffer) (Fisher Scientific).
15. Electroporation cuvettes, 0.1 cm (Bio-Rad).
16. LB agar plates (1.0% tryptone, 0.5% yeast extract, 1.0% NaCl. pH 7.0, 1.5% agar) supplemented with ampicillin (50-100 μg/mL).
17. LB (Luria-Bertani) medium (1.0% tryptone, 0.5% yeast extract, 1.0% NaCl, pH 7.0).
18. 96-Well cell culture cluster, flat bottom with low evaporation lid (Corning).
19. Glycerol.

Confirmation of Tester-Specific DNA Fragments by Dot-Blot and Southern Blot Hybridizations

Preparation of the Dot-Blot Membrane

1. QIAquick PCR Purification kit (QIAGEN).
2. Sterile ddH_2O.
3. Whatman 3 MM filter paper.
4. 1 M NaOH.

5. 0.4 *M* NaOH.
6. 200 mM EDTA, pH 8.2.
7. Hybond N+ charged nylon membrane (Amersham).
8. Dot-blot manifold such as the Bio-Dot Microfiltration Apparatus (Bio-Rad).
9. Clean glass tray.
10. 2× SSC.
11. UV transparent plastic wrap.

Preparation of the Southern Blot Membrane

1. QIAquick PCR Purification kit (QIAGEN).
2. Sterile ddH_2O.
3. Whatman 3 MM filter paper.
4. Agarose and ethidium bromide stain.
5. 0.25 *M* HCl.
6. Denaturation solution: 1.5 M NaCl, 0.5 M NaOH, store at room temperature.
7. Neutralization solution: 1.5 M NaCl, 0.5 *M* Tris•Cl, pH 7.0. Store at room temperature.
8. Two clean glass trays.
9. 20× SSC: 3 *M* NaCl, 0.3 M Na_3citrate.2H_2O, adjust to pH 7.0 with 1 M HC1. Store at room temperature.
10. 2× SSC. Store at room temperature.

Radiolabeling of DNA Probes for Dot and Southern Blots

1. Sterile ddH_2O.
2. Items supplied in the Random Primers DNA Labeling System (Invitrogen) dATP solution, dGTP solution, dTTP solution, random primers buffer mixture, Klenow fragment, stop buffer, store at –20°C.
3. [α-^{32}P]dCTP, 3000 Ci/mmol, 10 MCi/pL. Store at –20°C.

Purification of Radiolabeled DNA Probes for Dot and Southern Blots

1. Nensorb™ 20 Nucleic Acid Purification Cartridge (NEN Life Science Products, Boston, MA).
2. Disposable 5-mL syringe.
3. 100% Methanol.
4. Reagent A: 0. 1 M Tris•HCl, 10 mM triethylamine (TEA), 1 mM dipotassium or disodium EDTA, pH 7.7, stable for 1 month at 4°C. 5. 50% Methanol in water.
6. Molecular biology grade water. 7. 1.5-mL Eppendorf tube.

Hybridization of Dot and Southern Blot Membranes to Radiolabeled DNA Probes

1. Hybridization bottle.

2. PerfectHyb Plus Hybridization Buffer (Sigma-Aldrich).
3. Low-stringency wash solution: 2× SSC, 0.1% SDS. Store at room temperature.
4. Moderate stringency wash solution: 0.2× SSC, 0.1% SDS. Store at room temperature.
5. 2× SSC.
6. Phosphor imaging plates or X-ray film.
7. Phosphor imaging system such as BAS- 1800 II (Fujifilm) or X-ray film processor.

Stripping of Dot and Southern Blot Membranes

1. 0.1% SDS.
2. Glass dish.

Total RNA Extraction and 16S rRNA Dot-Blot Hybridization

Total RNA Extraction

1. 0.1 M sodium phosphate buffer, pH 6.0: to make 200 mL, mix 87.7 mL of solution A (27.6 g of $NaHPO_4$-H_2O/L [0.2 M]) with 12.3 mL of solution B (53.65 g of $Na_2HPO_4 \cdot 7H_2O$/L [0.2 M]), then dilute with H_2O to 200 mL.
2. Acetate buffer, pH 5.1 (50 mM sodium acetate, 10 mM EDTA).
3. 20% SDS (w/v).
4. Buffered phenol, pH 5.1. Store at 4°C in a dark glass bottle.
5. 0.1-mm zirconium beads, autoclaved. 6.2-mL Eppendorf tubes.
7. Phenol-chloroform-isoamyl alcohol (25:24:1).
8. Chloroform-isoamyl alcohol (24:1).
9. 3 M sodium acetate.
10. Absolute ethanol.
11. 80% Ethanol.
12. RNase-free water.
13. RQ1 RNase-free DNase (Promega).
14. Agarose and ethidium bromide stain.

Radiolabeling of DNA Probes for RNA Dot Blots

1. Polynucleotide kinase (Invitrogen).
2. 10× Kinase buffer (Invitrogen).
3. [β-^{32}P]ATP.
4. 1 % IGEPAL CA-630 (Sigma-Aldrich).
5. RNase-free water.
6. Nensorb™ 20 Nucleic Acid Purification Cartridge (NEN Life Science Products, Boston, MA).

7. Disposable 5-mL syringe.
8. 100% Methanol.
9. Reagent A: 0.1 M Tris·HCl, 10 mM triethylamine (TEA), 1 mM dipotassium or disodium EDTA. pH 7.7, stable for 1 month at 4°C.
10. 50% Methanol in water.
11. 1.5-mL Eppendorf tube.

Preparation of RNA Dot-Blot Membrane and Hybridization of Membrane to Radiolabeled DNA Probes

1. 2% Glutaraldehyde.
2. Hybond N+ charged nylon membrane (Amersham).
3. Whatman 3 MM filter paper.
4. Dot-blot manifold such as the Bio-Dot Microfiltration Apparatus (Bio-Rad).
5. Clean glass tray.
6. RNase free water.
7. UV transparent plastic wrap.
8. Hybridization bottle.
9. PerfectHyb Plus Hybridization Buffer (Sigma-Aldrich).
10. RNA dot-blot washing solution (1 × SSC, 1 % SDS).
11. Phosphor imaging plates or X-ray film.
12. Phosphor imaging system such as BAS-1800 11 (Fujifilm) or X-ray film processor.
13. ImageQuant (Fujifilm) or other densitometry software.

METHODS

Extraction of Total Genomic DNA from Environmental Samples

This procedure has been adapted from a method for the extraction of DNA from sediments and optimized for use with rumen samples. Before DNA isolation, fresh rumen contents can be strained through two layers of cheesecloth to remove large particulate matter. Alternate methods of DNA isolation may be preferable for other types of environmental samples.

1. Wash the sample by mixing 2 g of the environmental sample with 10 mL of 0.1 M sodium phosphate buffer, pH 8.0, shaking briefly. Centrifuge the sample at 6000g for 10 min at 4°C and discard the supernatant.
2. Suspend the sample in 10 mL of sodium phosphate buffer, pH 8.0, and repeat step 1.
3. Add 2 mL of lysis solution per gram of pelleted sample and incubate with gentle shaking (75 rpm on a rotating shaker) for 30 min at 37°C.
4. Add an equal volume of SDS solution to the sample and incubate with gentle shaking (75

rpm on a rotating shaker) for 30 min at 37°C.

5. Perform three cycles of freezing at –80°C and thawing at room temperature.
6. Add proteinase K to a final concentration of 50 μg/mL and incubate for 30 min at 37°C.
7. Add an equal volume of phenol and gently invert the sample tube for 4 min. Centrifuge at 10,000g for 30 min at 4°C. Pipet the aqueous phase into a clean tube.
8. Add an equal volume of phenol-chloroform-isoamyl alcohol mixture and gently invert the sample tube for 4 min. Centrifuge at 10,000g for 30 min at 4°C. Pipet the aqueous phase into a clean tube.
9. Add an equal volume of chloroform mixture and gently invert the sample tube for 4 min. Centrifuge at 10,000g for 15 Min at 4°C. Pipet the aqueous phase into a clean tube.
10. Add 10.5 *M* ammonium acetate to a final sample concentration of 2.5 *M*. Add an equal volume of cold isopropanol. Place the sample at –80°C for 30 min to precipitate the nucleic acids.
11. Centrifuge the sample at 12,000g for 30 min at 4°C. Discard the supernatant and wash the nucleic acid pellet with 2 mL of 70% EtOH.
12. Dry the pellet and suspend in 200 μL of TE buffer.
13. Add 1 μL of RNase OneTM to remove RNA from sample.
14. Store DNA at 4°C or -20°C.

Suppressive Subtractive Hybridization

The described protocol is based on the manufacturer's protocol for CLONTECH's PCR-Select™ Bacterial Genome Subtraction Kit (courtesy of Clontech Laboratories, Inc., a Takara Bio Co. [2006] Clontech Laboratories. Inc.). Specific steps have been amended due to metagenomic DNA being used instead of genomic DNA from pure cultures. Further information can additionally be obtained from the "Suppression Subtractive Hybridization" chapter in the *Methods in Molecular Biology* volume on "Gene Expression Profiling".

Restriction Digestion of Metagenomic DNA

Perform the following procedure with each experimental tester and driver metagenomic DNA, and with the control *E. coli* genomic DNA. The driver DNA will be used to "subtract" the DNA common with the tester DNA, thereby enriching for fragments unique to the tester DNA. Testing the efficacy of the restriction enzyme used is crucial to the success of the technique.

Specifically, it serves as an indicator for the presence of confounding inhibitors of latter steps if digestion does not occur. In our experiments we used A1uI as the restriction enzyme; for the E. *coli* control you must use the included *RsaI* restriction enzyme. If you do not, then you will not be able to conduct the associated control experiments using the *E. coli* DNA.

1. Label three 0.5-mL microcentrifuge tubes with a Sharpie as follows: T (this will contain the metagenornic tester DNA), D (this will contain the metagenomic driver DNA), and C (this will contain the *E. coli* control DNA). These designations will be used throughout the procedure.

2. Add the following reagents into the appropriately labeled 0.5-mL microcentrifuge tubes:

T-labeled tube (metagenomic tester DNA):

Metagenomic tester DNA (2 pg in X μL)	X μL
10× *AluI* restriction buffer	5.0 μL
AluI (10 U/μL)	1 .5 μL
Sterile, deionized H_2O	X μL
Total reaction volume:	50μL

D-labeled tube (metagenomic driver DNA):

Metagenomic driver DNA (2 μg in XμL)	XμL
10× *AluI* restriction buffer	5.0μL
AluI (10 U/μL)	1.5 μL
Sterile, deionized H_2O	X μL
Total reaction volume:	50μL

C-labeled tube (*E. coli* control DNA):

E. coli control DNA (2 μg in X μL)	XμL
10× *RsaI* restriction buffer	5.0μL
RsaI (10 U/μL)	1.5 μL
Sterile, deionized H_2O	X μL
Total reaction volume:	50μL

3. Mix each tube by gently flicking the end of the tube and centrifuge briefly. Vortex-mixing may give the appearance of digestion as opposed to specific restriction enzyme digestion of the DNA.
4. Incubate at 37°C for 5-16 h.
5. Use 5μL of the digest mixture to analyze the efficiency of digestion. Electrophorese 0.2 μg of uncut metagenomic DNA and 5μL of AluI-digested metagenomic DNA on a 1 % agarose gel run in 0.5× TBE buffer. Compare the results side by side. The undigested genomic DNA should appear as a high-molecular weight band at the top of the gel running higher than the 1 Kb DNA Ladder.

   (This will also be dependent on the method used to extract DNA from the environmental sample. Some methods are more prone to DNA shearing than others, e.g. those using bead-beating). The Alui-digested DNA, in contrast, will be drastically decreased in size.
6. Add 2.5μL of 0.2 M EDTA, pH 8.0, to terminate the reaction and transfer the volume to a 1.5-mL tube.
7. Add 450 μL of sterile, deionized H_2O to bring volume up to 500μL. Then add 500μL of phenol-chloroform-isoamyl alcohol (25:24:1). (This minimizes the loss of material associated with removing the aqueous layer in step 10.)

8. Vortex-mix thoroughly.
9. Centrifuge the tubes at 10,000 g for 10 min to separate phases.
10. Remove the top aqueous layer and place in a fresh 1.5-mL tube.
11. Add 500µL of chloroform-isoamyl alcohol (24:1) and vortex-mix thoroughly.
12. Centrifuge the tubes at 10,000 g for 10 min to separate phases.
13. Remove the top aqueous layer and place in a clean 1.5-mL tube.
14. Add 1µL of glycogen (20 mg/mL), 0.1 volume of 5 M sodium acetate, pH 5.2, and 2 volumes of 100% ice-cold ethanol. (Therefore, if the volume of the aqueous layer is 500µL, add 1µL of glycogen [20 mg/mL], 50µL of 5 M sodium acetate [pH 5.2], and 1000µL of 100% ice-cold ethanol.)
15. Vortex-mix the mixture thoroughly.
16. Centrifuge the tubes for 15 min at 10,000 g at room temperature. A small white pellet of precipitated DNA should appear at the bottom.
17. Remove the supernatant carefully.
18. Gently overlay the pellets with 200µL of 80% ethanol.
19. Centrifuge at 10,000 g for 5 min.
20. Carefully remove the supernatant.
21. Air-dry the pellets for 5-10 min.
22. Dissolve the pellet in 6.5 µL of H_2O and store at -20°C. The final concentration of driver DNA should be approx 300 ng/µL.

Adaptor Ligation

Three adaptor ligations must be performed for each metagenomic tester DNA, as well as for the control tester DNA. Each DNA is aliquotted into two separate tubes: one aliquot is ligated with adaptor 1 and the second is ligated with Adaptor 2R (tester T2 and C2). In a third tube, portions of each aliquot are combined so that the DNA is ligated with both adaptors (unsubtracted tester control T-c and C-c).

Each unsubtracted tester control DNA serves as a negative control for subtraction. An additional tester DNA for the positive control subtraction is prepared by mixing the RsaI-digested control *E. coli* genomic DNA with the Control DNA (ϕX174/*Hae*III digest) provided in the kit.

1. Label two tubes as "T-diluted" and "C-diluted". Dilute 1.2µL of each corre-sponding restriction enzyme-digested experimental DNA with 1.8µL of sterile, deionized H_2O. The digested metagenomic tester DNA, tube T from elsewhere in this chapter will go into "T-diluted," while digested control *E. coli* DNA, tube C from elsewhere in this chapter will go into "C-diluted."
2. Prepare positive control tester DNA:
   a. Briefly centrifuge the tube containing the control DNA (ϕX174/*Hae*III digest, provided in the kit).

b. Dilute 3 µL of the control DNA with *42µL* of sterile, deionized H_2O (up to 0.2 ng/µL).

c. Mix 1.2µL of the RsaI-digested control *E. coli* genomic DNA (300 ng/uL) with 1.8µL of the diluted ϕX 174/*Hae*III DNA. This is your control tester DNA.

3. Prepare a ligation master mix by combining the following reagents in a 0.5-ml-microcentrifuge tube. To ensure that you have sufficient master mix, prepare enough for all ligations plus one additional reaction.

	per reaction
Sterile, deionized H_2O	4µL
5× Ligation buffer	2µL
T4 DNA ligase (400 U/µL)	1µL

4. For the metagenomic tester DNA (two tubes labeled T1 and T2) and for the control tester DNA (two tubes labeled C1 and C2), combine the reagents below in 0.5-mL microcentrifuge tubes in the order shown. Pipet mixture up and down to mix thoroughly.

Tube no.:	T 1	T2	C 1	C2
Component	(µL)	(µL)	(µL)	(µL)
Diluted tester DNA*	1	1	1	1
Adaptor 1 (10 mM)	2	—	2	—
Adaptor 2R (10 mM)	—	2	—	2
Master mix	7	7	7	7
Final volume	10	10	10	10

5. In a fresh microcentrifuge tube labeled T-c, mix 1.5µL of tester T1 and 1.5µL of tester T2. This will be designated as the unsubtracted tester control T-c. Do the same for the control tester DNA (tester C 1 and C2; label this fresh tube as C-c). After ligation, approx 1/3 of the DNA molecules in each unsubtracted tester control tube will have two different adaptors on their ends.
6. Centrifuge tubes briefly, and incubate at 16°C overnight.
7. Terminate ligation reactions by adding 1µL of 0.2 M EDTA.
8. Heat samples at 72°C for 5 min to inactivate the ligase.
9. Briefly centrifuge the tubes. Preparation of the experimental and control Adaptor-ligated tester DNAs (T1, T2, C1, and C2) is now complete.
10. Dilute 1 µL of each unsubtracted tester control (T-c and C-c) in separate tubes into 1 mL of H_2O. These samples will be used for PCR amplification later.
11. Store samples at −20°C.

Analysis of Ligation

This subheading describes a quality control test for the ligation of adaptors and is specifically designed for the control *E. coli* genomic DNA included in the Clontech kit. The primers will amplify a 374-bp region of the 23S rRNA gene of E. *coli* and are used to verify that at least 25% of the control tester DNA fragments have adaptors on both ends. Amplified fragments span the adaptor/ DNA junctions of testers C1 and C2.

1. Dilute 1μL of each E. *coli* control-ligated DNA (e.g., testers Cl and C2) into 200 μL of H_2O.
2. Set up four PCR reactions with the following template and primer combinations (volumes are inμL):

	1	2	3	4
Tester C 1 (ligated to adaptor 1)	1	1	—	—
Tester *C2* (ligated to adaptor 2R)	—	—	1	1
23S RNA forward primer (10 mM)	1	1	1	1
23S RNA reverse primer (10 mM)	—	1	—	1
PCR primer 1 (10 mM)	1	—	1	—
Total volume	3	3	3	3

3. Prepare a master mix for all of the reaction tubes plus one additional tube. For each reaction planned, combine the reagents below in the order shown:

component	*Amount per reaction tube (μL)*	*Amount for 4 reactions (μL)*
Sterile H_2O	18.75	93.75
10×PCR reaction buffer	2.5	12.5
dNTP mix (10 mM)	0.5	2.5
Total volume	21.75	108.75

4. Mix well by vortex-mixing and then briefly centrifuge the tube.
5. Aliquot 21.75 μL of master mix into each of the reaction tubes from step 2 above.
6. Mix well by vortex-mixing and briefly centrifuge the tube.
7. Incubate the reaction mix in a thermal cycler (using a Applied Biosystems GeneAmp® System 9700) at 72°C. As soon as the temperature has reached 72°C, pause the thermal cycler and add 0.25 μL of TaKaRa Ex Taq DNA polymerase (1.25 U) to each tube. Incubate the tubes for 2 min to extend the adaptors. This step "fills in" the missing strand of the adaptors and thus creates binding sites for PCR primer 1. At this stage, do not remove the samples from the thermal cycler.
8. Immediately commence thermal cycling (using a Applied Biosystems GeneAmp® System 9700):

   · 94°C 30 s

   23 cycles:

   · 94°C 10 s

- 65°C 30 s
- 68°C 1 min

Final extension at 72°C for 7 min, then hold at 4°C indefinitely.

9. Examine the products by electrophoresis on a 2% agarose gel run in 0.5× TBE buffer. If you cannot see a product after 23 cycles, perform four more cycles, and again analyze the product by agarose gel electrophoresis.

First Hybridization

In the first hybridization, an excess of driver DNA is added to the tester DNA, and the samples are heat denatured and allowed to anneal. The concentration of fragments, present in single and multiple copies, is equalized among the type a molecules because reannealing is faster for the more abundant molecules due to the second-order kinetics of hybridization.

1. For each of the metagenomic (designated HT below) and control (HC) subtractions, combine the reagents in 0.2-mL microcentrifuge tubes in the order shown:

Component	*Hybridization sample HT 1 (μL)*	*Hybridization sample HT2 (μL)*
*Alu*I-digested driver DNA	2.0	2.0
Adaptor 1-ligated tester T1	1.0	—
Adaptor 2R-ligated tester T2	—	1.0
4× Hybridization buffer	1.0	1.0
Final volume	4.0	4.0

2. Incubate samples in a thermal cycler at 98°C for 1.5 min.
3. Incubate samples at 63°C for 1.5 h, then proceed immediately to the next section.

Second Hybridization

In the second hybridization the two samples from the first hybridization are mixed together and fresh denatured driver DNA is added to further enrich for tester-specific sequences. New hybrid molecules that correspond to tester-specific DNAs with different adaptors on each end are formed.

Do not denature the primary hybridization samples at this stage. Also, do not remove the hybridization samples from the thermal cycler for longer than is necessary to add fresh driver. Repeat the following steps for your experimental tester DNA and for the control tester DNA.

1. Dilute the 4× hybridization buffer with an equal volume of deionized H_2O to make 2× hybridization buffer.
2. Combine the following reagents in a sterile 0.2-mL microcentrifuge tube:

AluI-digested driver DNA	1 μL
2× Hybridization buffer	1 μL

3. Incubate in a thermal cycler at 98°C for 1.5 min.

4. Remove the tube of freshly denatured driver from the thermal cycler. Use the following procedure to simultaneously mix the driver with hybridization samples 1 and 2 (HT1 and HT2 for the metagenomic DNA, and HC1 and HC2 for the *E. coli* control DNA). This ensures that the two hybridization samples mix together only in the presence of driver.
   a. Set a micropipettor at 15µL.
   b. Gently touch the pipet tip to the sample surface of the tube containing hybridization sample 2 (HT2 for the metagenomic sample, or HC2 for the *E. coli* control).
   c. Carefully draw the entire sample part way into the pipet tip.
   d. Remove the pipet tip from the tube, and draw a small amount of air into the tip, creating a slight air space below the droplet of sample.
   e. Repeat steps b-d with the tube containing the freshly denatured driver. The pipet tip should now contain both samples separated by a small pocket of air.
   f. Transfer the entire mixture to the tube containing hybridization sample 1(HTI for the metagenomic sample, or HC1 for the *E. coli* control).
   g. Mix by pipetting up and down.
5. Briefly centrifuge the tube if necessary.
6. Incubate reaction at 63°C overnight.
7. Add 200 µL of dilution buffer to the tube and mix by pipetting.
8. Heat in a thermal cycler at 63°C for 7 min to eliminate nonspecific hybridization.
9. Store at –20°C.

PCR Amplification

In the following PCR, tester-specific DNA fragments are selectively amplified. The missing strands of the adaptors are first "filled in" again by a brief preincubation at 72°C which creates the binding sites for PCR primer 1. Following that, in the first round of PCR amplification, only dsDNA with different adaptor sequences on each end are exponentially amplified. Subsequently, in the second round, nested PCR is used to enrich for testerspecific sequences thereby reducing the background.

At least five reactions should be prepared:

1. The subtracted metagenomic DNA (mix of HT1/HT2).
2. The unsubtracted metagenomic tester control (T-c).
3. The subtracted control DNA from *E. coli* (mix of HC1/HC2).
4. The unsubtracted tester control for the control subtraction (C-c).
5. The PCR control subtracted DNA (provided in the kit).

The PCR control subtracted DNA from the kit provides a positive PCR control that can be compared directly with the subtracted control DNA from E. *coli* and contains a successfully subtracted mixture of *Hae*III-digested f*X 174 DNA fragments. In addition, you should also include a standard PCR control to ensure that the *Taq* DNA polymerase used in the PCR is performing efficiently.

1. Prepare the PCR templates as follows:
   a. Aliquot 1 µL of each diluted DNA (i.e., each subtracted sample and the corresponding diluted unsubtracted tester control) into an appropriately labeled tube. This is to be done for the prepared metagenomic subtracted sample (HT1/HT2), the diluted metagenomic unsubtracted sample (T-c), the prepared *E. coli* subtracted sample (HC1/HC2), and the diluted *E. coli* unsubtracted sample (C-c).
   b. Aliquot 1µL of the PCR control subtracted DNA (provided in the kit) into an appropriately labeled tube.
2. Prepare a master mix for all of the primary PCR tubes plus one additional tube. Combine the reagents in the order shown below:

Component	*Amount per reaction (µL)*	*For a 5-reaction experiment (µL)*
Sterile H_2O	19.75	118.5
10× PCR reaction buffer	2.5	15.0
dNTP mix (10 mM)	0.5	3.0
PCR primer 1 (10 mM)	1.0	6.0
Total volume	23.75	142.5

3. Vortex-mix well and briefly centrifuge the tube.
4. Aliquot 23.75 µL of master mix into each of the reaction tubes prepared in step 1.
5. Incubate the reaction mix in a thermal cycler (using a Applied Biosystems GeneAmp® System 9700) at 72°C. As soon as the temperature has reached 72°C, pause the thermal cycler and add 0.25 µL of TaKaRa Ex Taq DNA polymerase (1.25 U) to each tube. Incubate the tubes for 2 min to extend the adaptors. At this stage, do not remove the samples from the thermal cycler.
6. Immediately commence thermal cycling (using a Applied Biosystems GeneAm' System 9700):

   Initial hold at 94°C for 5 min

   25 cycles:

   - 94°C 10 s
   - 66°C 30 s
   - 72°C 1.5 min

   Final extension at 72°C for 7 min, then hold at 4°C indefinitely.
7. Three to five more cycles may be needed depending on the Taq DNA polymerise used. In our case, with TaKaRa Ex Taq DNA polymerase, we obtained optimal results with 5 extra cycles (total of 30).
8. Analyze 7 µL from each tube on a 2% agarose gel run in 0.5× TBE buffer. You can set these 7-µL aliquots aside and run them on the same gel used to analyze the secondary

PCR products (step 16 below).

9. Dilute 1 μL of each primary PCR mixture in 39 μL of H_2O
10. Aliquot 1 μL of each diluted primary PCR mixture from the previous step into an appropriately labeled tube.
11. Prepare master mix for the secondary PCRs plus one additional reaction by combining the reagents in the order shown:

Reagent	*Amount per reaction (μL)*	*For a 5-reaction experiment (μL)*
Sterile H_2O	18.75	112.5
10× PCR reaction buffer	2.5	15.0
Nested primer 1 (10 mM)	1.0	6.0
Nested primer 2R (10 mM)	1.0	6.0
dNTP mix (10 mM)	0.5	3.0
Total volume	23.75	142.5

12. Vortex-mix well and briefly centrifuge the tube.
13. Aliquot 23.75 μL of master mix into each reaction tube and incubate the reactions in a thermal cycler (using a Applied Biosystems GeneAmp" System 9700) at 94°C for 5 min. As soon as the temperature has reached 94°C, pause the thermal cycler and add 0.25 μL of TaKaRa Ex Taq DNA polymerase (1.25 U) to each tube.
14. Immediately commence thermal cycling (using a Applied Biosystems GeneAmp" System 9700):

    10-12 cycles:

    - 94°C 10 s
    - 68°C 30 s
    - 72°C 1.5 min

    Final extension at 72°C for 7 min, then hold at 4°C indefinitely.
15. Three to five more cycles may be needed depending on the *Taq* DNA polymerase used. In our case, with TaKaRa Ex *Taq* DNA polymerase, we obtained optimal results without extra cycles for the secondary PCR (total of 12 cycles only).
16. Analyze 7 μL from each reaction on a 2.0% agarose gel run in 0.5x TBE buffer.
17. Store reaction products at -20°C.

Analysis of PCR Products

1. Agarose get analysis of primary PCR. With the PCR control subtracted DNA, the major bands that appear after 25 cycles should correspond to the f*X174/HaeIII fragments. This result should look similar to the control subtraction performed. The f*X174/*Hae*III bands may appear only after the secondary PCR—this is normal. At this stage, success with the

control reactions is the only indication that the subtraction with metagenomic DNA may work.

The primary PCR products from the experimental subtraction with metagenomic DNA should appear as a smear of amplified products from 0.2-2 kb. Confirmation of successful subtraction with metagenomic DNA is described in the subsequent sections using Southern blots. What follows here are guidelines on how to interpret the results to this point now and troubleshoot some basic problems.

a. If you cannot see any products after 25 cycles, perform three to five more cycles, and analyze the products by gel electrophoresis. (As noted in elsewhere in this chapter our experiment required a total of 30 cycles). This variable is usually the most responsive and easiest to compare when products from increasing cycles are analyzed side-by-side on the same agarose gel.

b. If PCR products are not detected for either the subtracted (prepared E. *coli* subtracted sample [HC1/HC2]) or unsubtracted (unsubtracted tester control C-c) samples or the PCR control subtracted DNA, make sure that the *Taq* DNA polymerase is working. If the *Taq* DNA polymerase is working optimally, then try optimizing the PCR cycling parameters.

   Two variables to modify independently include decreasing the annealing or the extension temperature in small increments. Lowering these temperatures can dramatically increase the background. Initially, try reducing the annealing temperature from 66°C to 64°C. The extension temperature should be lowered in smaller increments (e.g., from 72°C to 71°C).

c. If PCR products are detected for the unsubtracted (unsubtracted tester control C-c) samples, but not in the subtracted sample, then dilute the primary PCR sample and perform more cycles of secondary PCR.

2. Agarose gel analysis of secondary PCR. The patterns of the secondary PCR products from the PCR control subtracted DNA and from the *E. coli* control subtraction should resemble each other (a few additional background bands may appear).

   Because of the increased complexity of metagenomes relative to pure cultures the experimental subtracted samples usually look like smears. In contrast, for subtraction experiments with DNA from pure cultures, a number of distinct bands may also appear in addition to the smeared products.

   a. If the expected *ϕX174/HaeIII* bands appear in the PCR Control Subtracted DNA, but not in the control subtraction (i.e., either a smear of products or randomly sized bands appear), then there was probably an issue during the subtraction. Poor ligation efficiency is likely. If the ligation efficiency is not a problem, then be sure that the protocol was followed correctly.

   b. If you cannot see a product after 12 cycles, perform three more cycles, and again check the product by gel electrophoresis. Be conservative when adding cycles; too many cycles will increase background.

   c. If you do not observe *ϕX174/HaeIII* bands in the PCR Control Subtracted DNA, you will need to optimize the PCR conditions. Again, because of the variation in *Tag* DNA polymerases and thermal cycler models, many of these conditions need to be determined empirically.

d. In a successful subtraction, the banding pattern of your unsubtracted DNA ligated with both adaptors (unsubtracted tester control T-c) should be different from the banding pattern of your experimental subtracted DNA samples. With metagenomic DNA this can sometimes be detected as a shift in the size of the smear of amplified products.

Clone Library of Enriched Fragments After Subtraction

In the following steps, 4µL of secondary PCR products from the metage-nomic subtraction will be cloned into the pCR® 4-TOPO® vector (Invitrogen) and transformed into E. *coli* TOP10 cells (Invitrogen) by electroporation. The described protocol is based on the manufacturers' protocol for the TOPO TA Cloning® Kit for Sequencing (courtesy of Invitrogen (D [2006]).

1. Set up the following reaction:

Secondary PCR product	4µL
Dilute salt solution	1µL
TOPO® vector	1µL

   Final volume: 6µL

   Addition of the dilute salt solution in the TOPO® cloning reaction brings the final concentration of NaCl and MgCl2 in the TOPO® cloning reaction to 50 mM and 2.5 mM, respectively.
2. Mix reaction gently and incubate for 5 min at room temperature (22-23°C).
3. Place the reaction on ice. Add 2 µL of the TOPO® Cloning reaction into a vial of One Shot" Electrocompetent E. *coli* and mix gently. Do not mix by pipetting up and down; gently stir with the pipet. Store remaining cloning reaction volume at –20°C when finished.
4. Carefully transfer solution to a 0.1-cm gapped cuvette, avoiding the formation of bubbles.
5. Electroporate your samples. Electroporation conditions are 2.5 kV, 25 µF, and 200 M* using a 0.1-cm gapped cuvet and a Bio-Rad GenePulser™.
6. Immediately add 250 µL of room temperature SOC medium.
7. Immediately plate transformants onto LB agar plates supplemented with 100 µg/mL of ampicillin and grow overnight at 37°C. Plate two different volumes to ensure that at least one plate will have well-spaced colonies. An efficient TOPO® cloning reaction should produce several hundred colonies.
8. Pick 96 random transformants and grow at 37°C in 200 µL of Luria-Bertani medium supplemented with ampicillin for 2 h.
9. Use the cycling conditions from the primary PCR, but the primers from secondary PCR to verify the presence of clone inserts and for use in dot and Southern blot hybridizations. Use 1 µL of each cell culture as a template for PCR screening.
10. Stock cultures of transformants should be maintained at –80°C in 96-well plates containing LB media with glycerol (1:1 ratio).

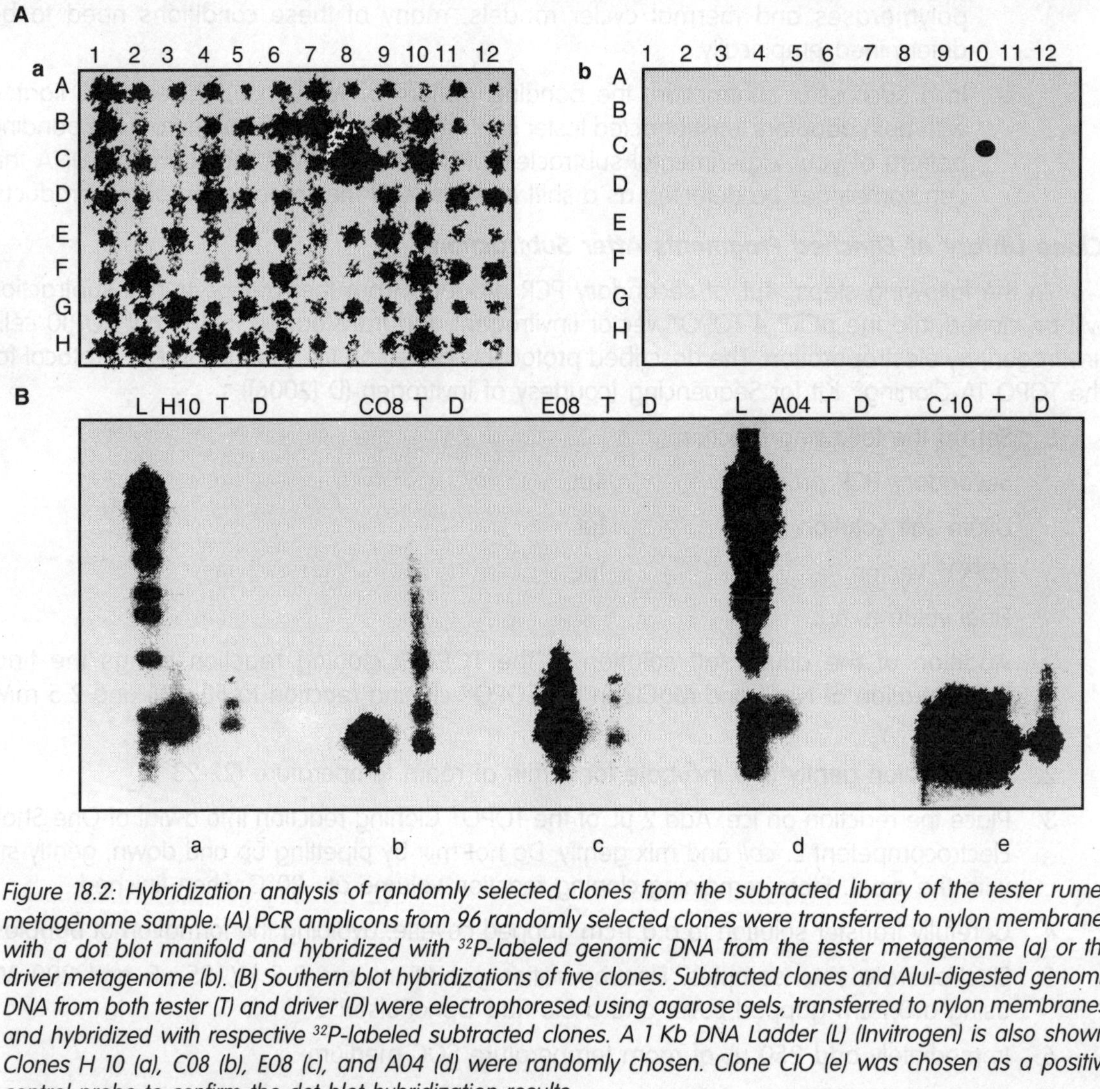

Figure 18.2: Hybridization analysis of randomly selected clones from the subtracted library of the tester rumen metagenome sample. (A) PCR amplicons from 96 randomly selected clones were transferred to nylon membranes with a dot blot manifold and hybridized with ^{32}P-labeled genomic DNA from the tester metagenome (a) or the driver metagenome (b). (B) Southern blot hybridizations of five clones. Subtracted clones and AluI-digested genomic DNA from both tester (T) and driver (D) were electrophoresed using agarose gels, transferred to nylon membranes, and hybridized with respective ^{32}P-labeled subtracted clones. A 1 Kb DNA Ladder (L) (Invitrogen) is also shown. Clones H 10 (a), C08 (b), E08 (c), and A04 (d) were randomly chosen. Clone CIO (e) was chosen as a positive control probe to confirm the dot blot hybridization results.

Confirmation of Tester-Specific DNA Fragments by Dot-Blot and Southern Blot Hybridizations

The efficiency of SSH can be tested by performing dot-blot and Southern blot hybridizations. In the dot-blot procedure, cloned subtraction fragments are spotted onto a membrane and hybridized to a tester metagenomic DNA probe and later to a driver metagenomic DNA probe. The Southern blot approach is similar, but uses the cloned subtraction fragments to probe a membrane containing the restriction digests of both tester and driver metage nomic DNA.

If the SSH procedure was successful, there should be little or no hybridization of subtraction fragments to the driver metagenomic DNA on either type of blot. The method described here utilizes radio-labeled DNA probes; however, other methods of labeling are also suitable. All membrane manipulations should be performed using sterile forceps. (For further reference information on Southern blotting consult *Current Protocols in Molecular Biology*).

Preparation of the Dot-Blot Membrane

1. Purify PCR products from randomly selected transformants using the QIAyuick PCR Purification kit (QIAGEN) as per the manufacturer's instructions. This results in purified DNA in a final volume of approx 50 µL per sample.
2. Dilute 10µL of each PCR product with 40µL of 1 M NaOH, 5 µL of 200 mm EDTA, pH 8.2, and 45µL of sterile water. Denature the samples by heating at 99.9°C for 10 min in a thermocycler.
3. Prepare a dot-blot manifold as per the manufacturer's instructions. A charged nylon membrane (Hybond N +) cut to fit the manifold is laid on the surface of a tray of distilled water to allow the membrane to wet by capillary action. The membrane is submerged and soaked for 10 min.
4. Cut filter paper to fit the manifold, soak the paper in sterile distilled water, and place it on the manifold.
5. Lay the wet membrane on top of the filter paper and assemble the manifold, ensuring that no bubbles are trapped in the apparatus.
6. Prewash the membrane by adding 500 µL of sterile ddH_2O per well. A small amount of bromphenol blue dye can be added to the prewash water if desired.
7. Turn on the vacuum and allow the water to be slowly pulled through the membrane. Then, turn the vacuum off and slowly break the vacuum seal.
8. Apply denatured DNA samples to the membrane (one sample per well).
9. Turn the vacuum on once more, allowing the DNA to be pulled onto the membrane and the liquid from the samples to be pulled through.
10. Rinse each well with 500 µL of 0.4 M NaOH and dismantle the manifold.
11. Rinse the membrane briefly in 2×SSC and air-dry for 10 min.
12. Crosslink the DNA to the membrane by wrapping the membrane in UV transparent plastic wrap and irradiating in a UV crosslinker (Stratalinker, Stratagene) using the Autocrosslink function as per the manufacturer's instructions.
13. Store the membrane dry at room temperature for up to 2-3 months, or at 4°C for longer periods of time.

Preparation of the Southern Blot Membrane

1. Purify PCR products from randomly selected transformants using the QIAquick PCR Purification kit (QIAGEN) as per the manufacturer's instructions. This results in purified DNA in a final volume of approx 50 µL per sample.
2. Load approx 50 ng of a purified subtraction fragment, 1 µg of tester metagenomic DNA digest (digested to completion with the same restriction enzyme used in the SSH) and 1 µg of driver metagenomic DNA digest (also digested to completion) in separate lanes on

a 0.8% agarose gel run in 0.5x TBE buffer. Electrophorese the gel for 3 h at 80 V, stain in 0.1 μg/mL of EtBr and visualize using a UV transilluminator.

3. Prepare the gel for DNA transfer by performing a series of washes at room temperature. All washes should be performed by slowly shaking the gel in a clean glass dish in approx 10 gel volumes of liquid.

   Between each of the washes, the gel should be rinsed with distilled water. Wash the gel once 0.25 M HCl for 30 min, twice in denaturation solution for 20 min, and twice in neutralization solution for 20 min.
4. Fill a glass dish with approx 1μL of 20x SSC. Balance a solid support large enough to hold the gel over the dish, leaving space on both sides of the support to allow a filter paper wick to hang down into the 20× SSC.
5. Cut a Whatman 3 MM filter paper wick so that its width is the same size as the length of the gel and its length is long enough to lay across the solid support and hang into the 20x SSC solution on both sides of the support. Wet this wick with 20x SSC and place it on the support.
6. Place the gel on the filter paper and squeeze out any air bubbles.
7. Cut a charged nylon membrane (Hybond N^+) to fit the agarose gel and lay it on the surface of a tray of distilled water to allow the membrane to wet by capillary action. The membrane is submerged and soaked for 10 min.
8. Place the wet membrane on the surface of the gel. Remove any air bubbles by slowly rolling a glass pipet over the surface. Flood the surface of the membrane with 20× SSC.
9. Cut five pieces of Whatman 3 MM filter paper to the size of the membrane, and stack on top of the membrane. Fold or cut paper towels to the size of the membrane and place on top of the filter paper.
10. Lay a solid support over the paper towels and place a weight on top of the support to hold the transfer pyramid in place.
11. Allow the transfer to proceed overnight at room temperature.
12. Disassemble the transfer pyramid, and using pencil, mark the position of gel wells and gel orientation on the membrane.
13. Rinse the membrane briefly in 2x SSC and air-dry for 10 min.
14. Cross link the DNA to the membrane by wrapping the membrane in UV transparent plastic wrap and irradiating in a UV crosslinker (Stratalinker, Stratagene) using the Autocrosslink function as per the manufacturer's instructions.
15. Store the membrane dry at room temperature for up to 2-3 months, or at 4°C for longer periods of time.

Radiolabeling of DNA Probes for Dot and Southern Blots

Labeling of several probes can be done simultaneously provided that the purified, labeled probes are stored at –20°C and used within a week. Dot blots should be probed with restriction enzyme digested metagenomic driver DNA first and subsequently with restriction enzyme digested metagenomic tester DNA.

Hybridizing the dot blot in this order reduces the likelihood of residual signal present on the membrane after stripping. The restriction enzymes used for digestion of metagenomic probes should be identical to those used in the subtraction procedure. For Southern blots, a probe should consist of a single purified subtraction fragment.

1. Dilute purified probe DNA to 10 ng/µL in distilled water. Denature 100 ng of purified probe DNA by heating for 5 min in a boiling water bath, then immediately cooling on ice.
2. To the denatured probe, make the following additions on ice:

   2 µL dATP solution

   2 µL dGTP solution 2 µL dTTP solution

   15 µL random primers buffer mixture

   5 µL (~50 µCi) [a*-^{3}2P]dCTP, 3000 Ci/mmol, 10 µCi/µL

   Distilled water to a total volume of 49 ML.

   Mix briefly.
3. Add 1 µL of Klenow fragment and mix thoroughly. Briefly centrifuge to deposit contents at the bottom of the tube.
4. Incubate at 25°C for 3 h.
5. Add 5 µL of stop buffer then proceed to probe purification or store probe at -20°C for up to 1 wk.

Purification of Radiolabeled DNA Probes for Dot and Southern Blots

Unincorporated dNTPs, salts, and other impurities should be removed from the DNA probes before hybridization. Purification using a Nensorb™ 20 Nucleic Acid Purification Cartridge (NEN Life Science Products, Boston, MA) is described here; however other purification methods may also be used.

In the following steps, whenever pressure is applied to the cartridge following the application of a reagent or a sample, the flow rate of the solvent should not exceed one drop per 2 s. In addition, drying of the cartridge resin should be avoided until the final elution step.

1. Clamp one Nensorb 20 cartridge to a support, remove the protective cap from the cartridge, and place a waste collection beaker below the cartridge.
2. Pipet 2 mL of 100% methanol into the top of the cartridge.
3. Securely attach an adapter to the top of the cartridge forming an airtight seal.
4. Attach a disposable 5-mL syringe to the adaptor and gently push the methanol through the cartridge until the meniscus reaches the top of the cartridge resin bed.
5. Disconnect the syringe and the adapter and add 2 ml- of reagent A to the top of the cartridge.

6. Reconnect the adapter and the syringe and gently push the reagent A through the cartridge until the meniscus reaches the top of the cartridge resin bed.
7. Disconnect the syringe and the adapter.
8. Add 200 µL of reagent A to the labeled probe and gently mix. Pipet this mixture into the top of the cartridge. Reattach the adapter and syringe and gently push the sample into the cartridge resin.
9. Disconnect the syringe and the adapter.
10. To wash the cartridge, pipet 3 mL of reagent A into the top of the cartridge, reattach the adapter and syringe, and gently push the liquid until the meniscus reaches the top of the cartridge resin bed.
11. Disconnect the syringe and the adapter and wash the cartridge with 3 mL of molecular biology grade water. Reattach the adapter and syringe and gently push the liquid until the meniscus reaches the top of the cartridge resin bed.
12. Remove the waste beaker and place a sterile 1.5-mL Eppendorf tube below the cartridge to collect the purified probe.
13. Disconnect the syringe and adapter.
14. Pipet 500 pL of 50% methanol in water into the top of the cartridge. Reattach the adapter and the syringe and elute the purified probe into the Eppendorf tube. (Continue eluting until no further liquid escapes the cartridge.)
15. Use the purified probe immediately for a dot blot or Southern blot hybridization or store at –20°C for up to 1 wk. All probes should have specific activity measure-ments of at least 10' cpm mg^{-1}.

Hybridization of Dot and Southern Blot Membranes to Radiolabeled DNA Probes

Hybridization of multiple membranes to a single probe can be performed simultaneously in a single hybridization bottle, provided that the quantity of probe and hybridization solution are adjusted accordingly and a mesh sheet is used to separate the membranes.

1. Carefully place the membrane in a hybridization bottle, add 7 mL of warm (68°C) PerfectHyb Plus Hybridization Buffer (Sigma-Aldrich), and place the bottle in a rotating hybridization oven for 15 min at 68°C. This is the prehybridization step.
2. Pour the prehybridization solution off of the membrane and discard.
3. Add 100 ng of radiolabeled probe to 7 mL of warm (68°C) PerfectHyb Plus and mix gently.
4. Add the PerfectHyb Plus containing the probe to the hybridization bottle containing the membrane.
5. Place the hybridization bottle in a rotating hybridization oven for 16 h at 68°C.
6. Pour off the hybridization solution and add 50 mL of low-stringency washing solution. Wash the membrane for 10 min at room temperature by gently shaking the hybridization bottle so that the wash solution flows gently over the membrane. Discard the wash solution

and repeat this step with fresh low-stringency wash solution.

7. Discard the low-stringency wash solution and add 50 mL of moderate strin-gency wash solution to the hybridization bottle. Place the bottle in the rotating hybridization oven for 15 min at 42°C. Discard the moderate-stringency wash solution and repeat this step.
8. Rinse the membrane briefly with 100 mL of 2x SSC.
9. Gently remove the membrane from the hybridization bottle, wrap it tightly in transparent plastic wrap, and expose the membrane to a phosphor imaging plate for 20 min. Longer exposures may be performed if desired.
10. Use a phosphor imaging system to capture the blot image from the phosphor imaging plate. Alternatively, X-ray film may be used for imaging, however exposure time must be increased to 1-3 h and an X-ray developer must be used. Examples of the results produced from both dot blots and Southern blots are shown in Figure elsewhere in this chapter

Stripping of Dot and Southern Blot Membranes

1. Place the wet membrane in a glass dish and pour 300-400 mL of boiling 0.1 % SDS over it. Gently shake the membrane until the 0.1 % SDS reaches room temperature.
2. Remove the membrane and wrap it tightly in plastic wrap. The membrane is now ready to be rehybridized, or it can be stored wet at 4°C for a few days.

Total RNA Extraction and 16S rRNA Dot-Blot Hybridization

After an SSH experiment, it can be useful to confirm the presence of specific microorganisms in the environmental sample and determine the proportion of their 16S rRNA relative to the total 16S rRNA in the tester and driver samples. For this purpose, DNA oligonucleotide probes are designed to target the populations of interest.

In the study exemplified here, it was desirable to determine the proportions of Bacteria and Archaea in the tester and driver rumen samples, and therefore domain-specific 16S rRNA probes were used. The rRNAs from reference cultures must be used as controls to ensure that each probe is hybridizing properly and to assist with quantification. Reference cultures should be chosen appropriately for the environmental metagenomes under scrutiny.

One reference culture should be obtained for each probe and defined quantities of RNA should be spotted on the blot. Measuring the intensity of hybridization signal from reference rRNA allows estimates to be made of the quantity of rRNA from environmental samples. The procedure for extraction of total RNA described below is slightly modified from the methods of Stahl et al. (1988) and Lin et al. (1997). (For further reference information on Northern blotting consult *Current Protocols in Molecular* Biology.

Total RNA Extraction

1. Combine 0.8 g of each environmental sample of interest with 400 μL of 0. 1 M sodium phosphate buffer, pH 6.0, 400 μL of acetate buffer pH 5. 1, 100 μL of 20% SDS (w/v), 500 μL of phenol (buffered to pH 5.1), and 0.1 g of sterile zirconium beads (0.1 mm diameter) in a 2.2-mL Eppendorf tube.
2. For reference microorganisms, centrifuge to obtain pellets from 40-mL cultures (E. *coli* BW

13711 or *M. acetivorans* C2A in this example) and combine 0.8 g from each with 400 µL of 0.1 M sodium phosphate buffer, pH 6.0, 400 µL of acetate buffer pH 5.1, 100 µL of 20% SDS (w/v), 500 µL of phenol (buffered to pH 5.1), and 0.1 g of sterile zirconium beads (0.1 min diameter) in a 2.2-mL Eppendorf tube.

3. Incubate all samples for 2 min at 60°C.
4. Homogenize all samples in a mini-beadbeater reciprocating shaker (Biospec Products, Bartlesville, OK).
5. Repeat steps 3 and 4 twice.
6. Centrifuge samples at 4°C at 10,000g for 5 min. Transfer supernatants to clean tubes.
7. To each sample, add 1 mL of phenol (buffered at pH 5.1), and shake in the mini-beadbeater for 2 min. Centrifuge samples for 5 min at 4°C and 10,000g. Transfer the aqueous phases to clean tubes.
8. Add 1 mL of phenol-chloroform-isoamyl alcohol (125:24:1) to each sample and shake in the mini-beadbeater for 2 min. Centrifuge samples for 5 min at 4°C and 10,000g. Transfer the aqueous phases to clean tubes. Repeat this step once.
9. Add 1 mL of chloroform-isoamyl alcohol (24:1) to each sample and shake in the mini-beadbeater for 2 min. Centrifuge samples for 5 min at 4°C and 10,000g. Transfer the supernatant from each sample to clean tubes.
10. Precipitate total nucleic acids from samples by addition of 50 µL of sodium acetate (3 M) and 1 mL of cold absolute ethanol to the supernatant followed by chilling at -80° C for 30 min.
11. Centrifuge samples at 4°C for 15 min at 10,000g to pellet nucleic acids.
12. Wash pellets with 100 µL of 80% ethanol, dry pellets, and suspend them in 100 µL of RNase-free water.
13. Treat samples with 1 U/µg of RQ 1 RNase-free DNase (Promega) for 30 min at 37°C followed by a second extraction with phenol, phenol-chloroform-isoamyl alcohol, and chloroform-isoamyl alcohol performed in the same manner as steps 7-9.
14. Nucleic acids were precipitated, pelleted, and washed again in the same manner as steps 10-12 but suspend in 40 µL of RNase-free water.
15. Estimate the concentration of RNA in each sample by measuring the absorbance at 260 nm and confirm the concentration by electrophoresing small aliquots on 1% agarose gels run in 0.5x TBE buffer. Adjust the concentration of each sample to 100 ng/uL^{-1}.

Radiolabeling of DNA Probes for RNA Dot Blots

End labeling of DNA oligonucleotide probes can be performed using polynucleotide kinase (Invitrogen) and [y*-^{32}P]ATP. The amount of polynucleotide kinase used in the labeling experiment should be calculated considering a 1:1 molar ratio of probe to [y*-^{32}P]ATP label.

1. In a 1.5-mL Eppendorf tube, combine the following:

3.0 µL	10× Kinase buffer

1.5 µL	1% IGEPAL
Calculated amount	Oligonucleotide probe
Calculated amount	[γ-^{32}P]ATP
1.0 µL	Polynucleotide kinase

Add RNase-free water to bring the total volume up to 30 µL.

2. Incubate the probe at 37°C for 30 min.
3. Inactivate polynucleotide kinase by heating probe to 65°C for 2 min.
4. Place probe on ice and purify using a Nensorb column as described for probes in elsewhere in this chapter
5. Use the purified probe immediately for a dot-blot hybridization or store at –20°C for up to 1 wk. All probes should have specific activity measurements of at least 107 cpm–1 mg^{-1}.

Preparation of RNA Dot-Blot Membrane and Hybridization of Membrane to Radiolabeled DNA Probes

Hybridization temperature and the temperature for the final membrane wash should be determined by the sequence and length of the probe used. Calcu-lations for determining T_m of probes and appropriate hybridization and wash temperatures can be found in *Current Protocols in Molecular Biology.* For hybridization of multiple membranes simultaneously to the same probe see elsewher in this chapter

1. Denature reference and environmental sample RNA by adding 3 volumes of 2% glutaraldehyde to 1 volume of RNA sample and incubate at room temperature for 10 min. Place samples on ice and dilute samples to desired concentrations with RNase- free water.
2. Prepare nylon membrane and dot blot manifold as described above in elsewher in this chapter. Prewash the membrane by adding 500 µL of sterile RNase free water to each well and allow the vacuum to pull the water slowly through the membrane.
3. Spot denatured RNA samples directly onto the nylon membrane, remove membrane from manifold, dry, and crosslink as described in elsewhere in this chapter RNA from each sample should be spotted onto the membrane in triplicate, in amounts of 100 ng, 50 ng, 20 ng, 5 ng, 2 ng, 0.5 ng, and 0.2 ng of RNA per spot (each spotted in a volume of 100 µL).
4. Carefully place the membrane in a hybridization bottle, add 7 mL of warm (40°C) PerfectHyb Plus Hybridization Buffer (Sigma-Aldrich), and place the bottle in a rotating hybridization oven for 2 h at 40°C. This is the prehybridization step.
5. During the last 10 min of prehybridization, denature probe by heating for 10 min to 100°C in a boiling water bath or a heat block.
6. Pour the prehybridization solution off of the membrane and discard.
7. Add 100 ng of denatured, radiolabeled probe to 7 mL of warm (40°C) Perfect-Hyb Plus and mix gently.
8. Add the PerfectHyb Plus containing the probe to the hybridization bottle containing the

membrane.

9. Place the hybridization bottle in a rotating hybridization oven for 16 h at 40°C or appropriate hybridization temperature calculated based on the sequence and length of the probe.
10. Pour off the hybridization solution and add 50 mL of warm (40°C) RNA dot blot washing solution. Place the bottle in the rotating hybridization oven for 1 h at 40°C. Discard the wash solution and repeat this step.
11. Pour off the wash solution and add 50 mL of fresh warm (40°C) wash solution. Place the bottle in the rotating hybridization oven for 30 min at an appropriate temperature calculated for the specific probe. Discard the wash solution, wrap wet membrane tightly in transparent plastic wrap, and expose the membrane to a phosphor imaging plate.
12. Capture an image of the blot using a phosphor imaging system or X-ray film processor as described for dot blots in elsewhere in this chapter.
13. Utilize a software program such as ImageQuant (Fujifilm) to determine densito-metry measurements on each dot blot.

DNA Sequencing and Bioinformatic Analsyses

Following Southern blot confirmation of the uniqueness of the cloned subtracted products relative to one metagenome versus another, the suppressive subtractive hybridization library should be sequenced.

Descriptive protocols for DNA sequencing and subsequent bioinformatic analyses is beyond the scope *of* this chapter and are better described in other volumesof the *Methods in Molecular Biology series.*

POINTS TO REMEMBER

1. To suspend precipitated DNA and mix reactions, gently pipet them up and down and then centrifuge the tube briefly to deposit contents at the bottom. For phenol-chloroform extractions vortex-mix the reactions. Always add enzymes to reaction mixtures last, and thoroughly incorporate the enzyme by gently pipetting the reaction mixture up and down.

   Do not increase the amount of enzyme added or the concentration of DNA in the reactions. The amounts and concentrations have been carefully optimized.

2. The choice of the appropriate restriction enzyme should depend on its ability to digest metagenomic DNA to produce fragments of optimal length. To determine if a sample is completely digested, remove small aliquots of DNA after 1 and 3 h of digestion and compare the samples on an agarose gel.

   If the DNA size distribution for both samples is the same (including relative to the E. *coli* positive control), then the digestion is complete. If the size distribution is not reduced after *AluI* digestion or is longer than 2 kb, your metagenomic DNA cannot be efficiently cut by (a) any restriction enzyme because of some inhibiting impurities in the DNA sample, or (b) the restriction enzyme because of the DNA's nucleotide composition (unlikely with metagenomic DNA).

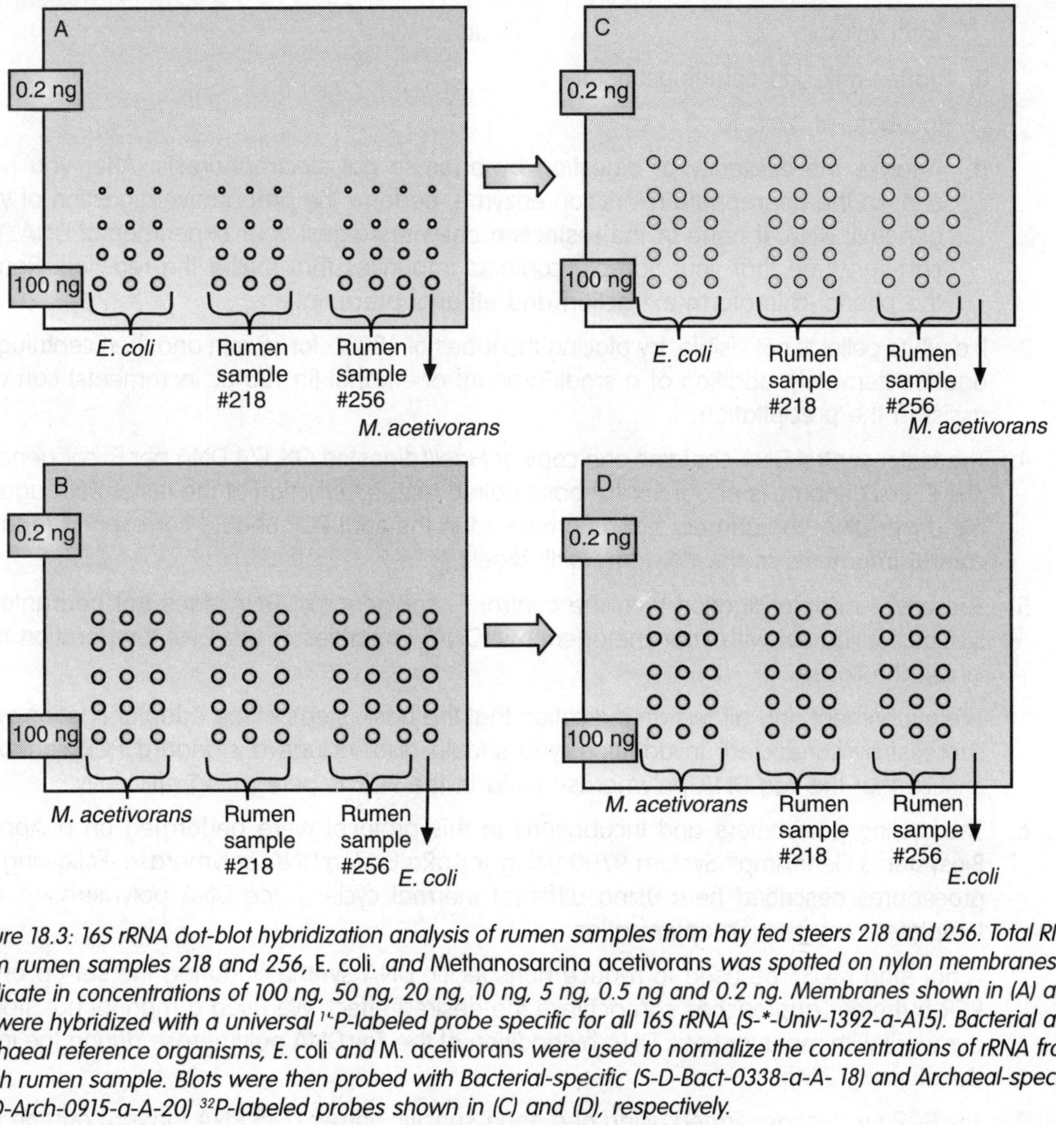

Figure 18.3: 16S rRNA dot-blot hybridization analysis of rumen samples from hay fed steers 218 and 256. Total RNA from rumen samples 218 and 256, E. coli. *and* Methanosarcina acetivorans *was spotted on nylon membranes in triplicate in concentrations of 100 ng, 50 ng, 20 ng, 10 ng, 5 ng, 0.5 ng and 0.2 ng. Membranes shown in (A) and (B) were hybridized with a universal ^{32}P-labeled probe specific for all 16S rRNA (S-*-Univ-1392-a-A15). Bacterial and Archaeal reference organisms, E.* coli *and* M. acetivorans *were used to normalize the concentrations of rRNA from each rumen sample. Blots were then probed with Bacterial-specific (S-D-Bact-0338-a-A- 18) and Archaeal-specific (S-D-Arch-0915-a-A-20) ^{32}P-labeled probes shown in (C) and (D), respectively.*

The described protocol here uses *AluI,* but there are other four-base, blunt-end producing restriction enzymes: *HaeIII* and *RsaI.* A combination of several six-base, blunt-cutting restriction enzymes can also be used to obtain the desired level of DNA digestion. In this case, perform several standard digestions with different restriction enzymes:

a. Add the following reagents into each reaction tube:

Metagenomic DNA (0.2 µg)	X µL
10× restriction enzyme buffer	1.0 µL
Restriction enzyme (1 U)	1.0 µL

Sterile, deionized H_2O	X µL
Total volume	10 µL

b. Vortex-mix and centrifuge briefly.

c. Incubate at 37°C for 3 h.

d. Analyze the efficiency of digestion by agarose gel electrophoresis. After you have chosen the appropriate restriction enzyme, perform the preparative digestion of your genomic DNA. If none of the restriction enzymes digest your experimental DNA, you can conclude that your sample contains impurities that inhibit the reaction. Repeat the phenol-chloroform extraction and ethanol precipitation.

3. If a white pellet is not visible, try placing the tubes at –80°C 'for 15 min and then centrifuging again. Alternately, addition of a small amount of ethanol (in 100-µL increments) can also assist in the precipitation.

4. The tester control DNA contains one copy of *HaeIII* digested OX 174 DNA per E. *coli* genome (the E. *coli* genome is about 5 x 10^6 base pairs). After subtraction of the tester DNA against the driver DNA, the primary bands produced in the final PCR should correspond to these control fragments of the f*X 174/*Hae*III digest.

5. Successful adaptor ligation with the control E. *coli* genomic DNA does not guarantee a successful ligation with your metagenomic DNA; impurities in the DNA preparation may inhibit the ligase.

   The experiment only allows an indication that the basic steps of the adaptor ligation were successfully conducted. In addition, you should also include a standard PCR control to ensure that the *Taq* DNA polymerase used in the PCR is performing efficiently.

6. The cycling parameters and incubations in this protocol were performed on a Applied Biosystems GeneAmp® System 9700 using TaKaRa Ex *Taq* DNA polymerase. Following the procedures described here using different thermal cyclers, *Taq* DNA polymerases, and templates may require optimization.

   A hot start MUST be used to reduce nonspecific DNA synthesis during the set-up of the PCR but other approaches can achieve the desired effect. We used a manual hot start in our work with metagenomic DNA by addition of the *Tay* DNA polymerase during the initial denaturation step.

7. The PCR product generated using one gene-specific primer (23S RNA forward primer) and PCR pr:mer 1 should be about the same intensity as the PCR product amplified using two gene specific primers (23S RNA forward and reverse primers).

   If the intensity of bands corresponding to these PCR products differs by more than 4-fold, then the efficiency of your ligation is less than 25%. This will significantly reduce your subt©raction efficiency, and we recommend that you repeat the ligation reaction.

   If the PCR product amplified using the 23S RNA forward primer and PCR Primer I contains a 574-bp band in addition to the expected 374-bp band, your DNA is not fully digested. If there is a significant amount of this 574-bp product, we recommend that you repeat the RsaI digestion. If your ligation did not work well, you should repeat the ligation before performing the hybridizations.

8. The CLONTECH PCR-Select"M Bacterial Genome Subtraction Kit was configured to work with genomic DNA obtained from pure cultures of microorganisms. Therefore, troubleshooting the procedure when working with pure cultures is much more straightforward, as in the case of optimizing the hybridization temperature used.

   This variable may warrant investigation if information is available as to the composition of the metagenome being studied. The hybridization temperature, for example, is dependent on the GC content of a particular genome. The optimal temperature for genomic DNA with average GC content (40-51%) is 63°C.

   If the GC content of the genomic DNA is unusually low (e.g., the genome of *Clostridium perfringens* has a GC content of 26.5%), then the hybridization temperature should be lowered.

9. The PCR mixture should now be enriched for metagenome tester-specific DNA. The products should be cloned to make a subtracted DNA library. The amplified products can be directly inserted into a T/A cloning vector.

10. If you experience arcing during electroporation then precipitate the remainder of the TOPO® Cloning reaction and suspend in water before trying again. Another factor in arcing of your samples during electroporation may be the volume of cells. They should be between 50 and 80 μL (0.1 cm cuvettes) or 100-200 μL (0.2 cm cuvettes).

11. A solution of 0.004% bromphenol blue in water can be used as the prewash water in dot blotting. The dye will slightly color the membrane in the areas where samples will be spotted. If the manifold is assembled and working properly the blue color will remain in areas where it is spotted. Bleeding of the color across the membrane indicates that the manifold is not assembled properly. In addition, the colored dye spots will allow you to easily discern which side of the membrane contains your samples after the dot blot manifold is disassembled.

12. When hybridizing multiple membranes simultaneously, increase the amount of probe and hybridization solution proportionally. To prevent the membranes from overlapping within the hybridization bottle, cut a piece of mesh such as Denville Scientific mesh (Denville Scientific, Metuchen, NJ) to the appropriate size and lay it between the membranes.

Index

A

B

C

D

Q

R

S

T